Springer Tracts in Advanced Robotics

Volume 143

The Springer Tracts in Advanced Robotics (STAR) publish new developments and advances in the fields of robotics research, rapidly and informally but with a high quality. The intent is to cover all the technical contents, applications, and multidisciplinary aspects of robotics, embedded in the fields of Mechanical Engineering, Computer Science, Electrical Engineering, Mechatronics, Control, and Life Sciences, as well as the methodologies behind them. Within the scope of the series are monographs, lecture notes, selected contributions from specialized conferences and workshops, as well as selected PhD theses.

Special offer: For all clients with a print standing order we offer free access to the electronic volumes of the Series published in the current year.

Indexed by SCOPUS, DBLP, EI Compendex, zbMATH, SCImago.

All books published in the series are submitted for consideration in Web of Science.

More information about this series at http://www.springer.com/series/5208

Bernd Henze

Whole-Body Control for Multi-Contact Balancing of Humanoid Robots

Design and Experiments

Springer

Bernd Henze
Deutsches Zentrum für Luft-und Raumfahrt
(DLR)
Intitut für Robotik und Mechatronik
Wessling, Germany

ISSN 1610-7438 ISSN 1610-742X (electronic)
Springer Tracts in Advanced Robotics
ISBN 978-3-030-87211-3 ISBN 978-3-030-87212-0 (eBook)
https://doi.org/10.1007/978-3-030-87212-0

This Springer imprint is published by the registered company Springer Nature Switzerland AG
The registered company address is: Gewerbestrasse 11, 6330 Cham, Switzerland

Series Editor's Foreword

At the dawn of the century's third decade, robotics is reaching an elevated level of maturity and continues to benefit from the advances and innovations in its enabling technologies. These all are contributing to an unprecedented effort to bringing robots to human environment in hospitals and homes, factories, and schools; in the field for robots fighting fires, making goods and products, picking fruits, and watering the farmland, saving time and lives. Robots today hold the promise for making a considerable impact in a wide range of real-world applications from industrial manufacturing to healthcare, transportation, and exploration of the deep space and sea. Tomorrow, robots will become pervasive and touch upon many aspects of modern life.

The *Springer Tracts in Advanced Robotics (STAR)* is devoted to bringing to the research community the latest advances in the robotics field on the basis of their significance and quality. Through a wide and timely dissemination of critical research developments in robotics, our objective with this series is to promote more exchanges and collaborations among the researchers in the community and contribute to further advancements in this rapidly growing field.

The monograph by Bernd Henze is based on the author's doctoral thesis. The contents are focused on control of legged, torque-controlled humanoid robots and are organized in ten chapters. The challenge of whole-body control is tackled to allow a humanoid robot not only to maintain multi-contact balance but also to manipulate objects while safely interacting with the environment.

Rich of examples developed by means of extensive experimentation on the humanoid robot TORO developed at DLR, this volume was the co-winner of the 2021 Georges Giralt PhD Award for the best doctoral thesis in Europe. A very fine addition to the STAR series!

Naples, Italy
October 2021

Bruno Siciliano
STAR Editor

Preface

This monograph is based on my dissertation written at the *Institute of Robotics and Mechatronics* of the *German Aerospace Center* (*DLR*). The defence took place at the *Technical University of Munich* (*TUM*) in April 2020.

This work aims at providing algorithms for balance control of legged, torque-controlled humanoid robots. A humanoid robot normally uses the feet for locomotion. This paradigm is extended by addressing the challenge of multi-contact balancing, which allows a humanoid robot to exploit an arbitrary number of contacts to support itself. Using multiple contacts increases the size of the support polygon, which in turn leads to an increased robustness of the stance and to an increased kinematic workspace of the robot. Both are important features for facilitating a transition of humanoid robots from research labs to real-world applications, where they are confronted with multiple challenging scenarios, such as climbing stairs and ladders, traversing debris, handling heavy loads, or working in confined spaces. The distribution of forces and torques among the multiple contacts is a challenging aspect of the problem, which arises from the closed kinematic chain given by the robot and its environment.

The developed framework also addresses the challenge of whole-body control by allowing a humanoid robot not only to maintain balance but also to interact with its environment, as, e.g. by carrying or manipulating objects. Of course, the forces and torques arising from the interaction task must be considered by the balancing and/or support task to avoid falling.

The whole-body control framework is generalized by combining the techniques for multi-contact balancing with multi-objective control, which allows for a more generic task definition. Kinematic and dynamic conflicts between the tasks are resolved via a prioritization. This generalization also allows the use of support contacts not only at the end-effectors (hands and feet) but at arbitrary locations on the body of the robot, such as the knees, pelvis, backpack, shoulders, or the elbows. This ability is essential for operating a humanoid robot in confined spaces, where there might be not enough space to completely "unfold" the limbs to rely on hands and feet only.

The developed control framework implements a compliant balancing behavior in order to ensure safety in case of accidental collisions between robot and environment or between robot and human. Furthermore, the framework employs passivity-based methods to achieve robustness with respect to external disturbances.

The performance and versatility of the developed whole-body control framework was demonstrated in numerous experiments and in the context of the EU-project *COMANOID*, which addressed the challenge of introducing humanoid robots into aircraft manufacturing.

I would like to take the opportunity to thank all persons who have contributed to the success of this work. Above all, I would like to thank Prof. Alin Albu-Schäffer and Dr. Christian Ott for their supervision and guidance during the past years. I also would like to thank all my co-authors of the previous works for the discussions, the contributions, and their help in obtaining and publishing the research results. In particular, I have to thank Dr. Máximo A. Roa, Alexander Werner, and Oliver Porges, who have been formidable colleagues and without whom this work would never have been possible. Great thanks also to Dr. Alexander Dietrich for introducing me to the field of multi-objective control. Thanks also to Ribin Balachandran for his insights into PO-PC. Furthermore, I would like to thank Dr. Firas Abi-Farraj, who visited *DLR* as a guest researcher for half a year, for the productive collaboration that resulted in two valuable publications.

Special thanks go to all the people who helped in developing and maintaining the humanoid robot *TORO*, including Florian Schmidt, Robert Burger, Benedikt Pleintinger, Martin Heumos, Ashok Meenakshi Sundaram, Alexander Beyer, Wieland Bertleff, Harald Wagner, and Michael Heumos. Great thanks also to Tilo Wüsthoff, who helped illustrating the thesis by providing the 3D renderings of *TORO*. Special thanks also to Samantha Stoneman for lending her voice for the audio commentary in many of the videos that are attached to the publications on which the thesis is based on.

I would like to thank Prof. Darius Burschka, Prof. Alin Albu-Schäffer, and Prof. Boris Lohmann for examining the Ph.D. thesis and administering the defense. I also have to thank Prof. Bruno Siciliano, the committee of the *Georges Girald Ph.D. Award*, and the team of *Springer* the honor of publishing my research in the *Springer Tracts in Advanced Robotics.*

The research activities that are incorporated in the dissertation have been partially funded by the *European Union Research and Innovation Programme Horizon 2020* under grant agreement number H2020-ICT-645097 (*COMANOID*) and by the *Initiative and Networking Fund of the Helmholtz Association* through a *Helmholtz Young Investigators Group* under grant agreement number VH-NG-808.

Munich, Germany
July 2021

Bernd Henze

Contents

Symbols and Abbreviations

Abbreviations

CoP	Center of Pressure
cQP	Constrained Quadratic Optimization Problem
DoF	Degree of Freedom
FTS	Force-Torque Sensor
HMCB	Hierarchical Multi-Contact Balancer
IMU	Inertial Measurement Unit
LIPM	Linear Inverted Pendulum Model
MCB	Multi-Contact Balancer
MCB-ACT	Multi-Contact Balancer with Automatic CoM Task
MCB-PC	Multi-Contact Balancer with Passivity Control
MCB+	Multi-Contact Balancer with PD+ control
SP	Support Polygon
TORO	TOrque-controlled humanoid RObot
WBC	Whole-Body Control
WDP	Wrench Distribution Problem
ZMP	Zero Moment Point

Sets

$\mathcal{S}_{\mathrm{SP}}$	Support Polygon
$\mathcal{S}_i$	Surface of contact i

Frames

$\mathcal{B}$	Base frame
$\mathcal{T}_i$	Task frame i
$\mathcal{W}$	World frame

Matrices

$\boldsymbol{\Lambda}$	Cartesian inertia matrix
$\boldsymbol{\mu}$	Cartesian Coriolis matrix
$\boldsymbol{\Xi}$	Mapping between the stacked task wrenches and $\boldsymbol{u}$
$\boldsymbol{Ad}$	Stacked adjoint matrix of all task frames $\mathcal{T}_i$
$\boldsymbol{Ad}(\boldsymbol{R}, \boldsymbol{x})$	Adjoint matrix according to Definition 3.2
$\boldsymbol{Ad}_{\text{int}}$	Stacked adjoint matrix of all interaction task frames $\mathcal{T}_i$
$\boldsymbol{Ad}_{\text{sup}}$	Stacked adjoint matrix of all support task frames $\mathcal{T}_i$
$\boldsymbol{A}_i$	Part of the contact model of contact i
$\boldsymbol{B}_\tau$	Part of the actuator model
$\boldsymbol{C}$	Coriolis matrix
$\boldsymbol{D}_{\text{c}}$	Damping matrix of the CoM compliance
$\boldsymbol{D}_{\text{int}}$	Combined damping matrix of the interaction compliances
$\boldsymbol{J}$	Stacked Jacobian matrix of all task frames $\mathcal{T}_i$
$\boldsymbol{J}'$	Stacked Sub-Jacobian matrix of all task frames $\mathcal{T}_i$
$\boldsymbol{J}_{\text{int}}$	Stacked Jacobian matrix of all interaction task frames $\mathcal{T}_i$
$\boldsymbol{J}'_{\text{int}}$	Stacked Sub-Jacobian matrix of all interaction task frames $\mathcal{T}_i$
$\boldsymbol{J}_{\text{sup}}$	Stacked Jacobian matrix of all support task frames $\mathcal{T}_i$
$\boldsymbol{J}'_{\text{sup}}$	Stacked Sub-Jacobian matrix of all support task frames $\mathcal{T}_i$
$\boldsymbol{J}_{\text{task}_k}$	Task Jacobian Matrix of task k
$\boldsymbol{M}$	Inertia matrix
$\boldsymbol{N}_{\text{null}}$	Null space projector of the null space controller (see Section 6.2)
$\boldsymbol{N}_{\text{task}_k}$	Null space projector of task k
$\boldsymbol{Q}_{\text{c}}$	Weighting matrix of the cQP for the CoM task
$\boldsymbol{Q}_{\text{int}}$	Weighting matrix of the cQP for the interaction task
$\boldsymbol{Q}_{\text{sup}}$	Weighting matrix of the cQP for the support task
$\boldsymbol{Q}_{\text{task}_k}$	Weighting matrix of the cQP for a general task k
$\boldsymbol{R}_{\text{c}}$	Orientation of the CoM frame C
$\boldsymbol{R}_{\text{c,d}}$	Desired orientation of the CoM frame C
$\boldsymbol{R}_i$	Orientation of task frame $\mathcal{T}_i$
$\boldsymbol{R}_{i,d}$	Desired orientation of the task frame $\mathcal{T}_i$
$\boldsymbol{S}(\boldsymbol{\lambda})$	Cross product matrix according to Definition 3.1
$\boldsymbol{Z}_{\text{task}_k}$	Null space base matrix of task k

Scalars

$\tilde{\mu}_i$	Friction coefficient of contact $\mathcal{T}_i$
ψ	Number of task frame
ψ_{int}	Number of interaction frames
ψ_{sup}	Number of support contacts
$f_{i,z}^{\text{min}}$	Minimum contact force of contact $\mathcal{T}_i$
m	Overall mass of the robot
m_k	Dimensionality of task k
n	Number of joints
$p_{i,x}^{\text{min/max}}$	Size of the contact area $\mathcal{S}_i$ in x-direction
$p_{i,y}^{\text{min/max}}$	Size of the contact area $\mathcal{S}_i$ in y-direction
r	Number of tasks
V_{c}^{K}	Potential energy of the CoM compliance
V_{int}^{K}	Potential energy of the interaction compliances

Vectors

$\boldsymbol{\delta}_{\text{c}}$	Residuum of the cQP for the CoM task
$\boldsymbol{\delta}_{\text{int}}$	Residuum of the cQP for the interaction task
$\boldsymbol{\delta}_{\text{sup}}$	Residuum of the cQP for the support task
$\boldsymbol{\delta}_{\text{task}_k}$	Residuum of the cQP for a general task k
$\boldsymbol{\nu}$	Generalized velocities
$\boldsymbol{\tau}$	Control torque
$\boldsymbol{\tau}^{\text{max}}$	Maximum joint torque
$\boldsymbol{\tau}_{\text{ext}}$	Vector of generalized external forces
$\boldsymbol{\tau}_{\text{pose}}^{\text{cpl}}$	Torque generated by a joint compliance
$\boldsymbol{\tau}_{\mu}$	Vector compensating for several coupling terms within $\boldsymbol{C}$
$\boldsymbol{\omega}_{\text{c}}$	Angular velocity of the CoM frame C
$\boldsymbol{\omega}_i$	Angular velocity of task frame $\mathcal{T}_i$
$\boldsymbol{b}_i$	Part of the contact model of contact i
$\boldsymbol{F}$	Stacked commanded wrenches of all task frames $\mathcal{T}_i$
$\boldsymbol{F}^{\text{ext}}$	Stacked external wrenches of all task frames $\mathcal{T}_i$
$\boldsymbol{F}^{\text{opt}}$	Optimal solution of the WDP
$\boldsymbol{F}_{\text{c}}^{\text{cpl}}$	Wrench generated by the CoM compliance
$\boldsymbol{F}_{\text{int}}$	Stacked commanded wrenches of the interaction tasks
$\boldsymbol{F}_{\text{int}}^{\text{cpl}}$	Stacked wrenches generated by the interaction compliances
$\boldsymbol{F}_{\text{int}}^{\text{ext}}$	Stacked external wrenches of the interaction tasks
$\boldsymbol{F}_{\text{sup}}$	Stacked commanded wrenches of the support contacts
$\boldsymbol{F}_{\text{sup}}^{\text{ext}}$	Stacked external wrenches of the support contacts
$\boldsymbol{F}_{\text{sup}}^{\text{res}}$	Commanded overall wrench resulting from the support contacts
$\boldsymbol{F}_{\text{sup}}^{\text{res,ext}}$	External overall wrench resulting from the support contacts

$\boldsymbol{F}_i$	Wrench applied at task frame $\mathcal{T}_i$
$\boldsymbol{g}$	Gravity vector
$\boldsymbol{g}_0$	Gravity vector acting at the coordinates of the CoM frame C
$\boldsymbol{u}$	Virtual control input
$\breve{v}_{\text{task}_k}$	Local velocity of task k
$\Delta\boldsymbol{v}_{\text{c}}$	Velocity error of the CoM frame C
$\Delta\boldsymbol{v}_{\text{int}}$	Stacked velocity errors of the interaction frames
$\Delta\boldsymbol{v}_{\text{int}}$	Stacked velocity errors of the support contacts
$\boldsymbol{v}$	Stacked Cartesian velocities of all task frames $\mathcal{T}_i$
$\boldsymbol{v}_{\text{c}}$	Cartesian velocities of the CoM frame C
$\boldsymbol{v}_{\text{c,d}}$	Desired velocities of the CoM frame C
$\boldsymbol{v}_{\text{int}}$	Stacked velocities of the interaction tasks
$\boldsymbol{v}_{\text{int,d}}$	Stacked desired velocities of the interaction frames
$\boldsymbol{v}_{\text{sup}}$	Stacked velocities of the support contacts
$\boldsymbol{v}_{\text{sup,d}}$	Stacked desired velocities of the support contacts
$\boldsymbol{v}_i$	Cartesian velocities of task frame $\mathcal{T}_i$
$\boldsymbol{v}_{i,d}$	Desired velocities of the task frame $\mathcal{T}_i$
$\dot{\boldsymbol{x}}_{\text{c}}$	Translational velocity of the CoM frame C
$\dot{\boldsymbol{x}}_i$	Translational velocity of task frame $\mathcal{T}_i$
$\boldsymbol{x}_{\text{c}}$	Position of the CoM frame C
$\boldsymbol{x}_{\text{c,d}}$	Desired position of the CoM frame C
$\boldsymbol{x}_i$	Position of task frame $\mathcal{T}_i$
$\boldsymbol{x}_{i,d}$	Desired position of the task frame $\mathcal{T}_i$

Chapter 1
Introduction

This chapter provides an introduction to the topic of whole-body control for multi-contact balancing of humanoid robots. Sections 1.1 and 1.2 describe the context and the challenges addressed in this work. Section 1.3 presents the achieved contributions with respect to the state of the art. The connection between the outline of this work and the previous publications is established in Sect. 1.4.

1.1 Motivation

Thanks to their versatility, humanoid robots can be employed in a large variety of applications, such as service robotics, industrial manufacturing, disaster scenarios, or space travel. In general, the use cases involve tasks that are monotone, exhausting, physically demanding, or dangerous for humans, which emphasizes the need for robotic solutions. The applications usually take place in environments explicitly designed for humans. Thus, it is a logical consequence to design the robots in a humanoid shape, such that they can use the same facilities and tools as humans. For this reason, science has taken an increasing interest in humanoid robotics over the past few years. For instance, the US *Defense Advanced Research Projects Agency* (DARPA) organized the *DARPA Robotics Challenge* (DRC), in which numerous international teams participated. The finals of the competition took place in 2015 and comprised several tasks from the context of a disaster scenario, such as driving a car, opening doors, traversing debris, and closing a valve (Spenko et al. 2018). In 2011, the US *National Aeronautics and Space Administration* (NASA) sent a humanoid robot, called *Robonaut 2*, to the *International Space Station* (ISS). The mission objective was to explore the possibility of using humanoid robots for supporting astronauts inside and outside of *ISS* (Diftler et al. 2012). The *German Aerospace Center* (*DLR*) investigates the potential of employing humanoid robots as service robots in the

B. Henze, *Whole-Body Control for Multi-Contact Balancing of Humanoid Robots*,
Springer Tracts in Advanced Robotics 143,
https://doi.org/10.1007/978-3-030-87212-0_1

health care sector in the form of the research project *SMiLE* (Smile 2019; Vogel et al. 2018). *DLR* also participated in the EU-project *COMANOID* (Comanoid 2019; Kheddar et al. 2019), which addressed the challenge of introducing humanoid robots into aircraft manufacturing. A more detailed overview of the use case and the research activities conducted in the context of *COMANOID* are provided in Sect. 9.3.

The field of humanoid robotics can be divided into robots with wheels and robots with legs. In general, humanoid robots with legs are considered to be more versatile and dexterous, as they can deal with obstacles and cluttered terrain, such as ladders, stairs, gravel, ditches, or debris. Another method of classification is offered by the low-level joint technology: In general, humanoid robots with position-controlled joints require less effort regarding the hardware design of the joint units. But humanoid robots with torque-controlled joints have the advantage of automatically adjusting their limbs to the environment, which is beneficial in challenging or unknown terrain. For instance, this feature is demonstrated in Sect. 6.1.3 with the torque-controlled humanoid robot *TORO*, developed by *DLR*, balancing on a pile of gym mats. The safety aspect is another advantage of torque-control: If a humanoid robot is sent into a completely or partially unknown environment, e.g. during a disaster scenario, then collisions between robot and environment are most likely to occur. If the robot is supposed to share the same workspace with a human co-worker, then there is also the potential of a collision between human and robot. In both cases, torque-control allows for the operation of a humanoid robot in a compliant and robust way to avoid damages or injuries to the robot, the environment, and the human.

Traditionally, a humanoid robot utilizes only the feet for locomotion for reasons of maneuverability. The field of multi-contact balancing generalizes the problem by allowing for an arbitrary number of contacts with the environment to gain a more robust stance in challenging terrain, such as rocks, ladders, or debris. Note that the contacts can be located arbitrarily on the body of the robot including the hands, elbows, shoulders, backpack, pelvis, knees, or feet. Using multiple contacts to support the robot features several advantages: The weight of the robot is distributed to a larger contact surface, which can reduce the sinking on soft terrain. Furthermore, the load can be (re-)distributed between the contacts if one of them is about to be overloaded. Multiple contacts also allow for the generation of internal stress, which is required for chimney climbing, for instance. But most important, using multiple contacts for balancing increases the size of the support polygon (SP) and therefore the kinematic workspace of the robot as well as the robustness of the stance in rough terrain.

For the reasons outlined above, this work focuses on torque-controlled humanoid robots with legs. The provided control algorithms implement a compliant and robust balancing behavior using multiple contacts for supporting the robot.

1.2 Problem Description

Ordinary industrial robots are usually mounted to the environment via a flange at their base. The forces and torques that are necessary to support and move the robot are transmitted via this flange to the environment. Thus, the kinematics and dynamics of an industrial robot are usually described via a so-called fixed-base model. In contrast to that, the field of humanoid robotics mainly utilizes a model description featuring a floating-base, which means that one central body of the robot structure (usually the pelvis or the trunk) is selected as the root or base of the kinematic chain. The base is only connected to the environment via six virtual Cartesian joints, which do not allow for any transmission of forces or torques. Thus, the robot must utilize its limbs to support itself by bringing them into contact with the environment. Figure 1.1 shows an exemplary scenario, where the robot exploits the right knee and the left foot for support.

As already mentioned, this work aims at providing algorithms for balance control of humanoid robots in multi-contact scenarios. From a technical point of view, using multiple contacts to support the robot creates a closed kinematic chain, as shown in Fig. 1.1. The consequence is the so-called Wrench Distribution Problem (WDP), which states that the overall wrench that is required to support the robot can be arbitrarily distributed to the contacts. Note that all feasible solutions of the WDP must respect the properties of each contact in order to avoid an unintentional lift-off,

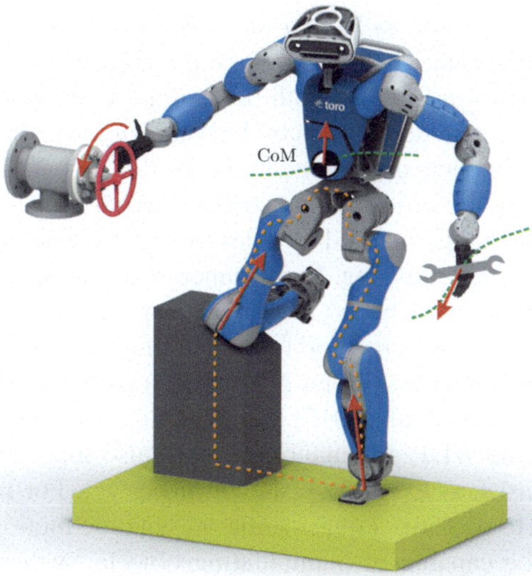

Fig. 1.1 Example of a humanoid robot performing several whole-body tasks while balancing on multiple contacts. The resulting closed kinematic chain is given in orange. The end-effector and CoM wrenches are represented by red arrows. The trajectories of the center of mass and the left hand are given in green

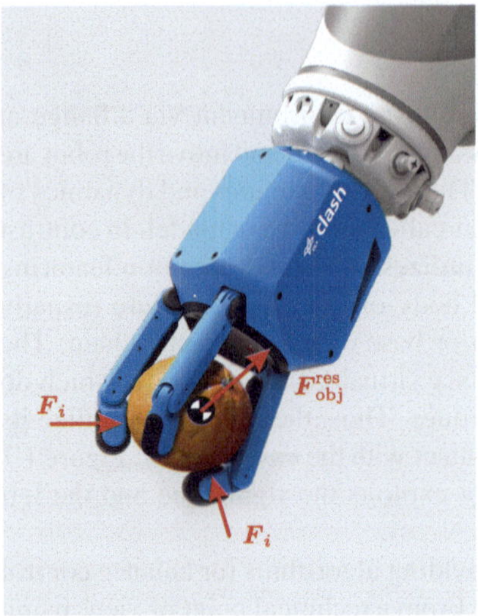

(a) The *CLASH* hand developed by *DLR* (Friedl et al., 2018).

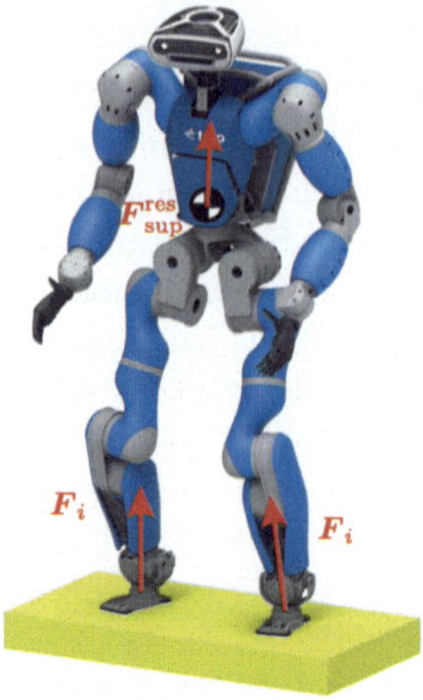

(b) The humanoid robot *TORO* developed by *DLR* (see Chapter 5).

Fig. 1.2 Analogy between the problem of multi-fingered grasping and multi-contact balancing: The overall wrench on the object ($\boldsymbol{F}^{\text{res}}_{\text{obj}}$) or respectively on the humanoid ($\boldsymbol{F}^{\text{res}}_{\text{sup}}$) is distributed to the contact wrenches $\boldsymbol{F}_i$

sliding, or tilting. According to Kumar and Waldron (1988) and Ott et al. (2011), the problem of multi-contact balancing and multi-fingered grasping are fundamentally similar: As shown in Fig. 1.2, both fields require the distribution of a desired overall wrench to the available contacts. In balancing, the desired overall wrench for supporting the robot gets distributed to the available contacts, while in grasping, the overall wrench on the object gets distributed to the finger contacts. Therefore, it is possible to interchange methods for describing and solving the WDP between the field of balancing and grasping. For instance, the concept of virtual linkage was introduced by Williams and Khatib (1993) for grasping and is used by Sentis (2010) for multi-contact balancing. A method for processing the contact properties was first formulated by Hirai (1991) and Borst et al. (2003) for multi-fingered grasping and later transferred by Caron et al. (2015) to humanoid balancing.

Besides solving the WDP, a humanoid robot is also supposed to perform several other meaningful tasks, as exemplary shown in Fig. 1.1. For instance, a humanoid robot must be able to perform a quasi-static or dynamic motion with its body. Furthermore, it must be capable of manipulating objects, such as carrying a tool or closing a valve. All these tasks require the generation of forces and torques, which must be considered in the WDP to properly counteract them. Note that there can be a kinematic or dynamic conflict between the tasks. For instance, the robot does not feature enough joints for fulfilling all tasks at the same time. Thus, this work incorporates methods from the field of multi-objective control (Dietrich 2016) to establish a prioritization of the tasks. The latter plays an important role if the support contacts are not only located at the end-effectors but also at arbitrary locations on the robot's body. From a different perspective, a humanoid robot can be interpreted as a tree of several serial kinematic chains on top of a closed kinematic loop. The latter is formed by the support contacts and features a redundancy in the space of the contact wrenches leading to the WDP. The serial kinematic chains can feature a kinematic redundancy, which is mitigated by prescribing multiple control objectives/tasks and prioritizing them. Thus, the control approaches presented in this work are from the field of Whole-Body Control (WBC), as they consider not only the support contacts but the complete body of the robot.

This works aims at providing control algorithms for robust and compliant balancing of humanoid robots. A compliant behavior is important to safely operate a humanoid robot in the case of accidental collisions with the environment or with humans. For this reason, most of the presented control approaches are passive or passivity-based. The passivity framework allows for modelling dynamic systems from an energetic point of view, where components can exchange energy via power ports. In this case, the robot (plus controller) can be regarded as a subsystem that is connected via a power port to the environment or to a human, respectively. If the subsystem is passive, then the robot can only emit a limited amount of energy, which implies a certain robustness with respect to external disturbances (see also Stramigioli 2015).

It is worth mentioning that the presented control approaches do not explicitly regulate the end-effector wrenches of the robot. This would not only require additional force-torque sensors at the end-effectors, but also introduce additional sensor noise into the control loop. Furthermore, the end-effector wrenches can be directly computed from the joint torques and vice versa, which would contradict the causality of the control loop. Instead, the presented control algorithms only regulate the state of the robot (position and velocity), whereas the commanded end-effector wrenches are used as a virtual control input. Of course, this can lead to a deviation between the commanded and actual end-effector wrenches. However, the presented experiments involving the humanoid robot *TORO* developed by *DLR* show that the deviations are insignificant, which emphasizes the validity and robustness of the proposed control approaches.

1.3 State of the Art and Contributions

A straightforward way to combine balancing with an interaction task is to consider separately the upper and the lower body of a humanoid robot. In general, the upper body is operated in a compliance mode to perform the intended manipulation task, while the lower body of the robot generates the required support. The modular design allows for a flexible combination of upper and lower body controllers. However, the forces and torques arising from the manipulation task must be taken into account by the lower body controller to generate sufficient support. For instance, Ibanez et al.

(2012) proposed a ZMP-based (Zero Moment Point) balancing controller in combination with a position-based controller for the upper body. The balancer utilizes preview control and is equipped with a disturbance input to account for the arising manipulation wrenches. We presented a similar approach in Ott et al. (2013) for kinesthetic teaching in combination with interaction-aware balancing (see Sect. 8.1). Here, the wrenches from the upper body are estimated via a momentum-based disturbance observer known from manipulator control (De Luca et al. 2006), which allows for a completely independent design of the upper and lower body control strategy.

More versatile approaches originate from the field of whole-body control by considering all the Degrees of Freedom (DoFs) of a humanoid robot. The field can be divided into control strategies employing inverse dynamics and passivity-based approaches. Let us first consider the branch based on inverse dynamics. Here, a vector of desired joint accelerations is obtained from a task formulation, which can also involve a prioritization, as discussed below. The desired accelerations are mapped to joint torques via the inverted dynamics equation of the robot. Often, the concept is combined with PD-control at the joint-level for stabilizing the desired joint positions and velocities. The desired torques computed from inverse dynamics are incorporated into the joint-level control loop in the form of a feedforward control action. Representative works from this field can be found in Mistry et al. (2010), Righetti et al. (2013), and Herzog et al. (2016), for instance.

Passivity-based approaches allow for the analysis and design of the robotic system from an energetic perspective. Here, the task formulation provides a set of desired task forces and/or torques, which are mapped to the commanded joint torques via their respective Jacobian matrices. A passivity-based approach for compliant balancing of humanoid robots was first proposed in Hyon et al. (2007) by computing suitable ground applied forces, which are mapped to joint torques. The concept was reused by Ott et al. (2011) exploiting the insight that the problem of balancing and grasping are conceptually similar (see Fig. 1.2). In Henze et al. (2016b), we extended the work of Ott et al. (2011) to a whole-body control framework by adding (a) the ability to handle multiple contacts, (b) interaction tasks for manipulation, and (c) a feedforward control. The latter allows for the execution of dynamic trajectories and results in a structure of the closed-loop system similar to classical PD+ control (Paden and Panja 1988; Whitcomb et al. 1993). Furthermore, we enriched the framework with a concept for passivity control on movable and deformable surfaces (Henze et al. 2018). Most of the works from the field of humanoid robotics assume that the floor is static, or that the dynamics of the floor is known (e.g. Sentis 2010; Mistry et al. 2010; Righetti et al. 2013; Herzog et al. 2016). But in order to operate the robot on a large variety of floors, we exploit the concept of passivity to make as few assumptions as possible on the floor dynamics. In particular, the framework is combined with methods from the field of Passivity Observer-Passivity Control (PO-PC) by Hannaford and Ryu (2002) and equipped with energy tanks (Franken et al. 2011). Besides this, we extended our framework in Abi-Farraj et al. (2019) with an automatic CoM tasks, which allows the robot to perform high-force interaction tasks. Recently, another passivity-based approach was proposed by Hirayama et al. (2018) for walking on uneven terrain.

Often, whole-body control (based on inverse dynamics as well as passivity-based) is combined with multi-objective control to achieve a prioritization of tasks. Methods for fixed-base manipulators are known from Nakamura et al. (1987), Hollerbach and Suh (1987), or Dietrich et al. (2015). In humanoid robotics, the approaches can be divided into the following categories: (a) orthogonal decomposition, (b) hierarchical quadratic programming, and (c) null space projection techniques. For instance, the approach by Mistry et al. (2010) utilizes a QR-decomposition of the constraint Jacobian matrix to achieve an analytically correct inverse dynamics formulation. The decomposition was later reused by Righetti et al. (2013) to minimize contact forces. Two representative works from the category of hierarchical programming are given by Escande et al. (2014) and Herzog et al. (2016). Here, numerical methods are employed to solve the inverse dynamics problem in a hierarchical way. Null space projection techniques are used in Sentis (2010)) and Henze et al. (2016a), for instance. Here, the hierarchy is implemented by projecting the forces and torques of one task onto the null space of all other tasks with a higher priority level. In Henze et al. (2016a), we were able to demonstrate a dynamic decoupling of the tasks after integrating multi-objective control (Dietrich 2016) into our passivity-based whole-body control framework. In Henze et al. (2017), we discovered that the combined framework can be exploited to generalize the concept of multi-contact balancing. In particular, the combined framework allows the robot to support itself not only with the hands and feet but also with arbitrary contacts on the robot's body, such as the knees, pelvis, backpack, shoulders, or the elbows. Apart from that, relatively little work has been presented to explicitly consider arbitrary contacts on the body of the robot. For instance, the approaches (Sentis 2010; Mistry et al. 2010; Righetti et al. 2013) control the constrained motion of the robot by accounting for the physical contact constraints, but the works do not explicitly consider contacts with arbitrary locations. A framework based on inverse dynamics combined with an orthogonal projection was presented by Aghili and Su (2016). The approach presented in Farnioli et al. (2016) is derived from a quasi-static analysis and tested on a humanoid robot with elastic joints. Although the latter two approaches are designed to theoretically handle contacts scattered all over the body of the robot, this particular aspect has not been validated in the demonstrated experiments.

By nature, the fields of multi-contact planning and multi-contact control are closely related, which allows for the exchange of knowledge and methods. For instance, a powerful multi-contact motion planner was presented by Bouyarmane and Kheddar (2012). The planner first computes a set of feasible contact configurations and afterwards generates the contact transitions and corresponding whole-body motions. In Bouyarmane et al. (2012), the planner was applied to several disaster response tasks using simulations of three different robot platforms. Another planning algorithm is presented by Murooka et al. (2015), which is capable of obtaining an optimal posture for interaction tasks that require large forces and torques. For instance, the robot can use its own body weight to push a heavy object by leaning against it, as shown in Fig. 6.16. However, Murooka et al. (2015) only consider the ZMP for evaluating the feasibility of the stance. A more general criterion for describing the feasible support wrenches originates from the field of grasping (Hirai 1991;

Borst et al. 2003) using polyhedral convex cones. The concept was then transfered by Caron et al. (2015) to humanoid balancing and used for multi-contact planning in Caron and Nakamura (2015). We applied the concept in Abi-Farraj et al. (2019) to instantaneous control by deriving an automatic task for the Center of Mass (CoM) that moves the robot into an optimal posture for generating the high interaction forces.

In summary, this work presents a framework for balance control of legged humanoid robots in multi-contact scenarios. It advances the state of the art by providing several theoretical contributions to the field as well as an extensive experimental evaluation of the presented methods. A list of the main contributions is given below:

- Extension of the work by Ott et al. (2011) into a passivity-based framework (see Henze et al. 2016b)
  - for whole-body control by incorporating interaction tasks,
  - for multi-contact balancing by allowing for an arbitrary number of contacts,
  - and for the tracking case by incorporating feedforward terms.
- Passivity control for balancing on movable and deformable surfaces (Henze et al. 2018).
- Automatic CoM task for performing high-force interaction tasks (Abi-Farraj et al. 2019).
- Generalization of the whole-body control framework by integrating hierarchical multi-objective control (Henze et al. 2016a). The latter features a task prioritization, which leads to a dynamic decoupling of the tasks.
- Transfer of the generalized framework to balancing in confined spaces by allowing for arbitrary contact locations on the body of the robot (Henze et al. 2017).
- Extensive experimental evaluation of the whole-body control framework.
- Demonstration of the performance and versatility of the whole-body control framework in various applications including planning (Werner et al. 2016; Sundaram et al. 2018), teleoperation (Porges et al. 2019; Abi-Farraj et al. 2018), and industrial manufacturing (Kheddar et al. 2019).

1.4 Publications and Outline

The research findings in this work resulted in five journal articles, ten conference publications, and two invited talks in the relevant robotic journals and conferences. The publications are decorated with three awards and two nominations as award finalists. A list of the publications that are incorporated into this monograph is given below:

Journal Publications

- Henze B, Dietrich A, Ott C (2016a) An approach to combine balancing with hierarchical whole-body control for legged humanoid robots. IEEE Robotics and Automation Letters 1(2):700–707. (**Award Winning**[1])
- Henze B, Roa MA, Ott C (2016b) Passivity-based whole-body balancing for torque-controlled humanoid robots in multi-contact scenarios. Int J of Robotics Research 35(12):1522–1543. (**Award Winning**[1])
- Henze B, Balachandran R, Roa MA, Ott C, Albu-Schäffer A (2018) Passivity analysis and control of humanoid robots on movable ground. IEEE Robotics and Automation Letters 3(4):3457–3464.
- Abi-Farraj F, Henze B, Ott C, Giordano PR, Roa MA (2019) Torque-based balancing for a humanoid robot performing high-force interaction tasks. IEEE Robotics and Automation Letters 4(2):2023–2030.
- Kheddar A, Caron S, Gergondet P, Comport A, Tanguy A, Ott C, Henze B, Mesesan G, Englsberger J, Roa MA, Wieber PB, Chaumette F, Spindler F, Oriolo G, Lanari L, Escande A, Chappellet K, Kanehiro F, Rabaté P (2019) Humanoid robots in aircraft manufacturing - the Airbus use-cases. IEEE Robotics and Automation Magazine 26(4):30–45. (**Award Winning**[2])

Conference Publications

- Ott C, Henze B, Lee D (2013) Kinesthetic teaching of humanoid motion based on whole-body compliance control with interaction-aware balancing. In: IEEE/RSJ Int. Conf. on Intelligent Robots and Systems, pp. 4615–4621.
- Henze B, Ott C, Roa MA (2014a) Posture and balance control for humanoid robots in multicontact scenarios based on model predictive control. In: IEEE/RSJ Int. Conf. on Intelligent Robots and Systems, pp. 3253–3258.
- Henze B, Werner A, Roa MA, Garofalo G, Englsberger J, Ott C (2014b) Control applications of TORO - a torque controlled humanoid robot. In: IEEE-RAS Int. Conf. on Humanoid Robots, pp. 841–841. (**Award Winning**[3])
- Englsberger J, Werner A, Ott C, Henze B, Roa MA, Garofalo G, Burger R, Beyer A, Eiberger O, Schmid K, Albu-Schäffer A (2014) Overview of the torque-controlled humanoid robot TORO. In: IEEE-RAS Int. Conf. on Humanoid Robots, pp. 916–923.
- Werner A, Henze B, Rodriguez DA, Gabaret J, Porges O, Roa MA (2016) Multi-contact planning and control for a torque-controlled humanoid robot. In: IEEE/RSJ Int. Conf. on Intelligent Robots and Systems, pp. 5708–5715.
- Henze B, Dietrich A, Roa MA, Ott C (2017) Multi-contact balancing of humanoid robots in confined spaces: Utilizing knee contacts. In: IEEE/RSJ Int. Conf. on Intelligent Robots and Systems, pp. 679–704. (**Award Finalist**[4])
- Abi-Farraj F, Henze B, Werner A, Panzirsch M, Ott C, Roa MA (2018) Humanoid teleoperation using task-relevant haptic feedback. In: IEEE/RSJ Int. Conf. on Intelligent Robots and Systems, pp. 5010–5017. (**Award Finalist**[5])
- Sundaram AM, Henze B, Porges O, Marton ZC, Roa MA (2018) Autonomous bipedal humanoid grasping with base repositioning and whole-body control. In: IEEE-RAS Int. Conf. on Humanoid Robots, pp. 395–402.

---

[1] Winner of the *Best WBC-Video Award 2016 (IEEE-RAS Technical Committee on Whole-Body Control)*.

[2] Winner of the *IEEE Robotics and Automation Magazine Best Paper Award 2020*.

[3] Winner of the *Best Video Award* at *IEEE-RAS Int. Conf. on Humnaoid Robots 2014*.

[4] Finalist for the *Best Paper Award on Safety, Security, and Rescue Robotics (in memory of Motohiro Kisoi)* at *IEEE/RSJ Int. Conf. on Intelligent Robots and Systems 2017*.

[5] Finalist for the *Best Paper Award on Safety, Security, and Rescue Robotics (in memory of Motohiro Kisoi)* at *IEEE/RSJ Int. Conf. on Intelligent Robots and Systems 2018*.

- Henze B, Roa MA, Werner A, Dietrich A, Ott C, Albu-Schäffer A (2019) Experiments with human-inspired behaviors in a humanoid robot: Quasi-static balancing using toe-off motion and stretched knees. In: IEEE Int. Conf. on Robotics and Automation, pp. 2510–2516.
- Porges O, Connan M, Henze B, Gigli A, Castellini C, Roa MA (2019) Awearable, ultralight interface for bimanual teleoperation of a compliant, whole-body-controlled humanoid robot. In: IEEE Int. Conf. on Robotics and Automation, URL www.youtube.com/watch?v=YLEUBFu5qgI, accessed September 13th, 2019.

Invited Talks

- Henze B (2018) Model-based posture control for a torque-controlled humanoid robot. In: The Int. Symposium on Wearable Robotics, Pisa, Italy, special session on *Biorobotics Approaches to Understand and Restore Human Balance*, (Invited talk).
- Henze B (2019) Hierarchical whole-body control for humanoid robots. In: IEEE Int. Conf. on Robotics and Automation, Montréal, Canada, workshop on *Continuous Management and Scheduling of Multiple Simultaneous Prioritized Tasks for Redundant Robots*, (Invited talk).

Besides the main contributions, two book chapters (Ott et al. 2017; Roa et al. 2018), two journal articles (Ott et al. 2015, 2016), and eight conference publications (Garofalo et al. 2015; Werner et al. 2017, 2018; Mesesan et al. 2017; Lakatos et al. 2019; Olivieri et al. 2019; Rossini et al. 2019; García-Haro et al. 2019) have been co-authored, which are related to the topic but not incorporated into this manuscript.

The connection between the publications and the outline of this work is established in Fig. 1.3. Chapter 3 provides two models: one for describing the floating-base dynamics of a humanoid robot and one for characterizing the properties of the support contacts. Chapter 4 details the design of Cartesian compliances, which are used throughout the work to implement a compliant balancing behavior. Chapter 5 introduces the reader to the hardware design of the humanoid robot *TORO*, which is used for evaluating the performance of the presented control concepts. The proposed framework for whole-body control is given in Chaps. 6 and 7. Section 6.1 introduces the basic concept of the framework as well as the numeric method that is used for solving the wrench distribution problem. Sections 6.3 to 6.5 present three extensions addressing the tracking case, passivity on movable ground, and interaction tasks with high forces. Chapter 7 generalizes the whole-body control framework by combining it with multi-objective control. The resulting prioritization facilitates a dynamic decoupling of the control tasks and allows humanoid robots to balance in confined spaces by exploiting arbitrary contacts on the robot's body. Chapter 8 presents two control approaches from the field of Model Predictive Control (MPC), which are both based on reduced dynamic models. The relevance and versatility of the developed framework for whole-body control is demonstrated in Chap. 9 in the form of several applications from the field of planning, teleoperation, and industrial manufacturing.

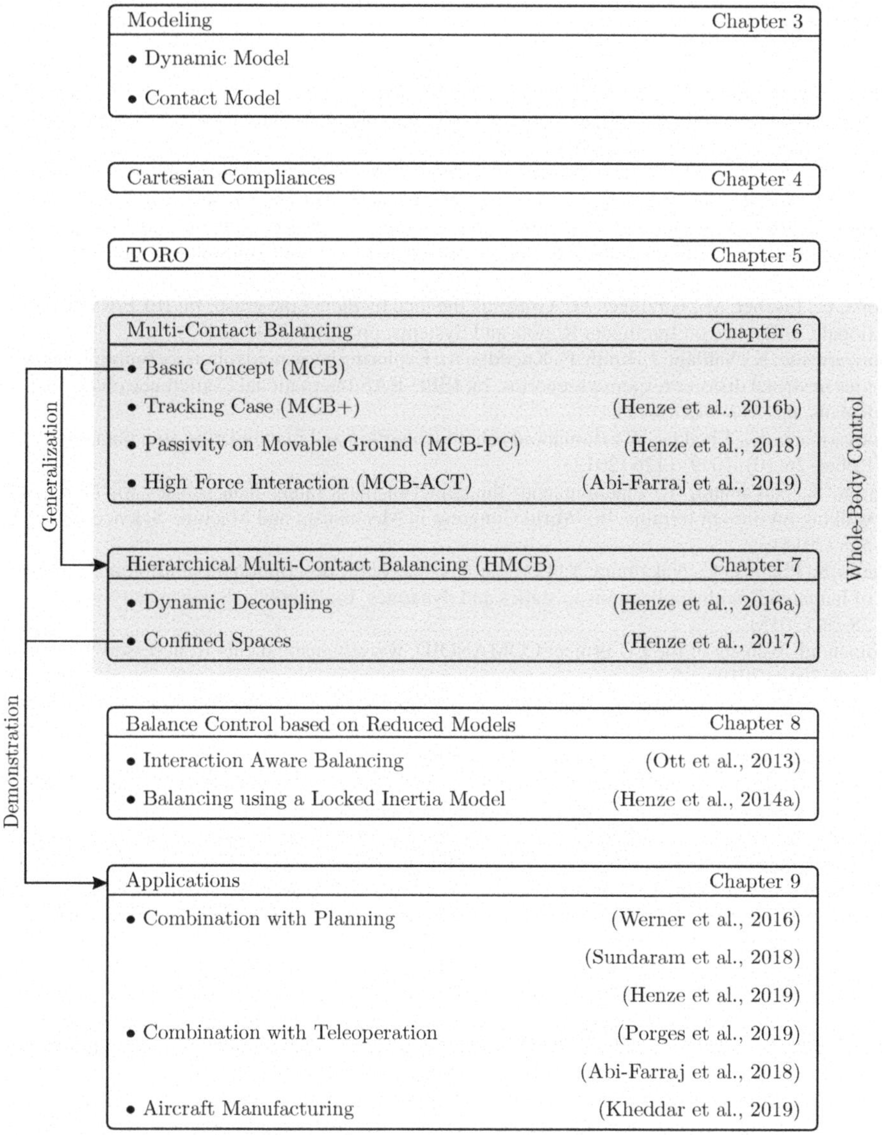

Fig. 1.3 Outline of the manuscript and relation of the main aspects to the publications

References

Abi-Farraj, F., Henze, B., Werner, A., Panzirsch, M., Ott, C., Roa, M.A.: Humanoid teleoperation using task-relevant haptic feedback. In: IEEE/RSJ International Conference on Intelligent Robots and Systems, pp. 5010–5017 (2018)

Abi-Farraj, F., Henze, B., Ott, C., Giordano, P.R., Roa, M.A.: Torque-based balancing for a humanoid robot performing high-force interaction tasks. IEEE Robot. Autom. Lett. **4**(2), 2023–2030 (2019)

Aghili, F., Su, C.Y.: Control of constrained robots subject to unilateral contacts and friction cone constraints. In: IEEE International Conference on Robotics and Automation, pp. 2347–2352 (2016)

Borst, C., Fischer, M., Hirzinger, G.: Grasping the dice by dicing the grasp. In: IEEE/RSJ International Conference on Intelligent Robots and Systems, pp. 3692–3697 (2003)

Bouyarmane, K., Vaillant, J., Keith, F., Kheddar, A.: Exploring humanoid robots locomotion capabilities in virtual disaster response scenarios. In: IEEE-RAS International Conference on Humanoid Robots, pp. 337–342 (2012)

Bouyarmane, K., Kheddar, A.: Humanoid robot locomotion and manipulation step planning. Adv. Robot. **26**(10), 1099–1126 (2012)

Caron, S., Nakamura, Y.: Planning with the center-of-mass rather than stances for humanoids walking on uneven terrains. In: World Congress in Mechanism and Machine Science, pp. 554–558 (2015)

Caron, S., Pham, Q.C., Nakamura, Y.: Leveraging cone double description for multi-contact stability of humanoids with applications to statics and dynamics. In: Robotics: Science and Systems, pp. 28–36 (2015)

Comanoid: Website of the EU-project COMANOID. www.comanoid.cnrs.fr, accessed September 13th, 2019 (2019)

De Luca, A., Albu-Schäffer, A., Haddadin, S., Hirzinger, G.: Collision detection and safe reaction with the DLR-III lightweight manipulator arm. In: IEEE/RSJ International Conference on Intelligent Robots and Systems, pp. 1623–1630 (2006)

Dietrich, A.: Whole-Body Impedance Control of Wheeled Humanoid Robots, Springer Tracts in Advanced Robotics, vol. 116, 1st edn. Springer, Switzerland (2016)

Dietrich, A., Ott, C., Albu-Schäffer, A.: An overview of null space projections for redundant, torque-controlled robots. Int. J. Robot. Res. **34**(11), 1385–1400 (2015)

Diftler, M.A., Ahlstrom, T.D., Ambrose, R.O., Radford, N.A., Joyce, C.A., De La Pena, N., Parsons, A.H., Noblitt, A.L.: Robonaut 2 - initial activities on-board the ISS. In: IEEE Aerospace Conference, pp. 1–12 (2012)

Englsberger, J., Werner, A., Ott, C., Henze, B., Roa, M.A., Garofalo, G., Burger, R., Beyer, A., Eiberger, O., Schmid, K., Albu-Schäffer, A.: Overview of the torque-controlled humanoid robot TORO. In: IEEE-RAS International Conference on Humanoid Robots, pp. 916–923 (2014)

Escande, A., Mansard, N., Wieber, P.B.: Hierarchical quadratic programming: fast online humanoid-robot motion generation. Int. J. Robot. Res. **33**(7), 1522–1543 (2014)

Farnioli, E., Gabiccini, M., Bicchi, A.: Toward whole-body loco-manipulation: Experimental results on multi-contact interaction with the walk-man robot. In: IEEE/RSJ International Conference on Intelligent Robots and Systems, pp. 2347–2352 (2016)

Franken, M., Stramigioli, S., Misra, S., Secchi, C., Macchelli, A.: Bilateral telemanipulation with time delays: a two-layer approach combining passivity and transparency. IEEE Trans. Robot. **27**(4), 741–756 (2011)

Friedl, W., Höppner, H., Schmidt, F., Roa, M.A., Grebenstein, M.: CLASH: compliant low cost antagonistic servo hands. In: IEEE/RSJ International Conference on Intelligent Robots and Systems, pp. 6469–6476 (2018)

García-Haro, J.M., Henze, B., Mesesan, G., Martínez, S., Ott, C.: Integration of dual-arm manipulation in a passivity based whole-body controller for torque-controlled humanoid robots. In: IEEE-RAS International Conference on Humanoid Robots, pp. 644–650 (2019)

Garofalo, G., Henze, B., Englsberger, J., Ott, C.: On the inertially decoupled structure of the floating base robot dynamics. In: Vienna International Conference on Mathematical Modelling, pp. 322–327 (2015)

Hannaford, B., Ryu, J.H.: Time-domain passivity control of haptic interfaces. IEEE Trans. Robot. Autom. **18**(1), 1–10 (2002)

Henze, B., Dietrich, A., Ott, C.: An approach to combine balancing with hierarchical whole-body control for legged humanoid robots. IEEE Robot. Autom. Lett. **1**(2), 700–707 (2016a)

Henze, B., Dietrich, A., Roa, M.A., Ott, C.: Multi-contact balancing of humanoid robots in confined spaces: utilizing knee contacts. In: IEEE/RSJ International Conference on Intelligent Robots and Systems, pp. 679–704 (2017)

Henze, B., Ott, C., Roa, M.A.: Posture and balance control for humanoid robots in multi-contact scenarios based on model predictive control. In: IEEE/RSJ International Conference on Intelligent Robots and Systems, pp. 3253–3258 (2014a)

Henze, B., Roa, M.A., Ott, C.: Passivity-based whole-body balancing for torque-controlled humanoid robots in multi-contact scenarios. Int. J. Robot. Res. **35**(12), 1522–1543 (2016b)

Henze, B., Roa, M.A., Werner, A., Dietrich, A., Ott, C., Albu-Schäffer, A.: Experiments with human-inspired behaviors in a humanoid robot: quasi-static balancing using toe-off motion and stretched knees. In: IEEE International Conference on Robotics and Automation, pp. 2510–2516 (2019)

Henze, B., Werner, A., Roa, M.A., Garofalo, G., Englsberger, J., Ott, C.: Control applications of TORO - a torque controlled humanoid robot. In: IEEE-RAS International Conference on Humanoid Robots, pp. 841–841 (2014b)

Henze, B.: Hierarchical whole-body control for humanoid robots. In: IEEE International Conference on Robotics and Automation, Montréal, Canada, Workshop on Continuous Management and Scheduling of Multiple Simultaneous Prioritized Tasks for Redundant Robots, (Invited talk) (2019)

Henze, B.: Model-based posture control for a torque-controlled humanoid robot. In: The International Symposium on Wearable Robotics, Pisa, Italy, special session on Biorobotics Approaches to Understand and Restore Human Balance, (Invited talk) (2018)

Henze, B., Balachandran, R., Roa, M.A., Ott, C., Albu-Schäffer, A.: Passivity analysis and control of humanoid robots on movable ground. IEEE Robot. Autom. Lett. **3**(4), 3457–3464 (2018)

Herzog, A., Rotella, N., Mason, S., Grimminger, F., Schaal, S., Righetti, L.: Momentum control with hierarchical inverse dynamics on a torque-controlled humanoid. Auton. Robots **40**(3), 473–491 (2016)

Hirai, S.: Analysis and planning of manipulation using the theory of polyhedral convex cones. Ph.D. thesis, Kyoto University (1991)

Hirayama, K., Hirosawa, N., Hyon, S.H.: Passivity-based compliant walking on torque-controlled hydraulic biped robot. In: IEEE-RAS International Conference on Humanoid Robots, pp. 552–557 (2018)

Hollerbach, J.M., Suh, K.: Redundancy resolution of manipulators through torque optimization. IEEE J. Robot. Autom. **3**(4), 308–316 (1987)

Hyon, S.H., Hale, J.G., Cheng, G.: Full-body compliant human-humanoid interaction: balancing in the presence of unknown external forces. IEEE Trans. Robot. **23**(5), 884–898 (2007)

Ibanez, A., Bidaud, P., Padois, V.: Unified preview control for humanoid postural stability and upper-limb interaction adaptation. In: IEEE/RSJ International Conference on Intelligent Robots and Systems, pp. 1801–1808 (2012)

Kheddar, A., Caron, S., Gergondet, P., Comport, A., Tanguy, A., Ott, C., Henze, B., Mesesan, G., Englsberger, J., Roa, M.A., Wieber, P.B., Chaumette, F., Spindler, F., Oriolo, G., Lanari, L., Escande, A., Chappellet, K., Kanehiro, F., Rabaté, P.: Humanoid robots in aircraft manufacturing - the Airbus use-cases. IEEE Robot. Autom. Mag. **26**(4), 30–45 (2019)

Kumar, V.R., Waldron, K.J.: Force distribution in closed kinematic chains. IEEE J. Robot. Autom. **4**(6), 657–664 (1988)

Lakatos, D., Federigi, Y., Gumpert, T., Henze, B., Hermann, M., Loeffl, F., Schmidt, F., Seidel, D., Albu-Schäffer, A.: A coordinate-based approach for static balancing and walking control of compliantly actuated legged robots. In: IEEE International Conference on Robotics and Automation, pp. 5909–9515 (2019)

Mesesan, G., Englsberger, J., Henze, B., Ott, C.: Dynamic multi-contact transitions for humanoid robots using divergent component of motion. In: IEEE International Conference on Robotics and Automation, pp. 4108–4115 (2017)

Mistry, M., Buchli, J., Schaal, S.: Inverse dynamics control of floating base systems using orthogonal decomposition. In: IEEE International Conference on Robotics and Automation, pp. 3406–3412 (2010)

Murooka, M., Nozawa, S., Kakiuchi, Y., Okada, K., Inaba, M.: Whole-body pushing manipulation with contact posture planning of large and heavy object for humanoid robot. In: IEEE International Conference on Robotics and Automation, pp. 5682–5689 (2015)

Nakamura, Y., Hanafusa, H., Yoshikawa, T.: Task-priority based redundancy control of robot manipulators. Int. J. Robot. Res. **6**(2), 3–15 (1987)

Olivieri, N., Henze, B., Braghin, F., Roa, M.A.: Experimental evaluation and modeling of passive falls in humanoid robots. In: IEEE-RAS International Conference on Humanoid Robots, pp. 1 – 7 (2019)

Ott, C., Henze, B., Lee, D.: Kinesthetic teaching of humanoid motion based on whole-body compliance control with interaction-aware balancing. In: IEEE/RSJ International Conference on Intelligent Robots and Systems, pp. 4615–4621 (2013)

Ott, C., Roa, M.A., Hirzinger, G.: Posture and balance control for biped robots based on contact force optimization. In: IEEE-RAS International Conference on Humanoid Robots, pp. 26–33 (2011)

Ott, C., Roa, M.A., Schmidt, F., Friedl, W., Englsberger, J., Burger, R., Werner, A., Dietrich, A., Leidner, D., Henze, B., Eiberger, O., Beyer, A., Bäuml, B., Borst, C., Albu-Schäffer, A.: Humanoid Robotics: A Reference, Springer Netherlands, Dordrecht, chap Mechanisms and Design of DLR Humanoid Robots, pp. 1–26 (2017)

Ott, C., Dietrich, A., Leidner, D., Werner, A., Englsberger, J., Henze, B., Wolf, S., Chalon, M., Friedl, W., Beyer, A., Eiberger, O., Albu-Schäffer, A.: From torque-controlled to intrinsically compliant humanoid robots. ASME Mech. Eng. Mag. **137**(6), 7–11 (2015)

Ott, C., Henze, B., Hettich, G., Seyde, T.N., Roa, M.A., Lippi, V., Mergner, T.: Good posture, good balance: comparison of bioinspired and model-based approaches for posture control of humanoid robots. IEEE Robot. Autom. Mag. **23**(1), 22–33 (2016)

Paden, B., Panja, R.: Globally asymptotically stable 'PD+' controller for robot manipulators. Int. J. Control **47**(6), 1697–1712 (1988)

Porges, O., Connan, M., Henze, B., Gigli, A., Castellini, C., Roa, M.A.: A wearable, ultralight interface for bimanual teleoperation of a compliant, whole-body-controlled humanoid robot. In: IEEE International Conference on Robotics and Automation, www.youtube.com/watch?v=YLEUBFu5qgI, accessed September 13th, 2019 (2019)

Righetti, L., Buchli, J., Mistry, M., Kalakrishnan, M., Schaal, S.: Optimal distribution of contact forces with inverse dynamics control. Int. J. Robot. Res. **32**(3), 280–298 (2013)

Roa, M.A., Henze, B., Ott, C.: Wearable Robotics: Challenges and Trends. Springer, Switzerland, chap Model-Based Posture Control for a Torque-Controlled Humanoid Robot, pp. 344–347 (2018)

Rossini, L., Henze, B., Braghin, F., Roa, M.A.: Optimal trajectory for active safe falls in humanoid robots. In: IEEE-RAS International Conference on Humanoid Robots, pp. 305–312 (2019)

Sentis, L.: Motion Planning for Humanoid Robots. Springer, London, chap Compliant Control of Whole-Body Multi-contact Behaviors in Humanoid Robots, pp. 29–66 (2010)

Smile: Website of the research project SMiLE. www.dlr.de/rm/smile, accessed September 13th, 2019 (2019)

Spenko, M., Buerger, S., Iagnemma, K. (eds.): The DARPA Robotics Challenge Finals: Humanoid Robots To The Rescue, Springer Tracts in Advanced Robotics, vol. 121. Springer, Cham (2018)

Stramigioli, S.: Mathematical Control Theory I (Lecture Notes in Control and Information Sciences 461). Springer, Cham, chap Energy-Aware Robotics, pp. 37–50 (2015)

Sundaram, A.M., Henze, B., Porges, O., Marton, Z.C., Roa, M.A.: Autonomous bipedal humanoid grasping with base repositioning and whole-body control. In: IEEE-RAS International Conference on Humanoid Robots, pp. 395–402 (2018)
Vogel, J., Leidner, D., Hagengruber, A., Panzirsch, M., Dietrich, A.: Das Projekt SMiLE. In: Cluster-Konferenz Zukunft der Pflege, pp. 212–216 (2018)
Werner, A., Henze, B., Keppler, M., Loeffl, F., Leyendecker, S., Ott, C.: Structure preserving multi-contact balance control for series-elastic and visco-elastic humanoid robots. In: IEEE/RSJ International Conference on Intelligent Robots and Systems, pp. 1233–1240 (2018)
Werner, A., Henze, B., Loeffl, F., Leyendecker, S., Ott, C.: Optimal and robust walking using intrinsic properties of a series-elastic robot. In: IEEE-RAS International Conference on Humanoid Robots, pp. 143–150 (2017)
Werner, A., Henze, B., Rodriguez, D.A., Gabaret, J., Porges, O., Roa, M.A.: Multi-contact planning and control for a torque-controlled humanoid robot. In: IEEE/RSJ International Conference on Intelligent Robots and Systems, pp. 5708–5715 (2016)
Whitcomb, L.L., Rizzi, A.A., Koditschek, D.E.: Comparative experiments with a new adaptive controller for robot arms. IEEE Trans. Robot. Autom. **9**(1), 59–70 (1993)
Williams, D., Khatib, O.: The virtual linkage: a model for internal forces in multi-grasp manipulation. In: IEEE International Conference on Robotics and Automation, pp. 1025–1030 (1993)

Chapter 2
Notation

This work is based on the following notation:

- Scalars are denoted with regular characters, such as a, α, V, or Ψ.
- Vectors are denoted by bold lower case characters, such as $\boldsymbol{v}_{\mathrm{c}}$ or $\boldsymbol{\tau}$. The only exception are wrenches $\boldsymbol{F} = (\,\boldsymbol{f}^T\ \boldsymbol{\tau}^T\,)^T$, which are a combination of a Cartesian force $\boldsymbol{f} \in \mathbb{R}^3$ and torque $\boldsymbol{\tau} \in \mathbb{R}^3$.
- Matrices are denoted by bold upper case characters, such as $\boldsymbol{M}$ or $\boldsymbol{\Lambda}$.
- Round and square brackets are used to combine symbols into vectors and matrices, respectively. For instance, a vector is given by $\boldsymbol{f} = (\,f_x\ f_y\ f_z\,)^T$ and a matrix by $\boldsymbol{J} = [\,\boldsymbol{Ad}\ \boldsymbol{J}'\,]$.
- The Euclidean norm of a vector is denoted by $\|\bullet\|$, as for instance $\|\boldsymbol{g}_0\| = \sqrt{\boldsymbol{g}_0^T\,\boldsymbol{g}_0}$.

In general, vectors and matrices can be expressed with respect to the base vectors of any arbitrary frame. In order to indicate the choice of base vectors, the notation is extended by a leading superscript: For instance, ${}^{A}\boldsymbol{\lambda}$ and ${}^{B}\boldsymbol{\lambda}$ indicate that the vector $\boldsymbol{\lambda}$ is given with respect to the base vectors of frame $\mathcal{F}_A$ or $\mathcal{F}_B$, respectively. For simplicity of notation, all position, velocity, force, and torque vectors are expressed in world coordinates, that is with respect to the base vectors of the world frame $\mathcal{W}$, unless otherwise mentioned. The superscript ${}^{\mathcal{W}}\bullet$ will be dropped.

B. Henze, *Whole-Body Control for Multi-Contact Balancing of Humanoid Robots*, Springer Tracts in Advanced Robotics 143,
https://doi.org/10.1007/978-3-030-87212-0_2

Chapter 3
Modeling

This chapter provides the underlying models for deriving the Whole-Body Control framework in Chaps. 6 and 7. Section 3.1 gives a brief overview on rigid body transformations in order to introduce the equations of motion of humanoid robots in Sect. 3.2. In multi-contact balancing, the distribution of forces and torques among the contacts is a challenging aspect of the problem, which arises from the closed kinematic chain given by the robot and its environment. Section 3.3 provides a mathematical description of this so-called Wrench Distribution Problem. The physical properties of the involved contacts are modeled and studied in Sects. 3.4 to 3.6.

3.1 Rigid Body Transformations

This section provides a brief summary on rigid body transformations, which will be used later to discuss the so-called wrench distribution problem (see Sect. 3.3). Let us consider two arbitrary frames $\mathcal{F}_A$ and $\mathcal{F}_B$ in six-dimensional Cartesian space attached to a rigid body (see Fig. 3.1). The position and orientation of both frames relative to a common world frame $\mathcal{W}$ are denoted by $\boldsymbol{x}_A \in \mathbb{R}^3$, $\boldsymbol{R}_A \in \mathcal{SO}(3)$ and $\boldsymbol{x}_B \in \mathbb{R}^3$, $\boldsymbol{R}_B \in \mathcal{SO}(3)$, respectively. The linear and angular velocities of frame $\mathcal{F}_A$ and $\mathcal{F}_B$ relative to the world frame $\mathcal{W}$ are given by $\boldsymbol{\chi}_A \in \mathbb{R}^3$, $\boldsymbol{\omega}_A \in \mathbb{R}^3$ and $\boldsymbol{\chi}_B \in \mathbb{R}^3$, $\boldsymbol{\omega}_B \in \mathbb{R}^3$. Both can be stacked into the six-dimensional Cartesian velocities

$$\boldsymbol{v}_A = \begin{pmatrix} \boldsymbol{\chi}_A \\ \boldsymbol{\omega}_A \end{pmatrix} \quad \text{and} \quad \boldsymbol{v}_B = \begin{pmatrix} \boldsymbol{\chi}_B \\ \boldsymbol{\omega}_B \end{pmatrix}. \tag{3.1}$$

The forces and torques acting on $\mathcal{F}_A$ and $\mathcal{F}_B$ are given by $\boldsymbol{f}_A \in \mathbb{R}^3$, $\boldsymbol{\tau}_A \in \mathbb{R}^3$ and $\boldsymbol{f}_B \in \mathbb{R}^3$, $\boldsymbol{\tau}_B \in \mathbb{R}^3$, respectively. Analogously to the velocities, the forces and

B. Henze, *Whole-Body Control for Multi-Contact Balancing of Humanoid Robots*,
Springer Tracts in Advanced Robotics 143,
https://doi.org/10.1007/978-3-030-87212-0_3

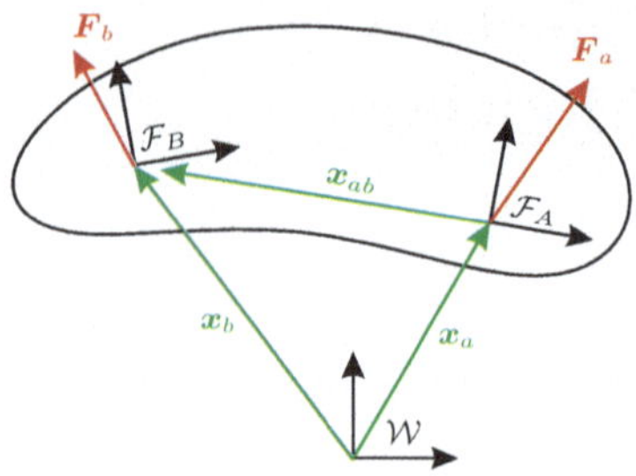

Fig. 3.1 Frames $\mathcal{F}_A$ and $\mathcal{F}_B$ located on a rigid body

torques can be stacked into the wrenches

$$\boldsymbol{F}_A = \begin{pmatrix} \boldsymbol{f}_A \\ \boldsymbol{\tau}_A \end{pmatrix} \quad \text{and} \quad \boldsymbol{F}_B = \begin{pmatrix} \boldsymbol{f}_B \\ \boldsymbol{\tau}_B \end{pmatrix}. \tag{3.2}$$

3.1.1 Body Coordinates

If the translational and rotational velocities are expressed in body frame, which means with respect to the base vectors of $\mathcal{F}_A$ and $\mathcal{F}_B$, then they can be computed via

$$\begin{aligned} {}^A\boldsymbol{\chi}_A &= \boldsymbol{R}_A^T({}^W\dot{\boldsymbol{x}}_A), & {}^B\boldsymbol{\chi}_B &= \boldsymbol{R}_B^T({}^W\dot{\boldsymbol{x}}_B), \\ \boldsymbol{S}({}^A\boldsymbol{\omega}_A) &= \boldsymbol{R}_A^T\dot{\boldsymbol{R}}_A, & \boldsymbol{S}({}^B\boldsymbol{\omega}_B) &= \boldsymbol{R}_B^T\dot{\boldsymbol{R}}_B. \end{aligned} \tag{3.3}$$

Here, the operator $\boldsymbol{S}(\bullet)$ denotes the cross product matrix according to Definition 3.1.

Definition 3.1 (*Cross product matrix* $\boldsymbol{S}(\boldsymbol{\lambda})$) Let $\boldsymbol{\lambda} = (\,\lambda_x^T\ \lambda_y^T\ \lambda_z^T\,)^T$ be an arbitrary vector in $\mathbb{R}^3$. The cross product matrix $\boldsymbol{S}(\boldsymbol{\lambda}) \in \mathbb{R}^{3\times 3}$ is defined as

$$\boldsymbol{S}(\boldsymbol{\lambda}) = \begin{bmatrix} 0 & -\boldsymbol{\lambda}_z & \boldsymbol{\lambda}_y \\ \boldsymbol{\lambda}_z & 0 & -\boldsymbol{\lambda}_x \\ -\boldsymbol{\lambda}_y & \boldsymbol{\lambda}_x & 0 \end{bmatrix}.$$

Note that the cross product matrix $\boldsymbol{S}(\boldsymbol{\lambda})$ is skew-symmetric, that is $\boldsymbol{S}(\boldsymbol{\lambda}) = -\boldsymbol{S}(\boldsymbol{\lambda})^T$ and $\boldsymbol{S}(-\boldsymbol{\lambda}) = \boldsymbol{S}(\boldsymbol{\lambda})^T$ hold.

The velocities of frame $\mathcal{F}_A$ can be mapped to frame $\mathcal{F}_B$ via

$$\underbrace{\begin{pmatrix} {}^A\boldsymbol{\chi}_A \\ {}^A\boldsymbol{\omega}_A \end{pmatrix}}_{{}^A\boldsymbol{v}_A} = \boldsymbol{Ad}(\boldsymbol{R}_{AB}, {}^A\boldsymbol{x}_{AB}) \underbrace{\begin{pmatrix} {}^B\boldsymbol{\chi}_A \\ {}^B\boldsymbol{\omega}_A \end{pmatrix}}_{{}^B\boldsymbol{v}_B} \tag{3.4}$$

with $\boldsymbol{Ad}(\bullet,\star)$ denoting the adjoint matrix according to Definition 3.2. The matrix $\boldsymbol{R}_{AB} = \boldsymbol{R}_A^T \boldsymbol{R}_B$ and the vector ${}^A\boldsymbol{x}_{AB} = \boldsymbol{R}_A^T({}^{\mathcal{W}}\boldsymbol{x}_B - {}^{\mathcal{W}}\boldsymbol{x}_A)$ describe the orientation and position of $\mathcal{F}_B$ relative to $\mathcal{F}_A$, expressed using the base vectors of $\mathcal{F}_A$.

Definition 3.2 (*Adjoint matrix* $\boldsymbol{Ad}(\boldsymbol{R},\boldsymbol{x})$) Let $\boldsymbol{R} \in \mathcal{SO}(3)$ and $\boldsymbol{x} \in \mathbb{R}^3$ be an arbitrary rotation matrix and an arbitrary position vector. Then, the adjoint matrix is defined as

$$\boldsymbol{Ad}(\boldsymbol{R},\boldsymbol{x}) = \begin{bmatrix} \boldsymbol{R} & \boldsymbol{S}(\boldsymbol{x})\boldsymbol{R} \\ \boldsymbol{0} & \boldsymbol{R} \end{bmatrix}$$

based on the cross product matrix $\boldsymbol{S}(\bullet)$ according to Definition 3.1. The inverse is given by $\boldsymbol{Ad}(\boldsymbol{R},\boldsymbol{x})^{-1} = \boldsymbol{Ad}(\boldsymbol{R}^T, -\boldsymbol{R}^T\boldsymbol{x})$.

The transpose of the adjoint matrix can be used to map forces and torques from frame $\mathcal{F}_A$ to frame $\mathcal{F}_B$ via

$$\underbrace{\begin{pmatrix} {}^B\boldsymbol{f}_B \\ {}^B\boldsymbol{\tau}_B \end{pmatrix}}_{{}^B\boldsymbol{F}_B} = \boldsymbol{Ad}(\boldsymbol{R}_{AB}, {}^A\boldsymbol{x}_{AB})^T \underbrace{\begin{pmatrix} {}^A\boldsymbol{f}_A \\ {}^A\boldsymbol{\tau}_A \end{pmatrix}}_{{}^A\boldsymbol{F}_A}. \tag{3.5}$$

Note that the wrenches are expressed in $\mathcal{F}_A$ and $\mathcal{F}_B$, respectively.

3.1.2 World Coordinates

If the translational and rotational velocities are expressed in world coordinates $\mathcal{W}$, then they can be computed via

$$\begin{aligned} {}^{\mathcal{W}}\boldsymbol{\chi}_A &= {}^{\mathcal{W}}\dot{\boldsymbol{x}}_A, & {}^{\mathcal{W}}\boldsymbol{\chi}_B &= {}^{\mathcal{W}}\dot{\boldsymbol{x}}_B, \\ \boldsymbol{S}({}^{\mathcal{W}}\boldsymbol{\omega}_A) &= \dot{\boldsymbol{R}}_A\boldsymbol{R}_A^T, & \boldsymbol{S}({}^{\mathcal{W}}\boldsymbol{\omega}_B) &= \dot{\boldsymbol{R}}_B\boldsymbol{R}_B^T. \end{aligned} \tag{3.6}$$

The velocities of frame $\mathcal{F}_A$ can be mapped to frame $\mathcal{F}_B$ via

$$\underbrace{\begin{pmatrix} {}^{\mathcal{W}}\boldsymbol{\chi}_A \\ {}^{\mathcal{W}}\boldsymbol{\omega}_A \end{pmatrix}}_{{}^{\mathcal{W}}\boldsymbol{v}_A} = \underbrace{\begin{bmatrix} \boldsymbol{I} & \boldsymbol{S}({}^{\mathcal{W}}\boldsymbol{x}_{AB}) \\ \boldsymbol{0} & \boldsymbol{I} \end{bmatrix}}_{\boldsymbol{Ad}(\boldsymbol{I},{}^{\mathcal{W}}\boldsymbol{x}_{AB})} \underbrace{\begin{pmatrix} {}^{\mathcal{W}}\boldsymbol{v}_B \\ {}^{\mathcal{W}}\boldsymbol{\omega}_B \end{pmatrix}}_{{}^{\mathcal{W}}\boldsymbol{v}_B}. \tag{3.7}$$

As the velocities are all in world coordinates, they do not need to be rotated. Consequently, the mapping can be considered a special case of the adjoint matrix with the identity matrix $\boldsymbol{I}$ as relative rotation. Note that the vector ${}^{\mathcal{W}}\boldsymbol{x}_{AB} = {}^{\mathcal{W}}\boldsymbol{x}_B - {}^{\mathcal{W}}\boldsymbol{x}_A$ is also given in world coordinates.

The transpose of the adjoint matrix can again be used to map forces and torques from frame $\mathcal{F}_A$ to frame $\mathcal{F}_B$ via

$$\underbrace{\begin{pmatrix} {}^{\mathcal{W}}\boldsymbol{f}_B \\ {}^{\mathcal{W}}\boldsymbol{\tau}_B \end{pmatrix}}_{{}^{\mathcal{W}}\boldsymbol{F}_B} = \boldsymbol{Ad}(\boldsymbol{I}, {}^{\mathcal{W}}\boldsymbol{x}_{AB})^T \underbrace{\begin{pmatrix} {}^{\mathcal{W}}\boldsymbol{f}_A \\ {}^{\mathcal{W}}\boldsymbol{\tau}_A \end{pmatrix}}_{{}^{\mathcal{W}}\boldsymbol{F}_A}. \tag{3.8}$$

Note that the forces and torques are all expressed in world frame $\mathcal{W}$.

For simplicity of notation, all position, velocity, force, and torque vectors will be expressed in world coordinates, that is with respect to the base vectors of the world fame $\mathcal{W}$, unless otherwise mentioned. The superscript ${}^{\mathcal{W}}\bullet$ will be dropped.

3.2 Robot Dynamics

Humanoid robots resemble the shape and appearance of a human, which usually leads to a design featuring a torso, arms, legs, and a head. The robot can use the limbs to support itself with respect to the environment but also to interact with it, e.g. for manipulating an object. From a mathematical point of view, a humanoid robot consists of a group of rigid bodies, which are linked to each other via joints. In order to be able to formulate the equations of motion, one of the bodies needs to be selected as the root or base of the kinematic chain. In the literature, there are two basic descriptions featuring either a fixed or a floating base. In the case of a fixed base, one of the bodies that is in contact with the environment to support the robot is defined as the root or base of the kinematic chain. Although it is easier to use a fixed-base model for motion planing (Werner et al. 2012), this has the disadvantage that the body serving as base is not allowed to move with respect to the environment. The latter makes it difficult to use the model for locomotion, which requires a repetitive attaching and detaching of the limbs. For this particular reason, models featuring a floating base are more common in literature (see e.g. Righetti et al. 2013; Sentis 2010; Herzog et al. 2016). Here, a central body of the humanoid robot, such as the hip or the torso, is selected as the root of the kinematic chains representing the limbs. This section introduces several floating-base models, which will be used in Chaps. 6, 7, and 8 to derive the discussed concepts for balance control.

3.2.1 Floating Base Dynamics

Considering the humanoid robot *TORO* (see Chap. 5), the body representing the hip is selected as the base of the kinematic chain. The state of the floating base is described by the position $\boldsymbol{x}_b \in \mathbb{R}^3$ and orientation $\boldsymbol{R}_b \in \mathcal{SO}(3)$ of the base frame $\mathcal{B}$ relative to the world frame $\mathcal{W}$ (see Fig. 3.2). The corresponding linear and angular velocities $\dot{\boldsymbol{x}}_b \in \mathbb{R}^3$ and $\boldsymbol{\omega}_b \in \mathbb{R}^3$ are stacked into the vector $\boldsymbol{v}_c = (\, \dot{\boldsymbol{x}}_b^T \; \boldsymbol{\omega}_b^T \,)^T$. Note that $\boldsymbol{x}_b$, $\dot{\boldsymbol{x}}_b$, and $\boldsymbol{\omega}_b$ are expressed in world coordinates. Based on the $n \in \mathbb{N}$ joint angles $\boldsymbol{q} \in \mathbb{R}^n$, the dynamics of the humanoid robot is given by

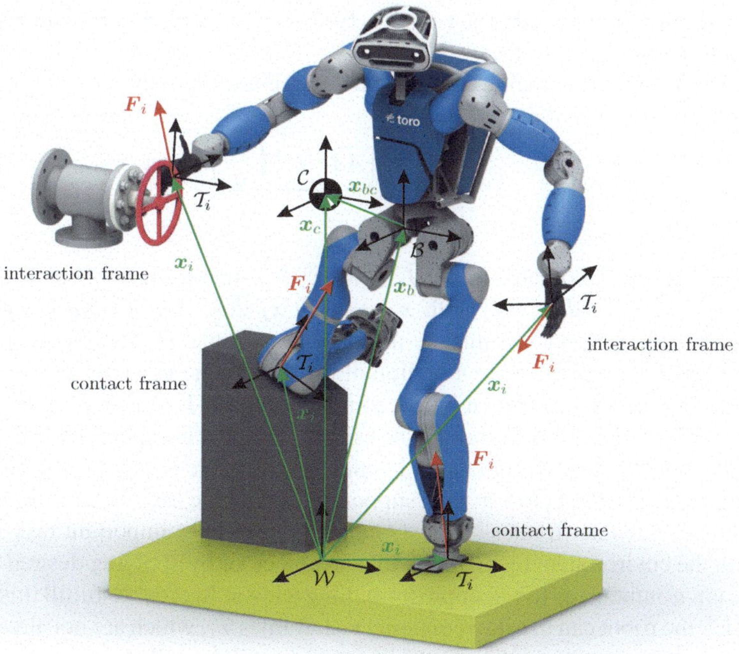

Fig. 3.2 Floating base model of a humanoid robot

$$\overline{M}\begin{pmatrix}\dot{v}_b\\ \ddot{q}\end{pmatrix}+\overline{C}\begin{pmatrix}v_b\\ \dot{q}\end{pmatrix}+\overline{g}=\begin{pmatrix}0\\ \tau\end{pmatrix}+\overline{\tau}_{\text{ext}} \tag{3.9}$$

with $\overline{M} \in \mathbb{R}^{(6+n)\times(6+n)}$ denoting the inertia matrix and $\overline{C} \in \mathbb{R}^{(6+n)\times(6+n)}$ the Coriolis/centrifugal matrix. Note that $\dot{\overline{M}} - 2\overline{C}$ is a skew-symmetric matrix (Ott 2008), which is strongly related to the passivity properties of the robot. The influence of gravity is represented by the vector $\overline{g} \in \mathbb{R}^{6+n}$. The joint torques generated by the actuators are denoted by $\tau \in \mathbb{R}^n$, while the generalized external forces are given by $\overline{\tau}_{\text{ext}} \in \mathbb{R}^{6+n}$.

Depending on the application, a legged humanoid robot has to fulfill several tasks simultaneously. An obvious task is to stabilize the CoM to maintain balance. Another two important tasks are to support itself or to interact with the environment. For these purposes, the robot features several task relevant points, which can be distributed all over the hull of the robot including feet, hands, knees, and elbows, for instance. Each of these points is characterized by a task frame $\mathcal{T}_i$ given by a position $x_i \in \mathbb{R}^3$ and an orientation $R_i \in \mathcal{SO}(3)$ with respect to the world frame $\mathcal{W}$. The overall number of task frames is given by $\psi \in \mathbb{N}$. The corresponding translational and rotational velocities are given by $\dot{x}_i \in \mathbb{R}^3$ and $\omega_i \in \mathbb{R}^3$, which can be stacked into $v_i = (\dot{x}_i^T\ \omega_i^T)^T$. The task wrench $F_i = (f_i^T\ \tau_i^T)^T$ combines the forces $f_i \in \mathbb{R}^3$

and torques $\boldsymbol{\tau}_i \in \mathbb{R}^3$ that act on $\mathcal{T}_i$. Note that $\boldsymbol{x}_i$ and $\boldsymbol{F}_i$ are expressed in world coordinates.

The Cartesian velocities of each task frame are given by

$$\boldsymbol{v}_i = \underbrace{\left[\overline{\boldsymbol{Ad}}_i \; \boldsymbol{J}_i'\right]}_{\bar{\boldsymbol{J}}_i} \begin{pmatrix} \boldsymbol{v}_b \\ \dot{\boldsymbol{q}} \end{pmatrix} . \tag{3.10}$$

based on the Jacobian matrix $\bar{\boldsymbol{J}}_i \in \mathbb{R}^{6\times(6+n)}$. According to Sect. 3.1, the adjoint matrix $\overline{\boldsymbol{Ad}}_i = \boldsymbol{Ad}(\boldsymbol{I}, \boldsymbol{x}_{ib})$ with $\boldsymbol{x}_{ib} = \boldsymbol{x}_b - \boldsymbol{x}_i$ describes a rigid body transformation between the task frame $\mathcal{T}_i$ and the base frame $\mathcal{B}$. Therefore, it correlates a motion of the base $\boldsymbol{v}_b$ with a motion of the task frame given by $\boldsymbol{v}_i$. A motion of the robot in joint space is taken into account by the Sub-Jacobian $\boldsymbol{J}_i' \in \mathbb{R}^{6\times n}$.

A subset of the task frames can be used by the robot to support itself by establishing contact with the environment and generating suitable contact wrenches. This task[1] will be referred to below as the "support task" and the associated frames as "support frames", "contact frames" or just "contacts". Another important task is to interact with the environment by manipulating an object, or by applying desired wrenches on the environment, as for example for pulling a lever. In order to fulfill this "interaction task", the robot can use the remaining task frames $\mathcal{T}_i$, which are not already occupied by the supporting task.[2] The frames concerned will be called "interaction frames" for the remainder of this work. The difference between the interaction and the support task is that the interaction wrenches are prescribed by the assignment of the robot, while the necessary support wrenches can be chosen freely to some extend by the controller. Note that the interaction frames can also be used to lift and reposition end-effectors of the robot in order to enable locomotion. To formalize this task assignment, let us consider the following numbering of the task frames $\mathcal{T}_i$: The first $\psi_{\text{sup}} \in \mathbb{N}$ task frames will be assigned to the supporting task with $1 \le i \le \psi_{\text{sup}}$. The remaining $\psi_{\text{int}} \in \mathbb{N}_0$ task frames will be associated with the interaction task by $\psi_{\text{sup}} + 1 \le i \le \psi$ with $\psi = \psi_{\text{sup}} + \psi_{\text{int}}$. Based on this notation, the task velocities can be combined into

$$\boldsymbol{v}_{\text{sup}} = \begin{pmatrix} \boldsymbol{v}_1 \\ \vdots \\ \boldsymbol{v}_{\psi_{\text{sup}}} \end{pmatrix} = \underbrace{\left[\overline{\boldsymbol{Ad}}_{\text{sup}} \; \boldsymbol{J}_{\text{sup}}'\right]}_{\bar{\boldsymbol{J}}_{\text{sup}}} \begin{pmatrix} \boldsymbol{v}_b \\ \dot{\boldsymbol{q}} \end{pmatrix} \tag{3.11}$$

and

[1] Note that this task was called "balancing task" (subscript "bal") in previous publications (Henze et al. 2016a, b, 2017).

[2] Note that it is also possible to operate the robot without any interaction task leading to $\psi_{\text{sup}} = \psi$. But this case will not be considered for simplicity of notation.

$$\boldsymbol{v}_{\text{int}} = \begin{pmatrix} \boldsymbol{v}_{\psi_{\text{sup}+1}} \\ \vdots \\ \boldsymbol{v}_{\psi} \end{pmatrix} = \underbrace{\left[\overline{\boldsymbol{Ad}}_{\text{int}}\ \bar{\boldsymbol{J}}'_{\text{int}}\right]}_{\bar{\boldsymbol{J}}_{\text{int}}} \begin{pmatrix} \boldsymbol{v}_b \\ \dot{\boldsymbol{q}} \end{pmatrix} \tag{3.12}$$

with

$$\overline{\boldsymbol{Ad}}_{\text{sup}} = \begin{bmatrix} \overline{\boldsymbol{Ad}}_1 \\ \vdots \\ \overline{\boldsymbol{Ad}}_{\psi_{\text{sup}}} \end{bmatrix}, \qquad \bar{\boldsymbol{J}}_{\text{sup}} = \begin{bmatrix} \bar{\boldsymbol{J}}'_1 \\ \vdots \\ \bar{\boldsymbol{J}}'_{\psi_{\text{sup}}} \end{bmatrix} \tag{3.13}$$

and

$$\overline{\boldsymbol{Ad}}_{\text{int}} = \begin{bmatrix} \overline{\boldsymbol{Ad}}_{\psi_{\text{sup}}+1} \\ \vdots \\ \overline{\boldsymbol{Ad}}_{\psi} \end{bmatrix}, \qquad \bar{\boldsymbol{J}}_{\text{int}} = \begin{bmatrix} \bar{\boldsymbol{J}}'_{\psi_{\text{sup}}+1} \\ \vdots \\ \bar{\boldsymbol{J}}'_{\psi} \end{bmatrix}. \tag{3.14}$$

If all generalized external forces exclusively act on the task frames $\mathcal{T}_i$, then $\boldsymbol{\tau}_{\text{ext}}$ becomes

$$\boldsymbol{\tau}_{\text{ext}} = \begin{bmatrix} \bar{\boldsymbol{J}}_{\text{sup}} \\ \bar{\boldsymbol{J}}_{\text{int}} \end{bmatrix}^T \begin{pmatrix} \boldsymbol{F}^{\text{ext}}_{\text{sup}} \\ \boldsymbol{F}^{\text{ext}}_{\text{int}} \end{pmatrix} \tag{3.15}$$

with the task wrenches $\boldsymbol{F}^{\text{ext}}_i \in \mathbb{R}^6$ stacked into

$$\boldsymbol{F}^{\text{ext}}_{\text{sup}} = \begin{pmatrix} \boldsymbol{F}^{\text{ext}}_1 \\ \vdots \\ \boldsymbol{F}^{\text{ext}}_{\psi_{\text{sup}}} \end{pmatrix} \quad \text{and} \quad \boldsymbol{F}^{\text{ext}}_{\text{int}} = \begin{pmatrix} \boldsymbol{F}^{\text{ext}}_{\psi_{\text{sup}+1}} \\ \vdots \\ \boldsymbol{F}^{\text{ext}}_{\psi} \end{pmatrix}. \tag{3.16}$$

Note that the controllers that will be presented in Chaps. 6, 7, and 8 utilize the commanded task wrenches $\boldsymbol{F}_i \in \mathbb{R}^6$ as a virtual control input instead of the control torque $\boldsymbol{\tau}$. The commanded task wrenches are analogously stacked into

$$\boldsymbol{F}_{\text{sup}} = \begin{pmatrix} \boldsymbol{F}_1 \\ \vdots \\ \boldsymbol{F}_{\psi_{\text{sup}}} \end{pmatrix} \quad \text{and} \quad \boldsymbol{F}_{\text{int}} = \begin{pmatrix} \boldsymbol{F}_{\psi_{\text{sup}+1}} \\ \vdots \\ \boldsymbol{F}_{\psi} \end{pmatrix}. \tag{3.17}$$

3.2.2 Centroidal Dynamics

As the location of the CoM is crucial for balancing, Hyon et al. (2007) proposed using the Center of Mass (CoM) of the robot instead of the hip as base for the dynamic model. Therefore, we introduced a frame C in Henze et al. (2016b), which has the same position $\boldsymbol{x}_{\mathrm{c}} \in \mathbb{R}^3$ as the CoM and the same orientation $\boldsymbol{R}_{\mathrm{c}} = \boldsymbol{R}_b$ as the hip of the robot. The linear and angular velocities $\dot{\boldsymbol{x}}_{\mathrm{c}} \in \mathbb{R}^3$ and $\boldsymbol{\omega}_{\mathrm{c}} = \boldsymbol{\omega}_b$ are stacked into $\boldsymbol{v}_{\mathrm{c}} = (\dot{\boldsymbol{x}}_{\mathrm{c}}^T \ \boldsymbol{\omega}_{\mathrm{c}}^T)^T$. Note that $\boldsymbol{x}_{\mathrm{c}}$, $\dot{\boldsymbol{x}}_{\mathrm{c}}$, and $\boldsymbol{\omega}_{\mathrm{c}}$ are expressed in world coordinates. The transformation obtained by replacing $\dot{\boldsymbol{x}}_b$ and $\boldsymbol{\omega}_b$ with $\dot{\boldsymbol{x}}_{\mathrm{c}}$ and $\boldsymbol{\omega}_{\mathrm{c}}$ is given by

$$\begin{pmatrix} \dot{\boldsymbol{x}}_{\mathrm{c}} \\ \boldsymbol{\omega}_{\mathrm{c}} \\ \dot{\boldsymbol{q}} \end{pmatrix} = \underbrace{\begin{bmatrix} \boldsymbol{I} & -\boldsymbol{S}(\boldsymbol{x}_{b\mathrm{c}}) & \boldsymbol{J}_{b\mathrm{c}} \\ \boldsymbol{0} & \boldsymbol{I} & \boldsymbol{0} \\ \boldsymbol{0} & \boldsymbol{0} & \boldsymbol{I} \end{bmatrix}}_{\boldsymbol{T}_{\mathrm{c}b}} \begin{pmatrix} \dot{\boldsymbol{x}}_b \\ \boldsymbol{\omega}_b \\ \dot{\boldsymbol{q}} \end{pmatrix}. \tag{3.18}$$

Here, $\boldsymbol{x}_{b\mathrm{c}} = \boldsymbol{x}_{\mathrm{c}} - \boldsymbol{x}_b$ denotes the lever arm between the CoM and the base frame $\mathcal{B}$, given in $\mathcal{W}$. The Jacobian $\boldsymbol{J}_{b\mathrm{c}} = \partial \boldsymbol{x}_{b\mathrm{c}} / \partial \boldsymbol{q}$ is needed to take into account the relative motion between frames $\mathcal{C}$ and $\mathcal{B}$. The inverse transformation $\boldsymbol{T}_{\mathrm{c}b} = \boldsymbol{T}_{b\mathrm{c}}^{-1}$ is given by

$$\begin{pmatrix} \dot{\boldsymbol{x}}_b \\ \boldsymbol{\omega}_b \\ \dot{\boldsymbol{q}} \end{pmatrix} = \underbrace{\begin{bmatrix} \boldsymbol{I} & \boldsymbol{S}(\boldsymbol{x}_{b\mathrm{c}}) & -\boldsymbol{J}_{b\mathrm{c}} \\ \boldsymbol{0} & \boldsymbol{I} & \boldsymbol{0} \\ \boldsymbol{0} & \boldsymbol{0} & \boldsymbol{I} \end{bmatrix}}_{\boldsymbol{T}_{b\mathrm{c}}} \begin{pmatrix} \dot{\boldsymbol{x}}_{\mathrm{c}} \\ \boldsymbol{\omega}_{\mathrm{c}} \\ \dot{\boldsymbol{q}} \end{pmatrix}. \tag{3.19}$$

Applying $\boldsymbol{T}_{b\mathrm{c}}$ to the dynamic model (3.9) leads to

$$\underbrace{\begin{bmatrix} \boldsymbol{M}_{\mathrm{cc}} & \boldsymbol{M}_{\mathrm{cq}} \\ \boldsymbol{M}_{\mathrm{cq}}^T & \boldsymbol{M}_{\mathrm{qq}} \end{bmatrix}}_{\boldsymbol{M}} \underbrace{\begin{pmatrix} \dot{\boldsymbol{v}}_{\mathrm{c}} \\ \ddot{\boldsymbol{q}} \end{pmatrix}}_{\dot{\boldsymbol{v}}} + \boldsymbol{C} \underbrace{\begin{pmatrix} \boldsymbol{v}_{\mathrm{c}} \\ \dot{\boldsymbol{q}} \end{pmatrix}}_{\boldsymbol{v}} + \underbrace{\begin{pmatrix} m\boldsymbol{g}_0 \\ \boldsymbol{0} \end{pmatrix}}_{\boldsymbol{g}} = \underbrace{\begin{pmatrix} \boldsymbol{0} \\ \boldsymbol{\tau} \end{pmatrix}}_{\boldsymbol{u}} + \boldsymbol{\tau}_{\mathrm{ext}} \tag{3.20}$$

with the transformed inertia and Coriolis matrix $\boldsymbol{M} = \boldsymbol{T}_{b\mathrm{c}}^T \overline{\boldsymbol{M}} \boldsymbol{T}_{b\mathrm{c}}$ and $\boldsymbol{C} = \boldsymbol{T}_{b\mathrm{c}}^T \overline{\boldsymbol{C}} \boldsymbol{T}_{b\mathrm{c}} + \boldsymbol{T}_{b\mathrm{c}}^T \overline{\boldsymbol{M}} \frac{d}{dt} \boldsymbol{T}_{b\mathrm{c}}$. Note that $\dot{\boldsymbol{M}} - 2\boldsymbol{C}$ is a skew-symmetric matrix (Ott 2008), which is strongly related to the passivity properties of the robot. The inertia matrix $\boldsymbol{M} \in \mathbb{R}^{(6+n)\times(6+n)}$ can be decomposed into $\boldsymbol{M}_{\mathrm{cc}} \in \mathbb{R}^{6\times 6}$, $\boldsymbol{M}_{\mathrm{qq}} \in \mathbb{R}^{n\times n}$, and $\boldsymbol{M}_{\mathrm{cq}} \in \mathbb{R}^{6\times n}$. The vector $\boldsymbol{v} \in \mathbb{R}^{6+n}$ denotes the generalized velocities of the robot combining $\boldsymbol{v}_{\mathrm{c}}$ and $\dot{\boldsymbol{q}}$. The scalar $m \in \mathbb{R}$ denotes the overall mass of the robot, while $\boldsymbol{g}_0 \in \mathbb{R}^6$ represents the vector of gravitational acceleration.[3] The structure of $\boldsymbol{g} = \boldsymbol{T}_{b\mathrm{c}}^T \bar{\boldsymbol{g}}$ and $\boldsymbol{M}$ results from the choice of $\mathcal{C}$ as base frame (Hyon et al. 2007; Ott et al. 2011; Henze et al. 2016b). The vector of generalized external forces $\bar{\boldsymbol{\tau}}_{\mathrm{ext}}$ transforms into $\boldsymbol{\tau}_{\mathrm{ext}} = \boldsymbol{T}_{b\mathrm{c}}^T \bar{\boldsymbol{\tau}}_{\mathrm{ext}}$. In Chap. 7, the vector $\boldsymbol{u} \in \mathbb{R}^{6+n}$ is used as a virtual control input.

Applying the transformation $\boldsymbol{T}_{b\mathrm{c}}$ to (3.11) and (3.12) yields

$$\boldsymbol{v}_{\mathrm{sup}} = \underbrace{\begin{bmatrix} \boldsymbol{Ad}_{\mathrm{sup}} & \boldsymbol{J}'_{\mathrm{sup}} \end{bmatrix}}_{\boldsymbol{J}_{\mathrm{sup}}} \underbrace{\begin{pmatrix} \boldsymbol{v}_{\mathrm{c}} \\ \dot{\boldsymbol{q}} \end{pmatrix}}_{\boldsymbol{v}} \tag{3.21}$$

[3] Note that $\boldsymbol{g}_0 \in \mathbb{R}^6$ is six-dimensional and comprises the translational and rotational DoFs of Cartesian space.

and

$$\boldsymbol{v}_{\text{int}} = \underbrace{\left[\boldsymbol{Ad}_{\text{int}} \; \boldsymbol{J}'_{\text{int}}\right]}_{\boldsymbol{J}_{\text{int}}} \underbrace{\begin{pmatrix} \boldsymbol{v}_c \\ \dot{\boldsymbol{q}} \end{pmatrix}}_{\boldsymbol{\nu}} \tag{3.22}$$

with $\boldsymbol{Ad}_{\text{sup}} = \overline{\boldsymbol{Ad}}_{\text{sup}} \boldsymbol{T}_{bc}^T$ and $\boldsymbol{J}'_{\text{sup}} = \bar{\boldsymbol{J}}'_{\text{sup}} \boldsymbol{T}_{bc}^T$. To simplify the notation even further, $\boldsymbol{v}_{\text{sup}}$ and $\boldsymbol{v}_{\text{int}}$ are combined into

$$\boldsymbol{v} = \begin{pmatrix} \boldsymbol{v}_{\text{sup}} \\ \boldsymbol{v}_{\text{int}} \end{pmatrix} = \underbrace{\left[\boldsymbol{Ad} \; \boldsymbol{J}'\right]}_{\boldsymbol{J}} \underbrace{\begin{pmatrix} \boldsymbol{v}_b \\ \dot{\boldsymbol{q}} \end{pmatrix}}_{\boldsymbol{\nu}} \tag{3.23}$$

leading to $\boldsymbol{Ad} = [\, (\boldsymbol{Ad}_{\text{sup}})^T \; (\boldsymbol{Ad}_{\text{int}})^T \,]^T$ and $\boldsymbol{J}' = [\, (\boldsymbol{J}'_{\text{sup}})^T \; (\boldsymbol{J}'_{\text{int}})^T \,]^T$. The commanded as well as the external task wrenches are analogously stacked into $\boldsymbol{F} = (\, \boldsymbol{F}_{\text{sup}}^T \; \boldsymbol{F}_{\text{int}}^T \,)^T$ and $\boldsymbol{F}^{\text{ext}} = (\, (\boldsymbol{F}_{\text{sup}}^{\text{ext}})^T \; (\boldsymbol{F}_{\text{int}}^{\text{ext}})^T \,)^T$, respectively. If all external forces and torques exclusively act on the task frame $\mathcal{T}_i$, then (3.15) becomes

$$\boldsymbol{\tau}_{\text{ext}} = \boldsymbol{J}^T \boldsymbol{F}^{\text{ext}} = \begin{bmatrix} \boldsymbol{Ad}_{\text{sup}}^T & \boldsymbol{Ad}_{\text{int}}^T \\ (\boldsymbol{J}'_{\text{sup}})^T & (\boldsymbol{J}'_{\text{int}})^T \end{bmatrix} \begin{pmatrix} \boldsymbol{F}_{\text{sup}}^{\text{ext}} \\ \boldsymbol{F}_{\text{int}}^{\text{ext}} \end{pmatrix}. \tag{3.24}$$

3.3 Wrench Distribution Problem

As can be seen in (3.24), the external task wrenches $\boldsymbol{F}_{\text{sup}}^{\text{ext}}$ and $\boldsymbol{F}_{\text{int}}^{\text{ext}}$ are mapped via the transpose of the stacked adjoint matrices $\boldsymbol{Ad}_{\text{sup}}$ and $\boldsymbol{Ad}_{\text{int}}$ onto the CoM wrench. The mapping onto the n joint torques is provided by the transpose of $\boldsymbol{J}'_{\text{sup}}$ and $\boldsymbol{J}'_{\text{int}}$.

The mapping of the external support wrenches via $\boldsymbol{Ad}_{\text{sup}}^T \in \mathbb{R}^{6 \times 6\psi_{\text{sup}}}$ is surjective but not injective for more than one contact (mapping from $6 + \psi_{\text{sup}}$ onto 6 coordinates). As a consequence, there exists a large set of external contact wrenches $\boldsymbol{F}_{\text{sup}}^{\text{ext}}$ that result in the same overall support wrench

$$\boldsymbol{F}_{\text{sup}}^{\text{res,ext}} = \boldsymbol{Ad}_{\text{sup}}^T \boldsymbol{F}_{\text{sup}}^{\text{ext}}. \tag{3.25}$$

Although the above equation is formulated for the external contact wrenches, the same phenomenon exists for the contact wrenches $\boldsymbol{F}_{\text{sup}}$ commanded by the controller

$$\boldsymbol{F}_{\text{sup}}^{\text{res}} = \boldsymbol{Ad}_{\text{sup}}^T \boldsymbol{F}_{\text{sup}}, \tag{3.26}$$

as detailed in Chaps. 6, 7, and 8. In order to generate a desired overall wrench $\boldsymbol{F}_{\text{sup}}^{\text{res}}$ at the CoM for balancing, the controller can choose the commanded contact wrenches $\boldsymbol{F}_{\text{sup}}$ from a set of feasible solutions, which is bounded by the contact model, as detailed below in Sect. 3.4. This challenge is commonly known as the so-called Wrench Distribution Problem (WDP). Note that the overall support wrench $\boldsymbol{F}_{\text{sup}}^{\text{res}} \in \mathbb{R}^6$ acts on the center of mass frame C. Therefore, the adjoint matrices stacked into $\boldsymbol{Ad}_{\text{sup}}^T = \left[\boldsymbol{Ad}_1^T \dots \boldsymbol{Ad}_{\psi_{\text{sup}}}^T \right]$ can be considered as a rigid body transformation shifting each contact wrench to the center of mass. In the literature, several names are used to refer to the mapping $\boldsymbol{Ad}_{\text{sup}}^T$ such as "Contact Map", "Graps Matrix", or "Stance Matrix". The term "Grasp Matrix" originates from the observation that balancing and grasping are fundamentally similar (see Fig. 1.2) both exhibiting the wrench distribution problem (Ott et al. 2011).

As the Contact Map $\boldsymbol{Ad}_{\text{sup}}^T$ is surjective but not injective for more than one contact, the mapping features a null space, which can be used by the robot to redistribute load from one contact to another and/or to generate internal stress between the contacts. Note that in mechanical engineering, the term "internal stress" is directly linked to the deformation of elastic objects (Gross et al. 2018). As the robot model does not feature any elastic elements and the contacts may also be rigid, it is difficult to provide a precise scientific definition of internal stress in this particular context. But in colloquial terms, we will consider internal stress as contact wrenches that do not contribute to the overall wrench at the center of mass. Therefore, our usage of the term "internal stress" is linked to the null space of the Contact Map $\boldsymbol{Ad}_{\text{sup}}^T$.

Generating internal stress appears to be counterproductive at first glance, as it increases the contact wrenches and therefore the joint torques. But the controller can use internal stress to ensure that the contact wrenches satisfy the contact model, which is detailed below in Sect. 3.4. As an example, Fig. 3.3 shows a scenario in which the robot uses internal stress to balance on a ridge. To ensure that the contact wrenches lie within their respective friction cone, the controller pushes the feet against the ridge. As the horizontal components of the contact wrenches cancel out, they can be considered as internal stress. By contrast, the vertical components of $\boldsymbol{F}_i$ add up to the overall wrench $\boldsymbol{F}_{\text{sup}}^{\text{res}}$ on the center of mass.

3.4 Contact Model

One essential problem in humanoid balancing is that the supporting contacts are not fixed to the environment as it is the case for robots with a fixed base. If a humanoid robot applies unsuitable wrenches at the supporting contacts, they can unintentionally lift off, tilt or start to slide. A common way of dealing with this problem is to provide the controller with a model of the contact properties, which restricts the commanded contact wrenches to an admissible set of values. This section presents the contact model which is used by the controllers in Chaps. 6, 7, and 8. Similar contact models have been formulated in literature, as for example in Audren et al. (2014), Lee and Goswami (2010), Kojima et al. (2015), Sentis (2010), Stephens and Atkeson (2010a), Wensing et al. (2013), Wieber (2006).

For convenience of notation, all quantities related to the contact model are expressed in $\mathcal{T}_i$ instead of the world frame $\mathcal{W}$, which will be indicated by the superscript $^i\bullet$.

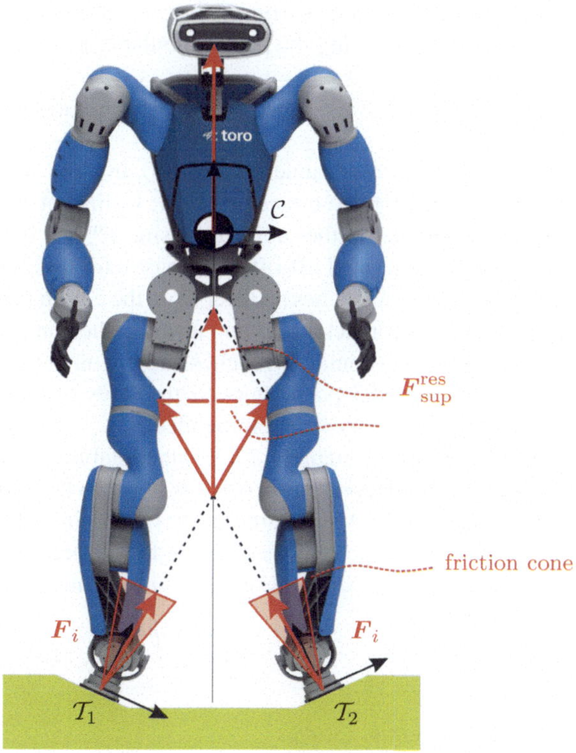

Fig. 3.3 Internal stress while balancing on a ridge

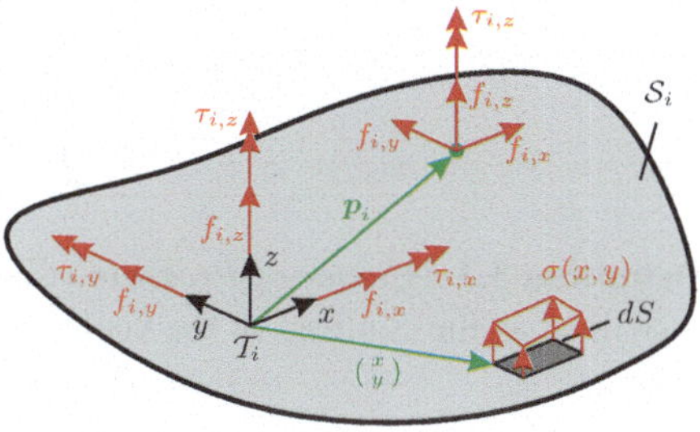

Fig. 3.4 Model of a flat contact

3.4.1 The General Case of a Flat and Unilateral Contact

For simplicity, let us consider a flat contact area $\mathcal{S}_i \subset \mathbb{R}^2$ (which is a subset of two-dimensional Cartesian space), as shown in Fig. 3.4. The contact frame $\mathcal{T}_i$ is defined such that the z-axis is perpendicular to the contact area $\mathcal{S}_i$. Furthermore, let us decompose the commanded contact wrench ${}^i\boldsymbol{F}_i$ into its elements ${}^i\boldsymbol{F}_i = (\, f_x \; f_y \; f_z \; \tau_x \; \tau_y \; \tau_z \,)^T$.

To prevent the contact from lifting off, the unilaterality of the contact must be considered by restricting the perpendicular contact force $f_{i,z}$ via

$$f_{i,z} \geq f_{i,z}^{\min} \quad \text{(unilaterality)} \tag{3.27}$$

to be above a minimum value of $f_{i,z}^{\min} \geq 0$. In most of the literature, $f_{i,z}^{\min}$ is set to zero (e.g. Kao et al. 2016; Audren et al. 2014; Kojima et al. 2015). However, based on the practical experience of the author, choosing $f_{i,z}^{\min} > 0$ leads to an increased robustness against an unintentional lift-off. One reason for this is joint friction, which can cause the actual contact wrenches to differ from the commanded values. To avoid an explicit measurement of contact forces and torques, the controller invokes the contact model on the commanded contact wrenches and not on the actual ones. Therefore, selection of the contact model parameters so that they are slightly more conservative than theoretically necessary is recommended.

Another relevant quantity for contact stability is the so-called Center of Pressure (CoP), which represents a point $\boldsymbol{p}_i \in \mathbb{R}^2$ in the same plane as the contact area $\mathcal{S}_i$ (Popovic et al. 2005; Vukobratović 2004; Sardain and Bessonnet 2004). The pressure inside a contact surface can be described by a vector field $\sigma(x, y) \in \mathbb{R}$ perpendicular to $\mathcal{S}_i$ with $(x, y) \in \mathcal{S}_i$ (see Fig. 3.4). Integrating over the contact area leads to the overall normal force

$$f_{i,z} = \int_{\mathcal{S}_i} \sigma(x, y)\, ds \tag{3.28}$$

and to the tangential torques

$$\begin{aligned} \tau_{i,x} &= \int_{\mathcal{S}_i} y\, \sigma(x, y)\, ds \\ \tau_{i,y} &= \int_{\mathcal{S}_i} -x\, \sigma(x, y)\, ds. \end{aligned} \tag{3.29}$$

Definition 3.3 (*Center of Pressure (CoP)*) The center of pressure ${}^i\boldsymbol{p}_i \in \mathbb{R}^2$ (CoP) is defined as the centroid of the vector field $\sigma(x, y) \in \mathbb{R}$ and is given by

$${}^i\boldsymbol{p}_i = \frac{\int_{\mathcal{S}_i} \sigma(x, y) \begin{pmatrix} x \\ y \end{pmatrix} ds}{\int_{\mathcal{S}_i} \sigma(x, y)\, ds}. \tag{3.30}$$

The center of pressure can also be regarded as the point ${}^i\boldsymbol{p}_i = (\, p_{i,x} \; p_{i,y} \,)^T$ in the same plane as $\mathcal{S}_i$, to which the contact wrench ${}^i\boldsymbol{F}_i$ must be shifted such that the tangential torques $\tau_{i,x}$ and $\tau_{i,y}$ vanish (see Fig. 3.4). Computing the sum of torques about the x- and y-axis of $\mathcal{T}_i$ leads to the definition

$$\begin{aligned} p_{i,x} &= +\tau_{i,y}/f_{i,z} \\ p_{i,y} &= -\tau_{i,x}/f_{i,z} \end{aligned} \tag{3.31}$$

of the CoP as a function of the contact wrench ${}^i\boldsymbol{F}_i$.

By definition, the center of pressure ${}^{i}\boldsymbol{p}_i \in \text{conv}(\mathcal{S}_i)$ cannot leave the convex hull of the contact area $\mathcal{S}_i$. If the CoP reaches the convex hull, the danger exists that the contact might start to tilt (Popovic et al. 2005). To prevent this, the contact wrenches commanded by the controller need to be restricted such that the commanded CoP lies within the convex hull of the contact area:

$$ {}^{i}\boldsymbol{p}_i \in \text{conv}(\mathcal{S}_i) \qquad \text{(center of pressure)}. \tag{3.32}$$

Slipping of the contacts can be avoided by considering a static friction model. For instance, translational friction can be described by the well-known Coulomb model (see e.g. Kao et al. 2016; Prattichizzo and Trinkle 2016), which limits the tangential forces via

$$\sqrt{f_{i,x}^2 + f_{i,y}^2} \leq \mu_{\text{trans},i}\, f_{i,z} \qquad \text{(trans. friction)} \tag{3.33}$$

using the non-negative friction coefficient $\mu_{\text{trans},i} \in \mathbb{R}$. There are several models originating from the field of multi-fingered grasping that account not only for translational but also for rotational friction constraining the perpendicular torque $\tau_{i,z}$ (Kao et al. 2016; Prattichizzo and Trinkle 2016; Howe et al. 1988). But these models are usually not applied to balancing control because they are often based on the assumption of a rather limited contact area such as a fingertip. Therefore, the models are not capable of transmitting tangential torques ($\tau_{i,x}$ and $\tau_{i,y}$) although this is a mandatory feature for considering the center of pressure given Definition 3.3.

3.4.2 Approximation for Rectangular Contacts

The contact model presented in Sect. 3.4.1 is nonlinear, which increases the effort for numerical optimization algorithms to solve the wrench distribution problem. As a consequence, most of the relevant literature approximates the contact area with polygons or rectangles (e.g. Audren et al. 2014; Lee and Goswami 2010; Kojima et al. 2015; Stephens and Atkeson 2010a; Wensing et al. 2013). Although polygons are more general, the assumption of rectangular contact surfaces is sufficient for most humanoid robots.

Let us consider a rectangular contact, as shown in Fig. 3.5, with the x-axis pointing to the front, the y-axis to the left, and the z-axis to the top of $\mathcal{S}_i$. The unilaterality constraint

$$f_{i,z} \geq f_{i,z}^{\min} \qquad \text{(unilaterality)} \tag{3.34}$$

can be adopted from the discussion on general contacts (3.27).

Due to the rectangular shape of $\mathcal{S}_i$, the constraint on the center of pressure (3.32) simplifies to

$$\begin{aligned} p_{i,x}^{\min} \leq \;& p_{i,x} \leq \; p_{i,x}^{\max} \\ p_{i,y}^{\min} \leq \;& p_{i,y} \leq \; p_{i,y}^{\max} \end{aligned} \tag{3.35}$$

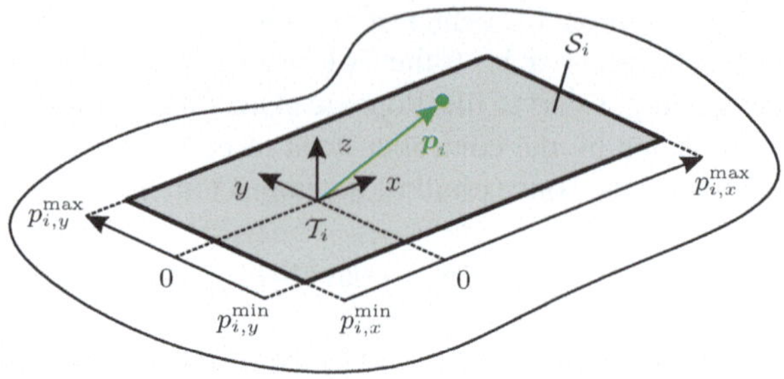

Fig. 3.5 Flat and rectangular contact

with $p_{i,x}^{\min}$, $p_{i,x}^{\max}$, $p_{i,y}^{\min}$, $p_{i,y}^{\max} \in \mathbb{R}$ specifying the physical dimensions. Inserting (3.31) leads to the formulation

$$\begin{aligned} p_{i,x}^{\min} f_{i,z} \leq \quad +\tau_{i,y} \leq \quad p_{i,x}^{\max} f_{i,z} \\ p_{i,y}^{\min} f_{i,z} \leq \quad -\tau_{i,x} \leq \quad p_{i,y}^{\max} f_{i,z} \end{aligned} \qquad \text{(center of pressure)} \qquad (3.36)$$

relating the tangential torques $\tau_{i,x}$ and $\tau_{i,y}$ with $f_{i,z}$.

Many other works (e.g. Audren et al. 2014; Lee and Goswami 2010; Kojima et al. 2015; Stephens and Atkeson 2010a; Wensing et al. 2013) only consider translational friction to avoid the problem of combining translational with rotational friction (see Sect. 3.4.1). To obtain a linear model for describing translational friction, (3.33) is usually approximated by a quadratic pyramid, which is inserted into the friction cone, as shown in Fig. 3.6. Choosing the friction parameter of the pyramid as $\tilde{\mu}_i = (1/\sqrt{2})\, \mu_{\text{trans},i}$ allows considering the x- and y-direction independently of each other (Henze et al. 2016b):

$$\begin{aligned} \left|f_{i,x}\right| \leq \tilde{\mu}_i\, f_{i,z} \\ \left|f_{i,y}\right| \leq \tilde{\mu}_i\, f_{i,z} \end{aligned} \qquad \text{(trans. friction)} \qquad (3.37)$$

This work extends the standard friction model (3.37) by also considering rotational friction. The extended model is derived by parameterizing the contact wrench ${}^{i}\boldsymbol{F}_i$ using the concept of base vectors (see Sect. 3.5 for a comparison of the different methods of parameterization). Despite (3.31), the center of pressure can also be parameterized via a set of normal force vectors $f_{i,z,j}$, which are located at the corners of the contact area, as shown in Fig. 3.7a. Here, the index $j \in \left[1 \ldots n_{i,\text{corners}}\right]$ is used to enumerate the n_{corners} corners of the contact area. To account for unilaterality and to maintain the CoP inside of $\mathcal{S}_i$, the forces are assumed to be non-negative via $f_{i,z,j} \geq 0$. Furthermore, it is assumed that each corner is subject to a friction pyramid limiting the tangential forces $f_{i,x,j}$ and $f_{i,y,j}$ at the corners. Therefore, the bounds on $f_{i,x,j}$ and $f_{i,y,j}$ limit the transmittable contact forces $f_{i,x}$ and $f_{i,y}$ as well as the contact torque $\tau_{i,z}$. Note that the mapping from the corner forces to the contact wrench is surjective but not injective for contacts featuring more than two corners. This means that the forces at the corners cannot be uniquely determined

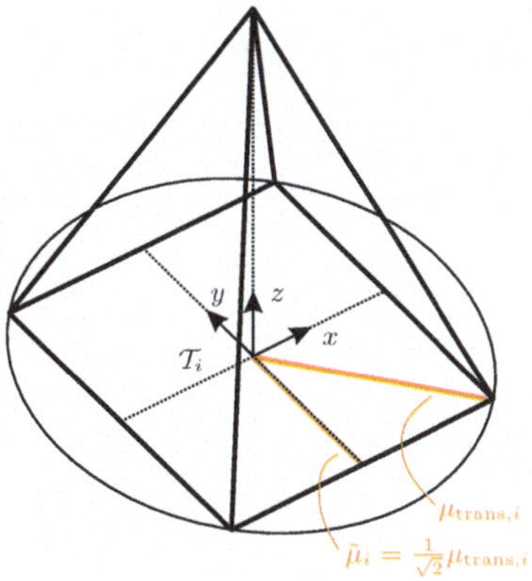

Fig. 3.6 Quadratic pyramid inscribed in the friction cone

from the contact wrench. To resolve the mapping, we will restrict our considerations to contacts whose length is greater than their width. In this case rotational friction will be dominated by the length of the contact, while the width can be neglected. Therefore, the contact is modeled with a friction pyramid located in the front and in the back of the contact as shown in Fig. 3.7b. This assumption is inspired by the geometry of the human foot, which is longer than wide. For instance, the feet of the humanoid robot *TORO* (see Chap. 5 and Fig. 3.7c) have a length of 21cm and a width of 9.5cm and thus fall into this particular category of contacts.

Computing the sum of torques about the front and the back of the contact area (see Fig. 3.7b) leads to

$$\begin{aligned} f_{i,z,\mathrm{front}}\,(p_{i,x}^{\max} - p_{i,x}^{\min}) &= f_{i,z}\,(-p_{i,x}^{\min}) - \tau_{i,y}, \\ f_{i,z,\mathrm{back}}\,(p_{i,x}^{\max} - p_{i,x}^{\min}) &= f_{i,z}\,(p_{i,x}^{\max}) + \tau_{i,y}, \end{aligned} \tag{3.38}$$

and

$$\begin{aligned} f_{i,y,\mathrm{front}}\,(p_{i,x}^{\max} - p_{i,x}^{\min}) &= f_{i,y}\,(-p_{i,x}^{\min}) + \tau_{i,z}, \\ f_{i,y,\mathrm{back}}\,(p_{i,x}^{\max} - p_{i,x}^{\min}) &= f_{i,y}\,(p_{i,x}^{\max}) - \tau_{i,z}. \end{aligned} \tag{3.39}$$

Solving for $f_{i,y,\mathrm{front}}$, $f_{i,z,\mathrm{front}}$, $f_{i,y,\mathrm{back}}$, $f_{i,z,\mathrm{back}}$ and inserting into

$$\begin{aligned} \left|f_{i,y,\mathrm{front}}\right| &\le \tilde{\mu}_i\, f_{i,z,\mathrm{front}} \\ \left|f_{i,y,\mathrm{back}}\right| &\le \tilde{\mu}_i\, f_{i,z,\mathrm{back}} \end{aligned} \tag{3.40}$$

yields

$$\begin{aligned} \left|f_{i,y}\,(-p_{i,x}^{\min}) + \tau_{i,z}\right| &\le \tilde{\mu}_i\, f_{i,z}\,(-p_{i,x}^{\min}) - \tilde{\mu}_i\, \tau_{i,y} \\ \left|f_{i,y}\,(p_{i,x}^{\max}) - \tau_{i,z}\right| &\le \tilde{\mu}_i\, f_{i,z}\,(p_{i,x}^{\max}) + \tilde{\mu}_i\, \tau_{i,y} \end{aligned} \qquad \text{(rot. friction)} \tag{3.41}$$

with $(p_{i,x}^{\max} - p_{i,x}^{\min}) > 0$ describing the length of the contact. Note that adding the two lines from (3.41) and exploiting the triangle inequality leads to the second line

Fig. 3.7 Friction model including rotation and translation

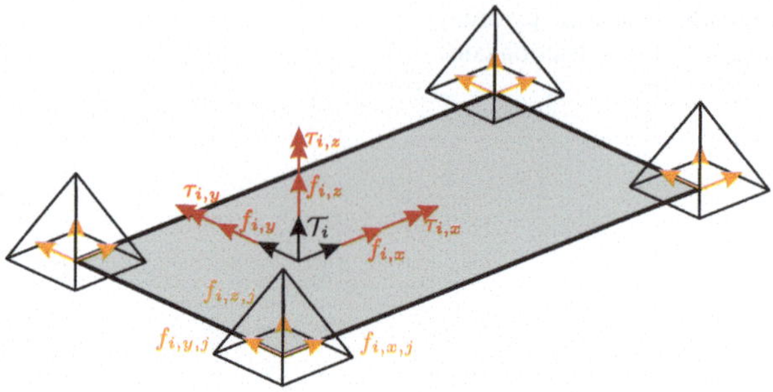

(a) Parameterization using base vectors.

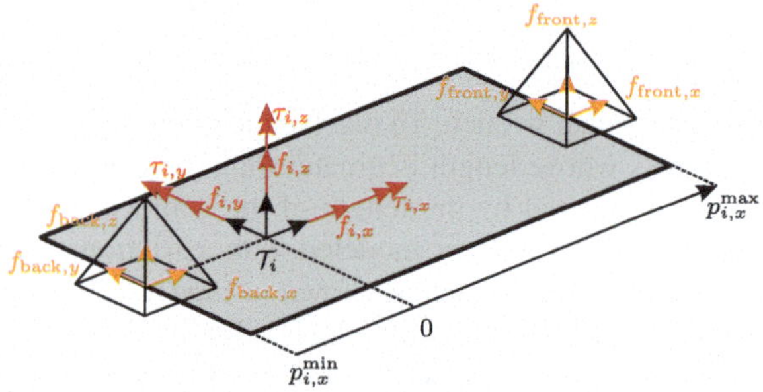

(b) Friction pyramids located at the front and in the back of the contact area $\mathcal{S}_i$.

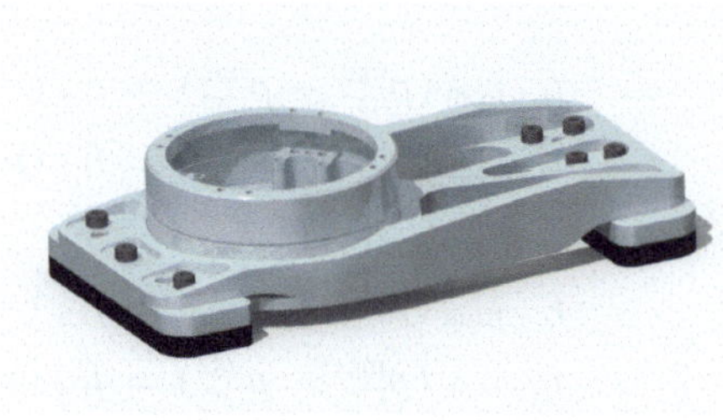

(c) Foot of the humanoid robot *TORO*.

of (3.37). Therefore, the second line of (3.37) may be omitted for the contact model, as it is redundant to (3.41).

In order to rewrite the contact model into a form that can be more easily processed by numerical algorithms, (3.34), (3.36), the first line of (3.37), and (3.41) are stacked into a linear contact model of the form ${}^{i}\boldsymbol{A}_i\,{}^{i}\boldsymbol{F}_i \leq \boldsymbol{b}_i$:

$$\underbrace{\begin{bmatrix} 0 & 0 & -1 & 0 & 0 & 0 \\ 0 & 0 & p_{i,x}^{\min} & 0 & -1 & 0 \\ 0 & 0 & -p_{i,x}^{\max} & 0 & 1 & 0 \\ 0 & 0 & p_{i,y}^{\min} & 0 & 1 & 0 \\ 0 & 0 & -p_{i,y}^{\max} & 0 & -1 & 0 \\ 1 & 0 & -\tilde{\mu}_i & 0 & 0 & 0 \\ -1 & 0 & -\tilde{\mu}_i & 0 & 0 & 0 \\ 0 & -p_{i,x}^{\min} & \tilde{\mu}_i\, p_{i,x}^{\min} & 0 & \tilde{\mu}_i & 1 \\ 0 & p_{i,x}^{\min} & \tilde{\mu}_i\, p_{i,x}^{\min} & 0 & \tilde{\mu}_i & -1 \\ 0 & p_{i,x}^{\max} & -\tilde{\mu}_i\, p_{i,x}^{\max} & 0 & -\tilde{\mu}_i & -1 \\ 0 & -p_{i,x}^{\max} & -\tilde{\mu}_i\, p_{i,x}^{\max} & 0 & -\tilde{\mu}_i & 1 \end{bmatrix}}_{{}^i\boldsymbol{A}_i} \underbrace{\begin{pmatrix} f_{i,x} \\ f_{i,y} \\ f_{i,z} \\ \tau_{i,x} \\ \tau_{i,y} \\ \tau_{i,z} \end{pmatrix}}_{{}^i\boldsymbol{F}_i} \leq \underbrace{\begin{pmatrix} -f_{i,z}^{\min} \\ 0 \\ 0 \\ 0 \\ 0 \\ 0 \\ 0 \\ 0 \\ 0 \\ 0 \\ 0 \end{pmatrix}}_{{}^i\boldsymbol{b}_i} . \tag{3.42}$$

Note that the superscript of ${}^i\boldsymbol{A}_i$ indicates that the corresponding wrench ${}^i\boldsymbol{F}_i$ is expressed in the local contact frame $\mathcal{T}_i$.

3.4.3 Approximation for Point Contacts with Fixed Orientation

A humanoid robot can also use contacts for support in situations where the model of a flat and rectangular contact from Sect. 3.4.2 does not apply. For example, Fig. 3.8a shows the hand of the robot *TORO*. As can be seen, the surface of the palm is rather buckled, which makes it difficult to model. The robot can also use the front edge of the feet to balance while performing a toe-off motion during stair climbing (see Fig. 3.8b). In the latter case, the rectangular contact of the foot degenerates to a line contact, which in theory could still transmit forces and torques in five Degrees of Freedom (DoFs). A toe-off motion can significantly increase the abilities of the robot, as discussed in Sect. 9.1. However, based on the practical experiences of the author, a toe-off motion is usually performed in situations with little perpendicular force $f_{i,z}$. As a result, the emerging line contact is rather unreliable with regard to transmitting forces and torques via the remaining DoFs.

For this reason, it is recommended that these contacts are modeled as a single point, which is located at the origin of the contact frame $\mathcal{T}_i$. The z-axis of the frame is still perpendicular to the contact surface, but the contact is only allowed to transmit forces in the direction of $f_{i,z}$. As $f_{i,x} = f_{i,y} = \tau_{i,x} = \tau_{i,y} = \tau_{i,z} = 0$ are restricted to zero, only unilaterality must be taken into account by the contacts model:

$$f_{i,z} \geq f_{i,z}^{\min} \quad \text{(unilaterality)}. \tag{3.43}$$

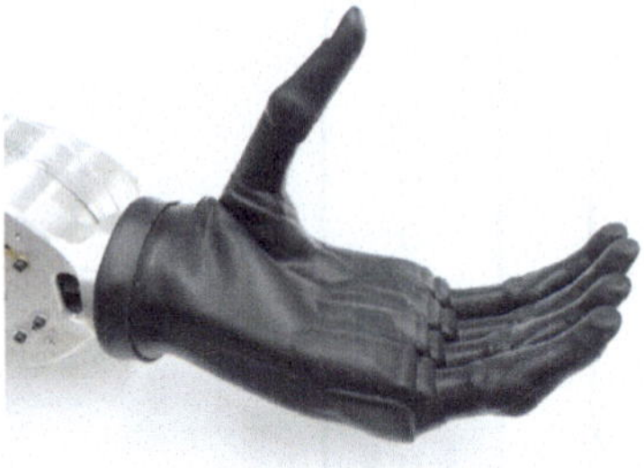

(a) Left hand of the humanoid robot *TORO*.

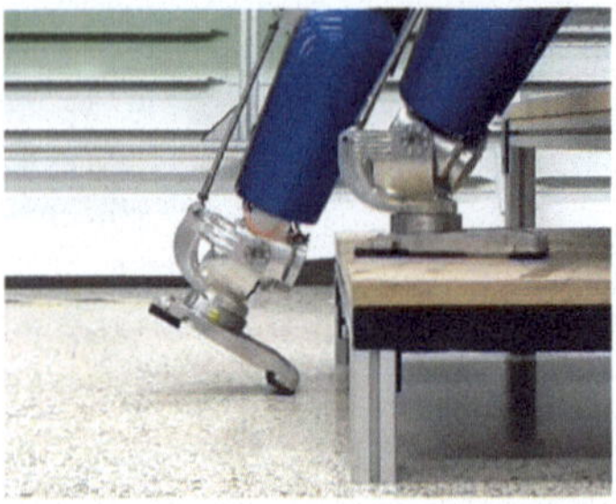

(b) Edge contact of a foot during toe-off motion.

Fig. 3.8 Examples of supporting contacts that are modeled as points

Therefore, the contact model can be rewritten as

$$\underbrace{\begin{bmatrix} 0 & 0 & -1 & 0 & 0 & 0 \\ 1 & 0 & 0 & 0 & 0 & 0 \\ -1 & 0 & 0 & 0 & 0 & 0 \\ 0 & 1 & 0 & 0 & 0 & 0 \\ 0 & -1 & 0 & 0 & 0 & 0 \\ 0 & 0 & 0 & 1 & 0 & 0 \\ 0 & 0 & 0 & -1 & 0 & 0 \\ 0 & 0 & 0 & 0 & 1 & 0 \\ 0 & 0 & 0 & 0 & -1 & 0 \\ 0 & 0 & 0 & 0 & 0 & 1 \\ 0 & 0 & 0 & 0 & 0 & -1 \end{bmatrix}}_{{}^i\boldsymbol{A}_i} \underbrace{\begin{pmatrix} f_{i,x} \\ f_{i,y} \\ f_{i,z} \\ \tau_{i,x} \\ \tau_{i,y} \\ \tau_{i,z} \end{pmatrix}}_{{}^i\boldsymbol{F}_i} \leq \underbrace{\begin{pmatrix} -f_{i,z}^{\min} \\ 0 \\ 0 \\ 0 \\ 0 \\ 0 \\ 0 \\ 0 \\ 0 \\ 0 \\ 0 \end{pmatrix}}_{{}^i\boldsymbol{b}_i} \tag{3.44}$$

again leading to a linear inequality of the form ${}^i\boldsymbol{A}_i\,{}^i\boldsymbol{F}_i \leq \boldsymbol{b}_i$.

3.5 Overview of the Parameterizations of the Wrench Distribution Problem

A review of the relevant literature reveals that there are several methods for parameterizing the wrench distribution problem. A comparison of the most common techniques is given in Table 3.1.

The straightforward way to parameterize the wrench distribution problem is to use the contact wrenches as decision variable in the optimization problem, as in Audren et al. (2014), Henze et al. (2016b), Kojima et al. (2015), Stephens and Atkeson (2010b). One advantage of this method is that the contact wrenches represent a minimal set of coordinates for describing the distribution of wrenches between the

Table 3.1 Methods for parameterizing the wrench distribution problem

Parameterization	Examples	Properties
Contact wrenches	(Audren et al. 2014) (Henze et al. 2016b) (Kojima et al. 2015) (Ott et al. 2011) (Stephens and Atkeson 2010b)	⊕ minimal representation ⊖ indirect param. of int. stress
Base vectors	(Hyon and Cheng 2006) (Lee and Goswami 2010)	⊖ redundant representation ⊖ indirect param. of int. stress ⊕ easy param. of constraints
Virtual linkage	(Williams and Khatib 1993) (Sentis 2010)	⊕ minimal representation ⊕ direct param. of int. stress ⊖ limited scalability

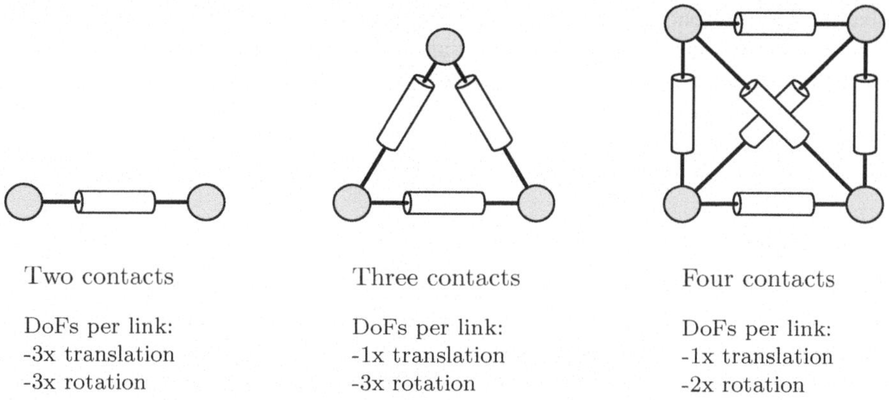

Fig. 3.9 Parameterization of the wrench distribution based on the concept of virtual linkage

contacts. However, the question which portion of the contact wrenches contributes to the overall wrench at the center of mass and which portion can be considered as internal stress can only be answered after analyzing the contact map $\boldsymbol{Ad}_{\text{sup}}^T$ from (3.24).

Another common method is to use base vectors to parameterize the wrench distribution problem (Hyon and Cheng 2006; Lee and Goswami 2010). Hyon and Cheng (2006) define a perpendicular force $f_{i,z,j}$ at each corner of the contact similar to Fig. 3.7a. The advantage of this formulation is that the constraints for unilaterality of the contact (3.34) and the center of pressure (3.36) can be easily ensured by requiring that $f_{i,z,j} \geq 0 \, \forall j = 1 \ldots n_{i,\text{corners}}$ (in the case of $f_{i,z}^{\text{min}} = 0$). However, the base vectors lead to a redundant or non-minimal set of coordinates for describing the wrench distribution. As before, internal stress can only be indirectly parameterized.

The concept of virtual linkage was introduced by Williams and Khatib (1993) for multi-fingered grasping and manipulation of objects. In Sentis (2010), it is used

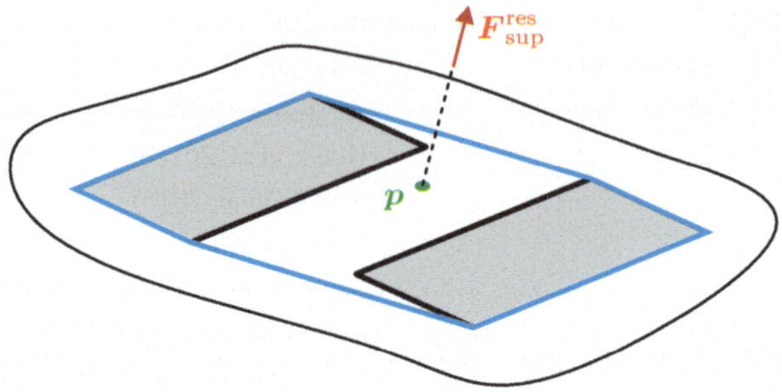

Fig. 3.10 Support polygon (blue) on flat and horizontal ground floor. The combined CoP $\boldsymbol{p}$ or ZMP is given in green

for balance control of humanoid robots. The concept proposes a net of virtual links that interconnects the contacts with each other, as shown in Fig. 3.9. The number of contacts determines the number of DoFs per link. For example in the case of three contacts, each virtual link provides one translational and three rotational DoFs. The advantage of the concept is that the forces and torques of these virtual joints directly relate to the internal stress, which allows an intuitive parameterization of the latter. Therefore, the contact wrenches can be parameterized via

$$\boldsymbol{F}_{\text{sup}} = \left[(\boldsymbol{Ad}_{\text{sup}}^T)^{\#}\ \boldsymbol{E}\right] \begin{pmatrix} \boldsymbol{F}_{\text{sup}}^{\text{res}} \\ \boldsymbol{\tau}_{\text{sup}}^{\text{internal}} \end{pmatrix} \tag{3.45}$$

as a function of the overall resulting wrench $\boldsymbol{F}_{\text{sup}}^{\text{resulting}} = \boldsymbol{Ad}_{\text{sup}}^T \boldsymbol{F}_{\text{sup}}$ and the generalized forces of the virtual links $\boldsymbol{\tau}_{\text{sup}}^{\text{internal}} \in \mathbb{R}^{6(\psi_{\text{sup}}-1)}$. Here, $(\boldsymbol{Ad}_{\text{sup}}^T)^{\#}$ denotes a pseudo-inverse of $\boldsymbol{Ad}_{\text{sup}}^T$ and $\boldsymbol{E} \in \mathbb{R}^{6\psi_{\text{sup}} \times 6(\psi_{\text{sup}}-1)}$ a mapping for the internal stress.[4] The disadvantage of the concept is that every contact change requires a modification of the connections inside the net (see Fig. 3.9), which makes it cumbersome to implement. Besides this, the complexity of the net grows with the number of contacts. Therefore, the advantage of an intuitive parameterization of internal stress only holds for a small number of contacts.

Considering the properties of the methods presented in Table 3.1, this work uses contact wrenches to parameterize the wrench distribution problem because it features a minimal representation and allows a simple implementation of the optimization problem.

3.6 Feasibility of the Overall Support Wrench

The forces and torques that each contact is able to transmit are limited by the contact model, as detailed in Sect. 3.4. As a consequence, the overall wrench that

[4] Note that in Williams and Khatib (1993), $\boldsymbol{E}$ is used to map only internal forces to the contacts, while the mapping including internal torques is more complicated. For simplicity of notation, $\boldsymbol{E}$ will be used for describing the mapping of both internal forces and torques.

the contacts can generate is limited and also poses an important quantity for planing and control. For balancing on flat and horizontal ground floors, the concept of the support polygon and the ZMP is introduced in Vukobratović (2004):

Definition 3.4 (*Zero Moment Point (ZMP)* (Popovic et al. 2005)) For balancing on flat and horizontal floor, the Zero Moment Point (ZMP) is defined as the point on the surface of the ground to which the overall contact wrench $\boldsymbol{F}_{\text{sup}}^{\text{res}}$ must be shifted such that the horizontal torques vanish. However, a torque about the vertical axis can still exist, which is why Sardain and Bessonnet (2004) introduced the more precise term "Zero tipping Moment Point".

Definition 3.5 (*Support Polygon (SP)*) Let $\mathcal{S}_{\text{SP}}$ denote the set of CoM locations $\boldsymbol{x}_{\text{c}}$ for which the given contact configuration allows static balancing without violating the contact model. The Static Support Area or Support Polygon (SP) is the projection of $\mathcal{S}_{\text{SP}}$ onto the horizontal plane of the world frame $\mathcal{W}$.

Popovic et al. (2005) showed that the ZMP coincides with the combined CoP for horizontal and flat ground surfaces. The combined CoP can be obtained analogously to (3.30) by computing the integral over all contact surfaces $\mathcal{S} = \mathcal{S}_1 \cup \ldots \cup \mathcal{S}_{\psi_{\text{sup}}}$. On a flat and horizontal floor, the support polygon is given by the convex hull of the contact areas $\mathcal{S}_i$ (for $f_{i,z}^{\min} = 0$), as shown in Fig. 3.10. The overall support wrench $\boldsymbol{F}_{\text{sup}}^{\text{res}}$ is said to be feasible[5] if the corresponding ZMP/ CoP is located inside the support polygon and the following assumptions are fulfilled:

- The horizontal forces within $\boldsymbol{F}_{\text{sup}}^{\text{res}}$ can be generated with the given contact configuration and contact model.
- The magnitude of the force within $\boldsymbol{F}_{\text{sup}}^{\text{res}}$ can be generated with the given contact configuration and contact model.
- The vertical torque within $\boldsymbol{F}_{\text{sup}}^{\text{res}}$ can be generated with the given contact configuration and contact model.

A more accurate criterion originates from the field of grasping (Hirai 1991; Borst et al. 2003) using polyhedral convex cones for describing the contact model. The concept was then transfered by Caron et al. (2015) to humanoid balancing. Note that the criterion addresses the general case of multiple contact surfaces with arbitrary locations and orientations.

According to Ziegler (2007), a polyhedron $\mathcal{P}$ in $\mathbb{R}^d$ can be defined as the intersection of $m \in \mathbb{N}$ closed half spaces. Each half space is characterized by an inequality $\boldsymbol{a}_{\mathcal{H},j}^T \boldsymbol{\lambda} \leq b_{\mathcal{H},j}$ with $\boldsymbol{\lambda}, \boldsymbol{a}_{\mathcal{H},j} \in \mathbb{R}^d$ and $b_{\mathcal{H},j} \in \mathbb{R}$ for $i = j \ldots m$. Note that each inequality represents a face of the polyhedron $\mathcal{P}$. Combining $\boldsymbol{a}_{\mathcal{H},j}$ and $b_{\mathcal{H},j}$ into the matrix $\boldsymbol{A}_{\mathcal{H}} \in \mathbb{R}^{m \times d}$ and the vector $\boldsymbol{b}_{\mathcal{H}} \in \mathbb{R}^m$ leads to the following definition:

Definition 3.6 (*$\mathcal{H}$-representation of a polyhedron* (Ziegler 2007)) A Polyhedron $\mathcal{P} \subseteq \mathbb{R}^d$ can be denoted as

---

[5] The term "stability" is often used to describe the feasibility of the contact wrenches. To avoid confusions, this work will use the term "stability" only in the sense of Lyapunov stability (Slotine and Lee 1991).

$$\mathcal{P}(\boldsymbol{A}_{\mathcal{H}}, \boldsymbol{b}_{\mathcal{H}}) = \{\boldsymbol{\lambda} : \boldsymbol{A}_{\mathcal{H}}\boldsymbol{\lambda} \leq \boldsymbol{b}_{\mathcal{H}}\} \tag{3.46}$$

for some $\boldsymbol{A}_{\mathcal{H}} \in \mathbb{R}^{m \times d}$ and $\boldsymbol{b}_{\mathcal{H}} \in \mathbb{R}^{m}$, which is called $\mathcal{H}$-representation.

Despite the $\mathcal{H}$-representation, a polyhedron can also be written as a combination of the convex hull of a finite point set $\boldsymbol{V}_{\text{finite}} \in \mathbb{R}^{d \times m'}$ and the cone generated by a finite set of vectors $\boldsymbol{V}_{\text{cone}} \in \mathbb{R}^{d \times m''}$:

Definition 3.7 (*$\mathcal{V}$-representation of a polyhedron* (Ziegler 2007)) A Polyhedron $\mathcal{P} \subseteq \mathbb{R}^d$ can be denoted as

$$\mathcal{P}(\mathcal{V}_{\text{finite}}, \mathcal{V}_{\text{cone}}) = \text{conv}(\mathcal{V}_{\text{finite}}) \oplus \text{cone}(\mathcal{V}_{\text{cone}}) \tag{3.47}$$

for some $\mathcal{V}_{\text{finite}} \in \mathbb{R}^{d \times m'}$ and $\mathcal{V}_{\text{cone}} \in \mathbb{R}^{d \times m''}$, which is called $\mathcal{V}$-representation.

The operator $\oplus$ denotes the so-called "Minkowsi-" or "vector sum", which allows the summation of two vector sets $\mathcal{V}_1, \mathcal{V}_2 \subseteq \mathbb{R}^d$ according to:

Definition 3.8 (*Minkowski-Sum* (Ziegler 2007)) The Minkowski-sum of two set of vectors $\mathcal{V}_1, \mathcal{V}_2 \subseteq \mathbb{R}^d$ is defined as

$$\mathcal{V}_1 \oplus \mathcal{V}_2 = \{\boldsymbol{\lambda}_1 + \boldsymbol{\lambda}_2 : \boldsymbol{\lambda}_1 \in \mathcal{V}_1, \boldsymbol{\lambda}_2 \in \mathcal{V}_2\}. \tag{3.48}$$

A powerful property of polyhedra is the Minkowski-Weyl-Theorem, which states that the $\mathcal{H}$-and $\mathcal{V}$-representation describe the same polyhedron:

Theorem 3.1 (Minkowski-Weyl-Duality (Ziegler 2007)) *A subset $\mathcal{P} \subseteq \mathbb{R}^d$ is a sum of a convex hull of a finite set of points plus a conical combination of vectors ($\mathcal{V}$-polyhedron)*

$$\mathcal{P}(\mathcal{V}_{\textit{finite}}, \mathcal{V}_{\textit{cone}}) = \text{conv}(\mathcal{V}_{\textit{finite}}) \oplus \text{cone}(\mathcal{V}_{\textit{cone}}) \tag{3.49}$$

for some $\mathcal{V}_{\textit{finite}} \in \mathbb{R}^{d \times m'}$ and $\mathcal{V}_{\textit{cone}} \in \mathbb{R}^{d \times m''}$ if and only if it is an intersection of closed halfspaces ($\mathcal{H}$-polyhedron)

$$\mathcal{P}(\boldsymbol{A}_{\mathcal{H}}, \boldsymbol{b}_{\mathcal{H}}) = \{\boldsymbol{\lambda} : \boldsymbol{A}_{\mathcal{H}}\boldsymbol{\lambda} \leq \boldsymbol{b}_{\mathcal{H}}\} \tag{3.50}$$

for some $\boldsymbol{A}_{\mathcal{H}} \in \mathbb{R}^{m \times d}$ and $\boldsymbol{b}_{\mathcal{H}} \in \mathbb{R}^{m}$.

With this knowledge on polyhedral geometry, the model of each contact (given by (3.42) or (3.44)) can be written as a polyhedron $\mathcal{P}_i({}^i\boldsymbol{A}_i, \boldsymbol{b}_i) = \{{}^i\boldsymbol{F}_i : {}^i\boldsymbol{A}_i{}^i\boldsymbol{F}_i \leq \boldsymbol{b}_i\}$ $\forall i = 1 \ldots \psi_{\text{sup}}$. The procedure for computing the overall wrench from the individual polyhedron $\mathcal{P}_i$ is schematically depicted in Fig. 3.11. The first step brings all contact models from their respective contact frames $\mathcal{T}_i$ to a common static frame as the world frame $\mathcal{W}$, for instance. This can be achieved by shifting each contact wrench ${}^i\boldsymbol{F}_i$ from $\mathcal{T}_i$ to $\mathcal{W}$ (see Sect. 3.1), which corresponds to multiplying ${}^i\boldsymbol{A}_i$ with $\boldsymbol{Ad}(\boldsymbol{I}, \boldsymbol{x}_i)^T$ from the right. The second step exploits the Minkowsi-Wyel-Duality (Theorem 3.1) to convert all contact models from the $\mathcal{H}$- to the $\mathcal{V}$-representation. The latter permits combination of the individual polyhedra $\mathcal{P}(\mathcal{V}_{i,\text{finite}}, \mathcal{V}_{i,\text{cone}})$ into

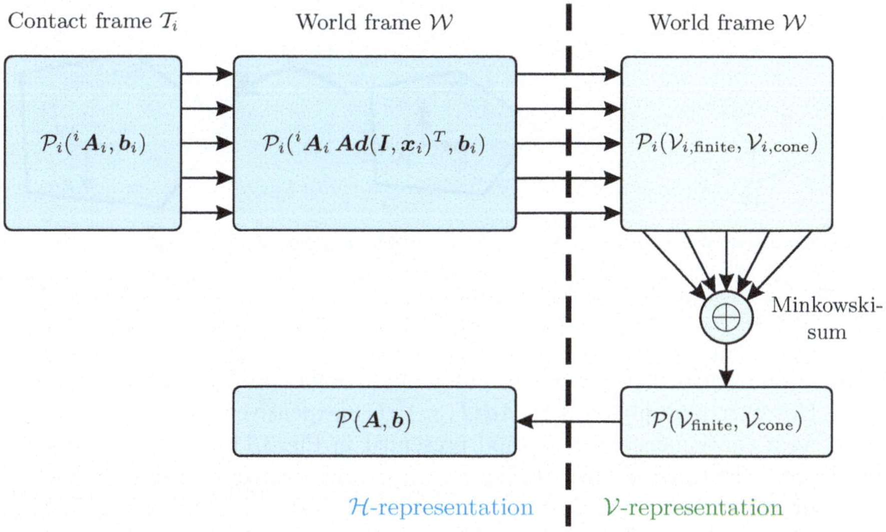

Fig. 3.11 Procedure for obtaining the set $\mathcal{P}$ of feasible vectors for the resulting support wrench $\boldsymbol{F}_{\text{sup}}^{\text{res}}$ from the contact models $\mathcal{P}_i$

one single polyhedron $\mathcal{P}(\mathcal{V}_{\text{finite}}, \mathcal{V}_{\text{cone}})$ via the Minkowsi-Sum (Definition 3.8). Note that $\mathcal{P}(\mathcal{V}_{\text{finite}}, \mathcal{V}_{\text{cone}})$ specifies the set of overall wrenches that are producible with the given contact configuration and contact model. To facilitate a simple and fast evaluation, the resulting polyhedron is converted back to the $\mathcal{H}$-representation $\mathcal{P}(\boldsymbol{A}, \boldsymbol{b})$.

A brute-force approach for testing if a desired overall wrench at the center of mass is producible with the supporting contacts is to try to solve the wrench distribution problem by the means detailed in Chaps. 6, 7, and 8. If the numerical optimization is able to find a solution, the wrench is producible. However, with this method it is only possible to check if one particular wrench is feasible. It does not provide the complete set of producible overall wrenches as the method described in Fig. 3.11 does. The latter method requires some numerical effort to convert the polyhedra from the $\mathcal{V}$- into the $\mathcal{H}$-representation and vice versa. For example, this can be done using the *cdd*-library by Fukuda and Prodon (1996), which is also available as part of the *Multi-Parametric Toolbox 3.0* (MPT3) for *Matlab/Simulink* (Herceg et al. 2013). Nevertheless, the numerical effort needs only to be invested once as long as all the involved frames (contact frames $\mathcal{T}_i$ and world frame $\mathcal{W}$) are stationary. This allows an offline computation of the resulting polyhedron $\mathcal{P}(\boldsymbol{A}, \boldsymbol{b})$ as soon as the contact configuration is known. Note that the obtained set of producible overall wrenches $\mathcal{P}(\boldsymbol{A}, \boldsymbol{b})$ is specified in the world frame. To check if a desired wrench given at the CoM is feasible, it must be mapped to $\mathcal{W}$ via the transposed inverse adjoint $\boldsymbol{Ad}(\boldsymbol{I}, \boldsymbol{x}_{\text{c}})^{-T}$, which yields

$$\boldsymbol{A}\ \boldsymbol{Ad}(\boldsymbol{I}, \boldsymbol{x}_{\text{c}})^{-T}\, \boldsymbol{F}_{\text{sup}}^{\text{res}} \leq \boldsymbol{b}. \tag{3.51}$$

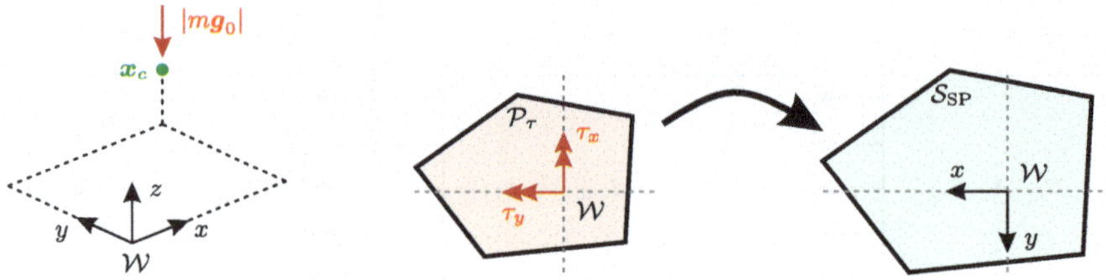

(a) Static balance of the CoM. (b) Converting $\mathcal{P}_\mathcal{T}$ into $\mathcal{S}_{\text{SP}}$.

Fig. 3.12 Computation of the Support Polygon

As the adjoint matrix depends on the configuration of the robot, it must be computed online. However, multiplying $\boldsymbol{A}$ by $\boldsymbol{Ad}(\boldsymbol{I}, \boldsymbol{x}_\text{c})^T$ is inexpensive in terms of numerical effort. As a consequence, the method presented in Fig. 3.11 is very efficient for performing a large number of checks as is for instance required for planning or model predictive control. In this work, the method is used to enable the humanoid robot *TORO* to lift and push heavy loads, as detailed in Sect. 6.5. Recently, Orsolino et al. (2018) have proposed an extension, which also includes torque limitations of the actuators. For the remainder of this work, we make the assumption that the desired overall wrench at the center of mass, which is required by the controllers presented in Chaps. 6, 7 and 8, can always be generated given the contact model (3.42):

Assumption 3.1 (*Feasibility of the desired overall support wrench*) The desired overall support wrench $\boldsymbol{F}_\text{sup}^\text{res}$ is always feasible with respect to the given contact configuration and contact model.

Furthermore, we make the assumption that the contact model is accurate enough to ensure that the contacts do not lift off, tilt, or slide unintentionally:

Assumption 3.2 (*Rigid contacts*) The contact frames $\mathcal{T}_i$ do not move relative to the contact surfaces $\mathcal{S}_i$ attached to the environment.

The procedure shown in Fig. 3.11 also enables computation of a static support polygon from $\mathcal{P}(\boldsymbol{A}, \boldsymbol{b})$ for multiple contact surfaces with arbitrary locations and orientations, as proposed in Caron and Nakamura (2015). Let us consider the case in which the x- and y-axis of the world frame $\mathcal{W}$ span the horizontal plane, as shown in Fig. 3.12a. According to Definition 3.5, the support polygon comprises all CoM locations $\boldsymbol{x}_\text{c} = (\, x_{\text{c},x} \; x_{\text{c},y} \; x_{\text{c},z} \,)^T$, projected onto the x-y-plane of $\mathcal{W}$, that enable static balancing without violating $\mathcal{P}(\boldsymbol{A}, \boldsymbol{b})$. The wrench that needs to be applied at $\mathcal{W}$ in order to compensate for the gravity force of the CoM is given by $\boldsymbol{F} = (\, 0 \; 0 \; m\|\boldsymbol{g}_0\| \; \tau_x \; \tau_y \; 0 \,)$. The torques

$$
\begin{aligned}
\tau_x &= -m \left\|\boldsymbol{g}_0\right\| x_{\text{c},y} \\
\tau_y &= +m \left\|\boldsymbol{g}_0\right\| x_{\text{c},x}
\end{aligned}
\qquad (3.52)
$$

arise due to the lever arms $x_{\text{c},x}$ and $x_{\text{c},y}$, respectively. As the torques τ_x and τ_y are limited by $\mathcal{P}(\boldsymbol{A}, \boldsymbol{b})$, the intersection

$$\mathcal{P}_\tau(\boldsymbol{A}_\tau, \boldsymbol{b}_\tau) = \mathcal{P}(\boldsymbol{A}, \boldsymbol{b}) \cup \left\{ \boldsymbol{F} : f_x = f_y = \tau_z = 0, f_z = m \left\| \boldsymbol{g}_0 \right\| \right\} \tag{3.53}$$

is computed. The resulting polyhedron $\mathcal{P}_\tau(\boldsymbol{A}_\tau, \boldsymbol{b}_\tau)$ is two-dimensional and specifies the feasible torques about the x- and y-axis of $\mathcal{W}$ in the form of

$$\boldsymbol{A}_\tau \begin{pmatrix} \tau_x \\ \tau_y \end{pmatrix} \leq \boldsymbol{b}_\tau . \tag{3.54}$$

Inserting (3.52) allows reformulating of $\mathcal{P}_\tau$ as a function of the CoM locations $x_{c,x}$ and $x_{c,y}$, as schematically depicted in Fig. 3.12b:

$$m \left\| \boldsymbol{g}_0 \right\| \boldsymbol{A}_\tau \begin{pmatrix} -x_{c,y} \\ +x_{c,x} \end{pmatrix} \leq \boldsymbol{b}_\tau . \tag{3.55}$$

Consequently, the static support polygon is given by

$$\mathcal{S}_{\mathrm{SP}} = \left\{ \begin{pmatrix} x \\ y \end{pmatrix} : m \left\| \boldsymbol{g}_0 \right\| \boldsymbol{A}_\tau \begin{pmatrix} -x_{c,y} \\ +x_{c,x} \end{pmatrix} \leq \boldsymbol{b}_\tau \right\} \tag{3.56}$$

limiting the projected CoM location $\boldsymbol{x}_c$ on the horizontal plane of $\mathcal{W}$.

References

Audren, H., Vaillant, J., Kheddar, A., Escande, A., Kaneko, K., Yoshida, E.: Model preview control in multi-contact motion-application to a humanoid robot. In: IEEE/RSJ International Conference on Intelligent Robots and Systems, pp. 4030–4035 (2014)

Borst, C., Fischer, M., Hirzinger, G.: Grasping the dice by dicing the grasp. In: IEEE/RSJ International Conference on Intelligent Robots and Systems, pp. 3692–3697 (2003)

Caron, S., Nakamura, Y.: Planning with the center-of-mass rather than stances for humanoids walking on uneven terrains. In: World Congress in Mechanism and Machine Science, pp. 554–558 (2015)

Caron, S., Pham, Q.C., Nakamura, Y.: Leveraging cone double description for multi-contact stability of humanoids with applications to statics and dynamics. In: Robotics: Science and Systems, pp. 28–36 (2015)

Fukuda, K., Prodon, A.: Combinatorics and Computer Science, Lecture Notes in Computer Science, vol. 1120. Springer, Berlin, Heidelberg, chap Double description method revisited, pp. 91–111 (1996)

Gross, D., Hauger, W., Schröder, J., Wall, W.A., Bonet, J.: Engineering Mechanics 2, 2nd edn. Springer, Berlin, Heidelberg (2018)

Henze, B., Dietrich, A., Ott, C.: An approach to combine balancing with hierarchical whole-body control for legged humanoid robots. IEEE Robot. Autom. Lett. **1**(2), 700–707 (2016a)

Henze, B., Dietrich, A., Roa, M.A., Ott, C.: Multi-contact balancing of humanoid robots in confined spaces: utilizing knee contacts. In: IEEE/RSJ International Conference on Intelligent Robots and Systems, pp. 679–704 (2017)

Henze, B., Roa, M.A., Ott, C.: Passivity-based whole-body balancing for torque-controlled humanoid robots in multi-contact scenarios. Int. J. Robot. Res. **35**(12), 1522–1543 (2016b)

Herceg, M., Kvasnica, M., Jones, C.N., Morari, M.: Multi-parametric toolbox 3.0. In: European Control Conference, pp. 502–510, http://control.ee.ethz.ch/~mpt, accessed September 13th, 2019 (2013)

Herzog, A., Rotella, N., Mason, S., Grimminger, F., Schaal, S., Righetti, L.: Momentum control with hierarchical inverse dynamics on a torque-controlled humanoid. Auton. Robots **40**(3), 473–491 (2016)

Hirai, S.: Analysis and planning of manipulation using the theory of polyhedral convex cones. Ph.D. thesis, Kyoto University (1991)

Howe, R.D., Kao, I., Cutkosky, M.R.: The sliding of robot fingers under combined torsion and shear loading. In: IEEE International Conference on Robotics and Automation, pp. 103–105 (1988)

Hyon, S.H., Cheng, G.: Passivity-based full-body force control for humanoids and application to dynamic balancing and locomotion. In: IEEE/RSJ International Conference on Intelligent Robots and Systems, pp. 4915–4922 (2006)

Hyon, S.H., Hale, J.G., Cheng, G.: Full-body compliant human-humanoid interaction: balancing in the presence of unknown external forces. IEEE Trans. Robot. **23**(5), 884–898 (2007)

Kao, I., Lynch, K.M., Burdick, J.W.: Springer Handbook of Robotics, 2nd edn. Springer, Berlin, Heidelberg, chap Contact Modeling and Manipulation, pp. 931–954 (2016)

Kojima, K., Nozawa, S., Okada, K., Inaba, M.: Shuffle motion for humanoid robot by sole load distribution and foot force control. In: IEEE/RSJ International Conference on Intelligent Robots and Systems, pp. 2187–2194 (2015)

Lee, S.H., Goswami, A.: Ground reaction force control at each foot: a momentum-based humanoid balance controller for non-level and non-stationary ground. In: IEEE/RSJ International Conference on Intelligent Robots and Systems, pp. 3157–3162 (2010)

Orsolino, R., Focchi, M., Mastalli, C., Dai, H., Caldwell, D.G., Semini, C.: Application of wrench based feasibility analysis to the online trajectory optimization of legged robots. In: IEEE/RSJ International Conference on Intelligent Robots and Systems, pp. 3363–3370 (2018)

Ott, C., Roa, M.A., Hirzinger, G.: Posture and balance control for biped robots based on contact force optimization. In: IEEE-RAS International Conference on Humanoid Robots, pp. 26–33 (2011)

Ott, C.: Cartesian Impedance Control of Redundant and Flexible-Joint Robots, Springer Tracts in Advanced Robotics, vol. 49, 1st edn. Springer, Berlin, Heidelberg (2008)

Popovic, M.B., Goswami, A., Herr, H.: Ground reference points in legged locomotion: definitions, biological trajectories and control implications. Int. J. Robot. Res. **24**(12), 1013–1032 (2005)

Prattichizzo, D., Trinkle, J.C.: Springer Handbook of Robotics, 2nd edn. Springer, Berlin, Heidelberg, chap Grasping, pp. 955–988 (2016)

Righetti, L., Buchli, J., Mistry, M., Kalakrishnan, M., Schaal, S.: Optimal distribution of contact forces with inverse dynamics control. Int. J. Robot. Res. **32**(3), 280–298 (2013)

Sardain, P., Bessonnet, G.: Forces acting on a biped robot. Center of Pressure - Zero Moment Point. IEEE Trans. Syst. Man Cybern. **34**(5), 630–637 (2004)

Sentis, L.: Motion Planning for Humanoid Robots. Springer, London, chap Compliant Control of Whole-body Multi-contact Behaviors in Humanoid Robots, pp. 29–66 (2010)

Slotine, J.J.E., Lee, W.: Applied Nonlinear Control. Prentice Hall International, Upper Saddle River (1991)

Stephens, B.J., Atkeson, C.G.: Dynamic balance force control for compliant humanoid robots. In: IEEE/RSJ International Conference on Intelligent Robots and Systems, pp. 1248–1255 (2010a)

Stephens, B.J., Atkeson, C.G.: Push recovery by stepping for humanoid robots with force controlled joints. In: IEEE-RAS International Conference on Humanoid Robots, pp. 52–59 (2010b)

Vukobratović, M.: Zero-moment point - thirty five years of its life. Int. J. Humanoid Robot. **1**(1), 157–173 (2004)

Wensing, P.M., Hammam, G.B., Dariush, B., Orin, D.E.: Optimizing foot centers of pressure through force distribution in a humanoid robot. Int. J. Humanoid Robot. **10**(03), 1–21 (2013)

Werner, A., Lampariello, R., Ott, C.: Optimization-based generation and experimental validation of optimal walking trajectories for biped robots. In: IEEE/RSJ International Conference on Intelligent Robots and Systems, pp. 4373–4379 (2012)
Wieber, P.B.: Trajectory free linear model predictive control for stable walking in the presence of strong perturbations. In: IEEE-RAS International Conference on Humanoid Robots, pp. 137–142 (2006)
Williams, D., Khatib, O.: The virtual linkage: a model for internal forces in multi-grasp manipulation. In: IEEE International Conference on Robotics and Automation, pp. 1025–1030 (1993)
Ziegler, G.M.: Lectures on Polytopes, Graduate Texts in Mathematics, vol. 152, 7th edn. Springer, New York (2007)

Chapter 4
Cartesian Compliance

This chapter defines a general Cartesian Compliance, which is used by the controllers presented in Chaps. 6 and 7. In general, a compliance combines a virtual spring and damper to stabilize a robot either in joint or Cartesian space. Although an implementation in Cartesian space is more cumbersome, a Cartesian compliance provides a more straightforward parameterization of the robot's motion with respect to the environment.

4.1 General Description

Let us consider an arbitrary frame $\mathcal{F}_A$ in six-dimensional Cartesian space, as shown in Fig. 4.1. The goal of a Cartesian compliance is to implement a desired stiffness and damping between both frames by applying the wrenches $\boldsymbol{F}_A^{\text{cpl}} \in \mathbb{R}^6$ at $\mathcal{F}_A$ and $\boldsymbol{F}_D^{\text{cpl}} \in \mathbb{R}^6$ at $\mathcal{F}_D$, respectively. The position and orientation of both frames relative to the world frame $\mathcal{W}$ are denoted by $\boldsymbol{x}_A \in \mathbb{R}^3$, $\boldsymbol{R}_A \in \mathcal{SO}(3)$ and $\boldsymbol{x}_D \in \mathbb{R}^3$, $\boldsymbol{R}_D \in \mathcal{SO}(3)$, respectively. The velocities of both frames are given by $\boldsymbol{v}_A = (\,\dot{\boldsymbol{x}}_A^T\ \boldsymbol{\omega}_A^T\,)^T$ and $\boldsymbol{v}_B = (\,\dot{\boldsymbol{x}}_D^T\ \boldsymbol{\omega}_D^T\,)^T$ with

$$\begin{aligned} \boldsymbol{S}(\boldsymbol{\omega}_A) &= \dot{\boldsymbol{R}}_A\, \boldsymbol{R}_A^T \\ \boldsymbol{S}(\boldsymbol{\omega}_D) &= \dot{\boldsymbol{R}}_D\, \boldsymbol{R}_D^T \end{aligned} \tag{4.1}$$

according to (3.6). Note that all position and velocity vectors are expressed in world coordinates as also are the wrenches $\boldsymbol{F}_A^{\text{cpl}}$ and $\boldsymbol{F}_D^{\text{cpl}}$. The control error is defined as

$$\begin{aligned} \boldsymbol{x}_{DA} &= \boldsymbol{x}_A - \boldsymbol{x}_D \\ \boldsymbol{R}_{DA} &= \boldsymbol{R}_D^T\, \boldsymbol{R}_A, \end{aligned} \tag{4.2}$$

B. Henze, *Whole-Body Control for Multi-Contact Balancing of Humanoid Robots*,
Springer Tracts in Advanced Robotics 143,
https://doi.org/10.1007/978-3-030-87212-0_4

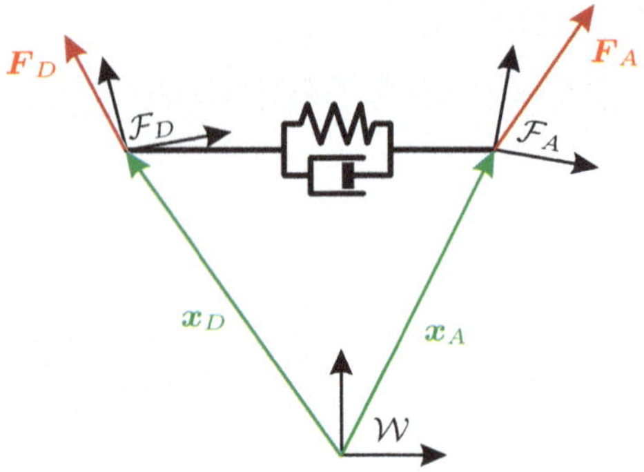

Fig. 4.1 Cartesian compliance connecting frame $\mathcal{F}_A$ with $\mathcal{F}_D$

which results in the Cartesian velocity

$$v_{DA} = \begin{pmatrix} \dot{x}_{DA} \\ \omega_{DA} \end{pmatrix} = v_A - v_D. \tag{4.3}$$

A Cartesian Compliance consists of a positive definite potential function $V^K(\boldsymbol{x}_{DA}, \boldsymbol{R}_{DA}) \in \mathbb{R}$ and a negative definite power dissipation $P^D(\boldsymbol{x}_{DA}, \boldsymbol{R}_{DA}, \dot{\boldsymbol{x}}_{DA}, \boldsymbol{\omega}_{DA}) \in \mathbb{R}$, which can be interpreted as the combination of a six-dimensional spring with a six-dimensional damper. The stiffness is deduced independently for translation and rotation in Sects. 4.2 and 4.3. The damping term is defined in Sect. 4.4 simultaneously governing translation and rotation. An overview of the architecture is given in Table 4.1. The respective stiffness matrices $\boldsymbol{K}_{\text{trans}}, \boldsymbol{K}_{\text{rot}} \in \mathbb{R}^{3\times3}$ as well as the damping matrix $\boldsymbol{D}' \in \mathbb{R}^{6\times6}$ are assumed to be symmetric, constant, and positive definite. The parameter matrices are given with respect to the axes of a reference frame $\mathcal{F}_{\text{ref}}$, whose orientation is given by $\boldsymbol{R}_{\text{ref}} \in \mathcal{SO}(3)$. Usually, the reference frame is selected to be identical to either $\mathcal{W}$, $\mathcal{F}_A$, or $\mathcal{F}_D$. If a moving frame is selected as reference ($\mathcal{F}_A$ or $\mathcal{F}_D$), then the parameter matrices become time variant as soon as they are rotated from $\mathcal{F}_{\text{ref}}$ into world frame $\mathcal{W}$ (see Table 4.2). The associated rotational velocity $\boldsymbol{\omega}_{\text{ref}} \in \mathbb{R}^3$ is given by

$$\boldsymbol{S}(\boldsymbol{\omega}_{\text{ref}}) = \dot{\boldsymbol{R}}_{\text{ref}} \boldsymbol{R}_{\text{ref}}^T. \tag{4.4}$$

Combining the wrenches of each element leads to the total wrench of the Cartesian compliance

Table 4.1 Architecture of a Cartesian compliance

	Derivation	Potential or Power	Wrench at $\mathcal{F}_A$	Wrench at $\mathcal{F}_D$
Trans. stiffness	Section 4.2	$V^{K,\text{trans}}$	$\boldsymbol{F}_A^{\text{cpl},K,\text{trans}}$	$\boldsymbol{F}_D^{\text{cpl},K,\text{trans}}$
Rot. stiffness	Section 4.3	$V^{K,\text{rot}}$	$\boldsymbol{F}_A^{\text{cpl},K,\text{rot}}$	$\boldsymbol{F}_D^{\text{cpl},K,\text{rot}}$
Damping	Section 4.4	P^D	$\boldsymbol{F}_A^{\text{cpl},D}$	$\boldsymbol{F}_D^{\text{cpl},D}$

Table 4.2 Overview of the Parameterization of a Cartesian compliance

	Size	Parameter in $\mathcal{F}_{\text{ref}}$	Parameter in $\mathcal{W}$
Trans. stiffness	$\boldsymbol{R}^{3\times3}$	$\boldsymbol{K}_{\text{trans}}$	$\boldsymbol{R}_{\text{ref}}\,\boldsymbol{K}_{\text{trans}}\,\boldsymbol{R}_{\text{ref}}^T$
Rot. stiffness	$\boldsymbol{R}^{3\times3}$	$\boldsymbol{K}_{\text{rot}}$	$\boldsymbol{R}_{\text{ref}}\,\boldsymbol{K}_{\text{rot}}\,\boldsymbol{R}_{\text{ref}}^T$
Damping	$\boldsymbol{R}^{6\times6}$	$\boldsymbol{D}'$	$\boldsymbol{D} = \text{diag}(\boldsymbol{R}_{\text{ref}}, \boldsymbol{R}_{\text{ref}})\,\boldsymbol{D}'\,\text{diag}(\boldsymbol{R}_{\text{ref}}, \boldsymbol{R}_{\text{ref}})^T$

$$\begin{aligned} \boldsymbol{F}_A^{\text{cpl}} &= \boldsymbol{F}_A^{\text{cpl},K,\text{trans}} + \boldsymbol{F}_A^{\text{cpl},K,\text{rot}} + \boldsymbol{F}_A^{\text{cpl},D} \\ \boldsymbol{F}_D^{\text{cpl}} &= \boldsymbol{F}_D^{\text{cpl},K,\text{trans}} + \boldsymbol{F}_D^{\text{cpl},K,\text{rot}} + \boldsymbol{F}_D^{\text{cpl},D} \end{aligned} \tag{4.5}$$

acting on frame $\mathcal{F}_A$ and $\mathcal{F}_D$, respectively. An overview of the generated wrenches is given in Table 4.1.

The translational and rotational stiffnesses are defined by two non-negative potentials $V^{K,\text{trans}}(\boldsymbol{x}_{DA}, \boldsymbol{R}_{DA}) \in \mathbb{R}$ and $V^{K,\text{rot}}(\boldsymbol{x}_{DA}, \boldsymbol{R}_{DA}) \in \mathbb{R}$ as a function of the control errors $\boldsymbol{x}_{DA}$ and $\boldsymbol{R}_{DA}$, which serve as a storage function for the elastic deformation energy. The generated wrenches are determined by the differential of $V^{K,\text{trans}}(\boldsymbol{x}_{DA}, \boldsymbol{R}_{DA})$ and $V^{K,\text{rot}}(\boldsymbol{x}_{DA}, \boldsymbol{R}_{DA})$ with respect to the position and orientation of the frames $\mathcal{F}_A$ and $\mathcal{F}_D$. To circumvent the problem of deriving the potential functions with respect to a rotation matrix, the wrenches can also be expressed as the derivatives of $\dot{V}^{K,\text{trans}}$ and $\dot{V}^{K,\text{rot}}$ with respect to the velocities $\boldsymbol{v}_A$, $\boldsymbol{v}_D$, and $\boldsymbol{v}_W$:

$$\begin{aligned} \boldsymbol{F}_A^{\text{cpl},K,\text{trans}} &= -\partial\dot{V}^{K,\text{trans}}/\partial\boldsymbol{v}_A & \boldsymbol{F}_A^{\text{cpl},K,\text{rot}} &= -\partial\dot{V}^{K,\text{rot}}/\partial\boldsymbol{v}_A \\ \boldsymbol{F}_D^{\text{cpl},K,\text{trans}} &= -\partial\dot{V}^{K,\text{trans}}/\partial\boldsymbol{v}_D & \boldsymbol{F}_D^{\text{cpl},K,\text{rot}} &= -\partial\dot{V}^{K,\text{rot}}/\partial\boldsymbol{v}_D. \end{aligned} \tag{4.6}$$

Consequently, computing the time derivatives

$$\begin{aligned} \dot{V}^{K,\text{trans}} &= -\boldsymbol{v}_A^T\,\boldsymbol{F}_A^{\text{cpl},K,\text{trans}} - \boldsymbol{v}_D^T\,\boldsymbol{F}_D^{\text{cpl},K,\text{trans}} \\ \dot{V}^{K,\text{rot}} &= -\boldsymbol{v}_A^T\,\boldsymbol{F}_A^{\text{cpl},K,\text{rot}} - \boldsymbol{v}_D^T\,\boldsymbol{F}_D^{\text{cpl},K,\text{rot}} \end{aligned} \tag{4.7}$$

allows identification of the wrenches acting on $\mathcal{F}_A$ and $\mathcal{F}_D$.

Various approaches for designing the damping of a Cartesian compliance can be found, as for instance methods based on a diagonalization or factorization of the system dynamics (Albu-Schäffer et al. 2004). For simplicity, this work focuses on a damping design featuring constant damping matrices $\boldsymbol{D}' \in \mathbb{R}^{6\times6}$ (with respect to $\mathcal{F}_{\text{ref}}$) implementing viscous damping. The damping elements are derived in Sect. 4.4 by defining a non-positive rate of dissipated energy

$$P^D = \boldsymbol{v}_A^T\,\boldsymbol{F}_A^{\text{cpl},D} + \boldsymbol{v}_D^T\,\boldsymbol{F}_D^{\text{cpl},D} \tag{4.8}$$

leading to the associated wrenches on $\mathcal{F}_A$ and $\mathcal{F}_D$.

The wrenches generated by the stiffness elements are deduced in Sects. 4.2 and 4.3. The overview in Table 4.3 reveals that the wrenches $\boldsymbol{F}_D^{\text{cpl},K,\text{trans}}$ and $\boldsymbol{F}_D^{\text{cpl},K,\text{rot}}$ acting on $\mathcal{F}_D$ are connected via a mapping $\boldsymbol{\Omega}_{AD} \in \mathbb{R}^{6\times6}$ to the wrenches $\boldsymbol{F}_A^{\text{cpl},K,\text{trans}}$

Table 4.3 Overview of the wrenches generated by a Cartesian compliance

Wrenches	Parameters	Trans. stiffness $\boldsymbol{F}_{\bullet}^{\mathrm{cpl},K,\mathrm{trans}}$	Rot. stiffness $S(\boldsymbol{\tau}_{\bullet}^{\mathrm{cpl},K,\mathrm{rot}})$	Damping $\boldsymbol{F}_{\bullet}^{\mathrm{cpl},D}$
$\boldsymbol{F}_A^{\mathrm{cpl}}$	... in frame $\mathcal{W}$	$\begin{pmatrix} -\boldsymbol{K}_{\mathrm{trans}}\,\boldsymbol{x}_{DA} \\ \boldsymbol{0} \end{pmatrix}$	$+2\,\mathrm{as}(\tilde{\boldsymbol{K}}\,\boldsymbol{R}_D\,\boldsymbol{R}_A^T)$	$-\boldsymbol{D}'\,(\boldsymbol{v}_A - \boldsymbol{\Omega}_{AD}\,\boldsymbol{v}_D)$
	... in frame $\mathcal{F}_A$	$\begin{pmatrix} -(\boldsymbol{R}_A\,\boldsymbol{K}_{\mathrm{trans}}\,\boldsymbol{R}_A^T)\,\boldsymbol{x}_{DA} \\ S(\boldsymbol{x}_{DA})\,(\boldsymbol{R}_A\,\boldsymbol{K}_{\mathrm{trans}}\,\boldsymbol{R}_A^T)\,\boldsymbol{x}_{DA} \end{pmatrix}$	$+2\,\mathrm{as}(\boldsymbol{R}_D\,\tilde{\boldsymbol{K}}\,\boldsymbol{R}_A^T)$	$-\begin{bmatrix} \boldsymbol{R}_A & \boldsymbol{0} \\ \boldsymbol{0} & \boldsymbol{R}_A \end{bmatrix} \boldsymbol{D}' \begin{bmatrix} \boldsymbol{R}_A & \boldsymbol{0} \\ \boldsymbol{0} & \boldsymbol{R}_A \end{bmatrix}^T (\boldsymbol{v}_A - \boldsymbol{\Omega}_{AD}\,\boldsymbol{v}_D)$
	... in frame $\mathcal{F}_D$	$\begin{pmatrix} \boldsymbol{0} \\ -(\boldsymbol{R}_D\,\boldsymbol{K}_{\mathrm{trans}}\,\boldsymbol{R}_D^T)\,\boldsymbol{x}_{DA} \end{pmatrix}$	$+2\,\mathrm{as}(\boldsymbol{R}_D\,\tilde{\boldsymbol{K}}\,\boldsymbol{R}_A^T)$	$-\begin{bmatrix} \boldsymbol{R}_D & \boldsymbol{0} \\ \boldsymbol{0} & \boldsymbol{R}_D \end{bmatrix} \boldsymbol{D}' \begin{bmatrix} \boldsymbol{R}_D & \boldsymbol{0} \\ \boldsymbol{0} & \boldsymbol{R}_D \end{bmatrix}^T (\boldsymbol{v}_A - \boldsymbol{\Omega}_{AD}\,\boldsymbol{v}_D)$
$\boldsymbol{F}_D^{\mathrm{cpl}}$	... in frame $\mathcal{W}$	$\begin{pmatrix} \boldsymbol{K}_{\mathrm{trans}}\,\boldsymbol{x}_{DA} \\ \boldsymbol{0} \end{pmatrix}$	$-2\,\mathrm{as}(\boldsymbol{R}_D\,\boldsymbol{R}_A^T\,\tilde{\boldsymbol{K}})$	$\boldsymbol{\Omega}_{AD}^T\,\boldsymbol{D}'\,(\boldsymbol{v}_A - \boldsymbol{\Omega}_{AD}\,\boldsymbol{v}_D)$
	... in frame $\mathcal{F}_A$	$\begin{pmatrix} (\boldsymbol{R}_A\,\boldsymbol{K}_{\mathrm{trans}}\,\boldsymbol{R}_A^T)\,\boldsymbol{x}_{DA} \\ \boldsymbol{0} \end{pmatrix}$	$-2\,\mathrm{as}(\boldsymbol{R}_D\,\tilde{\boldsymbol{K}}\,\boldsymbol{R}_A^T)$	$\boldsymbol{\Omega}_{AD}^T \begin{bmatrix} \boldsymbol{R}_A & \boldsymbol{0} \\ \boldsymbol{0} & \boldsymbol{R}_A \end{bmatrix} \boldsymbol{D}' \begin{bmatrix} \boldsymbol{R}_A & \boldsymbol{0} \\ \boldsymbol{0} & \boldsymbol{R}_A \end{bmatrix}^T (\boldsymbol{v}_A - \boldsymbol{\Omega}_{AD}\,\boldsymbol{v}_D)$
	... in frame $\mathcal{F}_D$	$\begin{pmatrix} (\boldsymbol{R}_D\,\boldsymbol{K}_{\mathrm{trans}}\,\boldsymbol{R}_D^T)\,\boldsymbol{x}_{DA} \\ S(\boldsymbol{x}_{DA})\,(\boldsymbol{R}_D\,\boldsymbol{K}_{\mathrm{trans}}\,\boldsymbol{R}_D^T)\,\boldsymbol{x}_{DA} \end{pmatrix}$	$-2\,\mathrm{as}(\boldsymbol{R}_D\,\tilde{\boldsymbol{K}}\,\boldsymbol{R}_A^T)$	$\boldsymbol{\Omega}_{AD}^T \begin{bmatrix} \boldsymbol{R}_D & \boldsymbol{0} \\ \boldsymbol{0} & \boldsymbol{R}_D \end{bmatrix} \boldsymbol{D}' \begin{bmatrix} \boldsymbol{R}_D & \boldsymbol{0} \\ \boldsymbol{0} & \boldsymbol{R}_D \end{bmatrix}^T (\boldsymbol{v}_A - \boldsymbol{\Omega}_{AD}\,\boldsymbol{v}_D)$

Table 4.4 Mapping between the wrenches generated by a Cartesian compliance

Parameters in ...	frame $\mathcal{W}$	frame $\mathcal{F}_A$	frame $\mathcal{F}_D$
$\boldsymbol{\Omega}_{AD} = \ldots$	$\begin{bmatrix} \boldsymbol{I} & \boldsymbol{0} \\ \boldsymbol{0} & \boldsymbol{R}_A \boldsymbol{R}_D^T \end{bmatrix}$	$\boldsymbol{Ad}(\boldsymbol{I}, \boldsymbol{x}_{AD})$	$\boldsymbol{Ad}(\boldsymbol{I}, \boldsymbol{x}_{AD})$

and $\boldsymbol{F}_A^{\text{cpl},K,\text{rot}}$ acting on frame $\mathcal{F}_A$. The damping design in Sect. 4.4 facilitates the same mapping, which yields

$$\boldsymbol{F}_D^{\text{cpl}} = -\boldsymbol{\Omega}_{AD}^T \boldsymbol{F}_A^{\text{cpl}} \tag{4.9}$$

for the combined wrench on $\mathcal{F}_A$ and $\mathcal{F}_D$. The mapping $\boldsymbol{\Omega}_{AD} \in \mathbb{R}^{6\times6}$ is specified in Table 4.4 depending on the selected reference frame. Consequently, a Cartesian compliance can be formulated in more general terms as

$$\begin{aligned} \boldsymbol{F}_A^{\text{cpl}} &= -(\nabla V^K)^T - \boldsymbol{D}\,(\boldsymbol{v}_A - \boldsymbol{\Omega}_{AD}\boldsymbol{v}_D) \\ \boldsymbol{F}_D^{\text{cpl}} &= +\boldsymbol{\Omega}_{AD}^T\,(\nabla V^K)^T + \boldsymbol{\Omega}_{AD}^T \boldsymbol{D}\,(\boldsymbol{v}_A - \boldsymbol{\Omega}_{AD}\boldsymbol{v}_D) \end{aligned} \tag{4.10}$$

with ∇V^K denoting the differential $\nabla V^K = \partial \dot{V}^K / \partial \boldsymbol{v}_A$ of the overall (positive definite) potential $V^K = V^{K,\text{trans}} + V^{K,\text{rot}}$. The damping term is abbreviated with the positive definite damping matrix $\boldsymbol{D} = \text{diag}(\boldsymbol{R}_{\text{ref}}, \boldsymbol{R}_{\text{ref}})\,\boldsymbol{D}'\,\text{diag}(\boldsymbol{R}_{\text{ref}}, \boldsymbol{R}_{\text{ref}})^T$, which consists of the constant matrix $\boldsymbol{D}'$ rotated from the reference frame $\mathcal{F}_{\text{ref}}$ to the world frame $\mathcal{W}$. The power balance of the compliance simplifies to

$$\dot{V}^K = -\underbrace{\boldsymbol{v}_A^T \boldsymbol{F}_A^{\text{cpl}}}_{\text{port at } \mathcal{F}_A} - \underbrace{\boldsymbol{v}_D^T \boldsymbol{F}_D^{\text{cpl}}}_{\text{port at } \mathcal{F}_D} - \underbrace{(\boldsymbol{v}_A - \boldsymbol{\Omega}_{AD}\boldsymbol{v}_D)^T \boldsymbol{D}\,(\boldsymbol{v}_A - \boldsymbol{\Omega}_{AD}\boldsymbol{v}_D)}_{\text{dissipation}}, \tag{4.11}$$

by inserting (4.5) and (4.10) into (4.7). The first two terms represent the power ports of the compliance, which allow the exchange of energy with both frames $\mathcal{F}_A$ and $\mathcal{F}_D$. The third term represents the power $P^D = -(\boldsymbol{v}_A - \boldsymbol{\Omega}_{AD}\boldsymbol{v}_D)^T \boldsymbol{D}\,(\boldsymbol{v}_A - \boldsymbol{\Omega}_{AD}\boldsymbol{v}_D) \leq 0$, which is dissipated by damping.

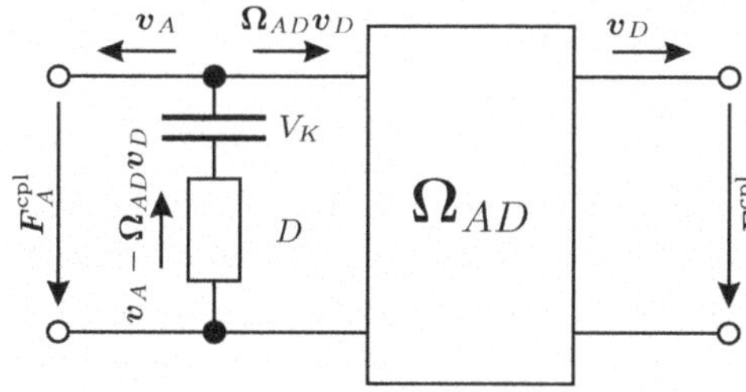

(a) Two-port system.

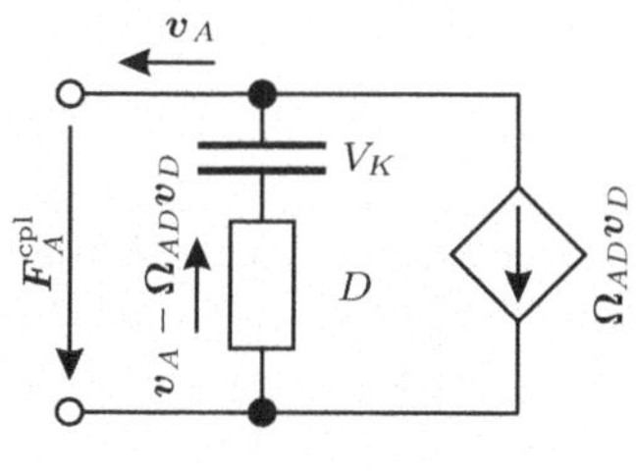

(b) One-port system.

Fig. 4.2 Cartesian compliance represented as an electrical network

A Cartesian compliance can be visualized, for instance, in the form of a bond-graph (Duindam et al. 2009) or an electrical circuit, as shown in Fig. 4.2. If the compliance is considered as a two-port system (Fig. 4.2a), then the power ports in (4.11) are given on the left- and right-hand side of the circuit. The deformation energy is stored in the capacitor representing the potential function V^K. The resistor represents the power dissipation P^D via the damping term. In most cases, the desired frame or set point is specified via software, which means that only the wrench $\boldsymbol{F}_A^{\text{cpl}}$ needs to be implemented, while $\boldsymbol{F}_D^{\text{cpl}}$ can be omitted. This simplification means that the two-port system of Fig. 4.2a can be reduced to a single-port system, as shown in Fig. 4.2b. The only power port that remains is the one at frame $\mathcal{F}_A$. A motion of the set point/desired frame is represented by the current source on the right-hand side of the circuit. As a consequence, the set point, represented by the only active component inside the circuit, can be used to inject or remove energy from the Cartesian compliance.

4.2 Translational Stiffness

Let us define the elastic potential for deriving the translational stiffness as

$$V^{K,\text{trans}} = \frac{1}{2}\boldsymbol{x}_{DA}^T \left(\boldsymbol{R}_{\text{ref}} \boldsymbol{K}_{\text{trans}} \boldsymbol{R}_{\text{ref}}^T\right) \boldsymbol{x}_{DA} \tag{4.12}$$

with the time derivative

$$\begin{aligned}\dot{V}^{K,\text{trans}} &= \dot{\boldsymbol{x}}_{DA}^T \left(\boldsymbol{R}_{\text{ref}} \boldsymbol{K}_{\text{trans}} \boldsymbol{R}_{\text{ref}}^T\right) \boldsymbol{x}_{DA} + \boldsymbol{x}_{DA}^T \left(\dot{\boldsymbol{R}}_{\text{ref}} \boldsymbol{K}_{\text{trans}} \boldsymbol{R}_{\text{ref}}^T\right) \boldsymbol{x}_{DA} \\ &= \dot{\boldsymbol{x}}_{DA}^T \left(\boldsymbol{R}_{\text{ref}} \boldsymbol{K}_{\text{trans}} \boldsymbol{R}_{\text{ref}}^T\right) \boldsymbol{x}_{DA} + \boldsymbol{x}_{DA}^T \boldsymbol{S}(\boldsymbol{\omega}_{\text{ref}}) \left(\boldsymbol{R}_{\text{ref}} \boldsymbol{K}_{\text{trans}} \boldsymbol{R}_{\text{ref}}^T\right) \boldsymbol{x}_{DA} \\ &= \dot{\boldsymbol{x}}_{DA}^T \left(\boldsymbol{R}_{\text{ref}} \boldsymbol{K}_{\text{trans}} \boldsymbol{R}_{\text{ref}}^T\right) \boldsymbol{x}_{DA} - \boldsymbol{\omega}_{\text{ref}}^T \boldsymbol{S}(\boldsymbol{x}_{DA}) \left(\boldsymbol{R}_{\text{ref}} \boldsymbol{K}_{\text{trans}} \boldsymbol{R}_{\text{ref}}^T\right) \boldsymbol{x}_{DA}\end{aligned} \tag{4.13}$$

considering (4.4). Inserting (4.3) yields

$$\begin{aligned}\dot{V}^{K,\text{trans}} &= \dot{\boldsymbol{x}}_A^T \left(\boldsymbol{R}_{\text{ref}} \boldsymbol{K}_{\text{trans}} \boldsymbol{R}_{\text{ref}}^T\right) \boldsymbol{x}_{DA} \\ &\quad - \dot{\boldsymbol{x}}_D^T \left(\boldsymbol{R}_{\text{ref}} \boldsymbol{K}_{\text{trans}} \boldsymbol{R}_{\text{ref}}^T\right) \boldsymbol{x}_{DA} \\ &\quad - \boldsymbol{\omega}_{\text{ref}}^T \boldsymbol{S}(\boldsymbol{x}_{DA}) \left(\boldsymbol{R}_{\text{ref}} \boldsymbol{K}_{\text{trans}} \boldsymbol{R}_{\text{ref}}^T\right) \boldsymbol{x}_{DA}.\end{aligned} \tag{4.14}$$

Comparing (4.14) with (4.7) reveals that the translational stiffness generates a force acting on the frames $\mathcal{F}_A$ and $\mathcal{F}_D$, but also on the reference frame. The latter vanishes if $\boldsymbol{K}_{\text{trans}}$ equals a scaled identity matrix because of $\boldsymbol{S}(\boldsymbol{x}_{DA}) \left(\boldsymbol{R}_{\text{ref}} \boldsymbol{K}_{\text{trans}} \boldsymbol{R}_{\text{ref}}^T\right) \boldsymbol{x}_{DA} = \boldsymbol{S}(\boldsymbol{x}_{DA}) \boldsymbol{x}_{DA} = \boldsymbol{0}$. This effect can be explained by Fig. 4.3. If $\boldsymbol{K}_{\text{trans}}$ is chosen differently, the contour lines of the potential $V^{K,\text{trans}}$ no longer represent a sphere in three dimensional space. Rotating the reference frame also rotates the contour lines, which results in a change of $V^{K,\text{trans}}$ for a constant control error $\boldsymbol{x}_{DA}$. This change of the energy stored inside the spring is transferred via the power port $\boldsymbol{\omega}_{\text{ref}}^T \boldsymbol{\tau}_{\text{ref}}^{\text{cpl},K,\text{trans}}$ resulting in a torque $\boldsymbol{\tau}_{\text{ref}}^{\text{cpl},K,\text{trans}}$ on the reference frame.

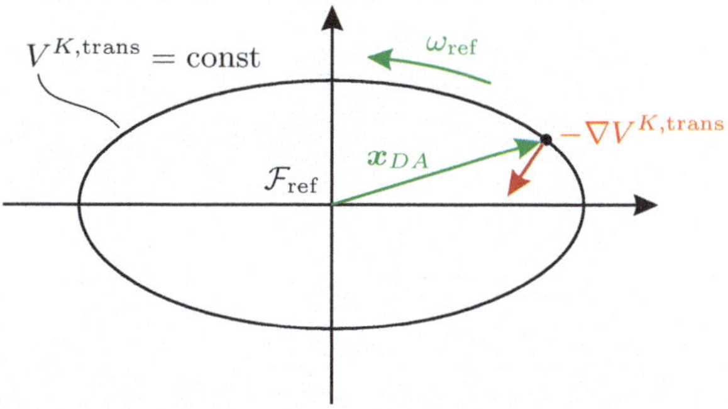

Fig. 4.3 Translational stiffness with an elliptical potential

If the reference frame is chosen to be $\mathcal{W}$, then $\boldsymbol{R}_{\mathrm{ref}} = \boldsymbol{I}$ holds and $\boldsymbol{\omega}_{\mathrm{ref}} = \boldsymbol{0}$ because of the world frame being stationary. The wrenches can be identified as

$$\begin{aligned} \boldsymbol{F}_A^{\mathrm{cpl},K,\mathrm{trans}} &= \begin{pmatrix} -\boldsymbol{K}_{\mathrm{trans}}\,\boldsymbol{x}_{DA} \\ \boldsymbol{0} \end{pmatrix} \\ \boldsymbol{F}_D^{\mathrm{cpl},K,\mathrm{trans}} &= \begin{pmatrix} \boldsymbol{K}_{\mathrm{trans}}\,\boldsymbol{x}_{DA} \\ \boldsymbol{0} \end{pmatrix} \end{aligned} \tag{4.15}$$

by comparing (4.14) with (4.7). If the reference frame is chosen to be $\mathcal{F}_A$ ($\boldsymbol{R}_{\mathrm{ref}} = \boldsymbol{R}_A$, $\boldsymbol{\omega}_{\mathrm{ref}} = \boldsymbol{\omega}_A$), then the wrenches can be identified as

$$\begin{aligned} \boldsymbol{F}_A^{\mathrm{cpl},K,\mathrm{trans}} &= \begin{pmatrix} -\left(\boldsymbol{R}_A\,\boldsymbol{K}_{\mathrm{trans}}\,\boldsymbol{R}_A^T\right)\boldsymbol{x}_{DA} \\ \boldsymbol{S}(\boldsymbol{x}_{DA})\left(\boldsymbol{R}_A\,\boldsymbol{K}_{\mathrm{trans}}\,\boldsymbol{R}_A^T\right)\boldsymbol{x}_{DA} \end{pmatrix} \\ \boldsymbol{F}_D^{\mathrm{cpl},K,\mathrm{trans}} &= \begin{pmatrix} \left(\boldsymbol{R}_A\,\boldsymbol{K}_{\mathrm{trans}}\,\boldsymbol{R}_A^T\right)\boldsymbol{x}_{DA} \\ \boldsymbol{0} \end{pmatrix} \end{aligned} \tag{4.16}$$

by comparing (4.14) with (4.7). If the reference frame is chosen to be $\mathcal{F}_D$ ($\boldsymbol{R}_{\mathrm{ref}} = \boldsymbol{R}_D$, $\boldsymbol{\omega}_{\mathrm{ref}} = \boldsymbol{\omega}_D$), then the wrenches can be identified as

$$\begin{aligned} \boldsymbol{F}_A^{\mathrm{cpl},K,\mathrm{trans}} &= \begin{pmatrix} -\left(\boldsymbol{R}_D\,\boldsymbol{K}_{\mathrm{trans}}\,\boldsymbol{R}_D^T\right)\boldsymbol{x}_{DA} \\ \boldsymbol{0} \end{pmatrix} \\ \boldsymbol{F}_D^{\mathrm{cpl},K,\mathrm{trans}} &= \begin{pmatrix} \left(\boldsymbol{R}_D\,\boldsymbol{K}_{\mathrm{trans}}\,\boldsymbol{R}_D^T\right)\boldsymbol{x}_{DA} \\ \boldsymbol{S}(\boldsymbol{x}_{DA})\left(\boldsymbol{R}_D\,\boldsymbol{K}_{\mathrm{trans}}\,\boldsymbol{R}_D^T\right)\boldsymbol{x}_{DA} \end{pmatrix} \end{aligned} \tag{4.17}$$

by comparing (4.14) with (4.7).

A physical interpretation of these results is given in Fig. 4.4. In the case of a diagonal stiffness matrix $\boldsymbol{K}_{\mathrm{trans}}$, both frames are connected via a series of uniaxial springs representing the diagonal entries (or eigenvalues) of $\boldsymbol{K}_{\mathrm{trans}}$. If $\boldsymbol{K}_{\mathrm{trans}}$ is specified in frame $\mathcal{F}_A$ (see Fig. 4.4a), then the axes of the springs are aligned with the base vectors of $\mathcal{F}_A$. This can be interpreted as a mechanism holding the uniaxial springs, which is attached to a virtual body carrying frame $\mathcal{F}_A$. Frame $\mathcal{F}_D$ is connected to

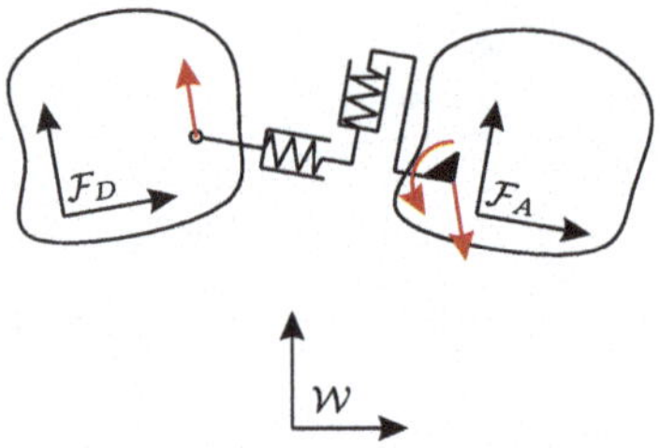

(a) $\boldsymbol{K}_{\text{trans}}$ in $\mathcal{F}_A$.

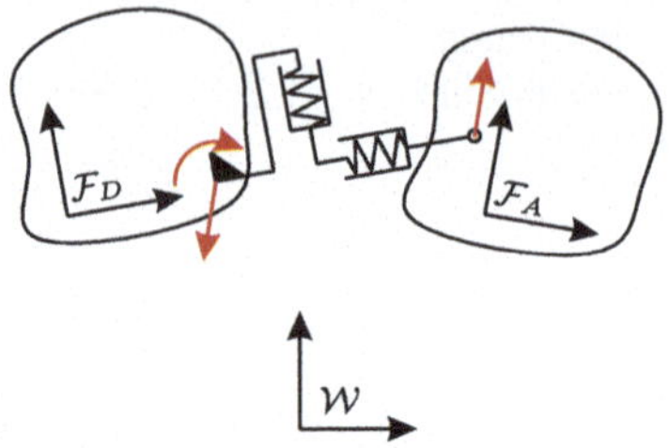

(b) $\boldsymbol{K}_{\text{trans}}$ in $\mathcal{F}_D$.

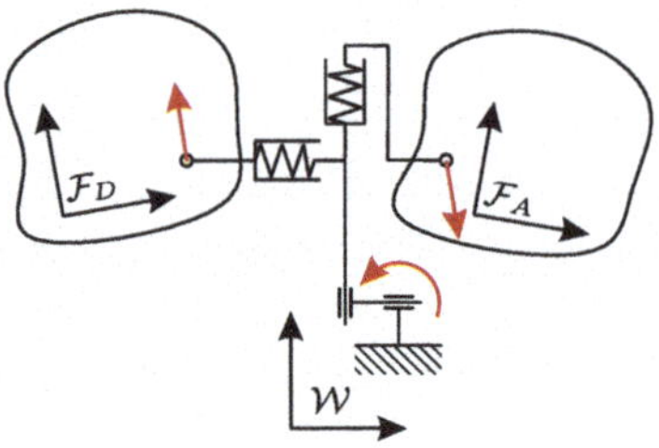

(c) $\boldsymbol{K}_{\text{trans}}$ in $\mathcal{W}$.

Fig. 4.4 Physical Interpretation of a translational stiffness featuring a diagonal $\boldsymbol{K}_{\text{trans}}$

the mechanism via a rotational joint. Therefore, all forces acting on $\mathcal{F}_D$ are balanced by a force and a torque on $\mathcal{F}_A$, as indicated by (4.16). The case of $\boldsymbol{K}_{\text{trans}}$ being specified in frame $\mathcal{F}_D$ can be analogously explained using the mechanism shown in Fig. 4.4b. If $\boldsymbol{K}_{\text{trans}}$ is expressed in the world frame, the mechanism becomes more complicated, as shown in Fig. 4.4c. The set of uniaxial springs is connected to the frames $\mathcal{F}_A$ and $\mathcal{F}_D$ via rotational joints. The orientation of the springs is ensured via a connection with the world frame blocking a rotational motion. Of course, the connection with the world frame can transmit loads in the form of a torque. However, this torque does not contribute to the power balance of the translational stiffness because the world frame is not moving.

4.3 Rotational Stiffness

In Ott et al. (2011) and Henze et al. (2016), the rotational stiffness is defined via a positive definite potential function based on the quaternion representation of the rotation error $\boldsymbol{R}_{DA}$. Adapting the potential energy to our requirements and notation yields

$$V^{K,\text{rot}} = 2\boldsymbol{\eta}^T \boldsymbol{R}_D^T \boldsymbol{R}_{\text{ref}} \boldsymbol{K}_{\text{rot}} \boldsymbol{R}_{\text{ref}}^T \boldsymbol{R}_D \boldsymbol{\eta} \tag{4.18}$$

with η and $\boldsymbol{\eta}$ denoting the scalar and the vector part of the unit quaternion η representing $\boldsymbol{R}_{DA}$. It was shown by Zhang and Fasse (2000) that the potential can also be formulated by using the trace of the rotation matrix $\boldsymbol{R}_{DA}$ directly. Although both formulations lead to the same wrench, a numerical analysis conducted by the author of this work revealed that the matrix potential features a higher numeric precision for small rotation errors $\boldsymbol{R}_{DA} \approx \boldsymbol{I}$. In preparation of the next steps, let us first introduce the following identities:

Theorem 4.1 *Let $\boldsymbol{A} \in \mathbb{R}^{n\times m}$ and $\boldsymbol{B} \in \mathbb{R}^{m\times n}$ be two arbitrary matrices. Then the following identity holds (Gruber 2013):*

$$\mathrm{tr}(\boldsymbol{A}\boldsymbol{B}) = \mathrm{tr}(\boldsymbol{B}\boldsymbol{A}). \tag{4.19}$$

Theorem 4.2 *Let $\boldsymbol{A} = [\, \boldsymbol{a}_1 \ldots \boldsymbol{a}_n \,] \in \mathbb{R}^{n\times n}$ and $\boldsymbol{B} = [\, \boldsymbol{b}_1 \ldots \boldsymbol{b}_n \,] \in \mathbb{R}^{n\times n}$ be two arbitrary matrices. Then the following identity holds (Gruber 2013):*

$$\mathrm{tr}(\boldsymbol{A}^T\boldsymbol{B}) = \sum_{j=1}^{n} \boldsymbol{a}_j^T \boldsymbol{b}_j. \tag{4.20}$$

Theorem 4.3 *Let $\boldsymbol{A} \in \mathbb{R}^{n\times n}$ and $\boldsymbol{B} \in \mathbb{R}^{n\times n}$ be two arbitrary matrices. Then the following identity holds (Zhang and Fasse 2000):*

$$\mathrm{tr}(\boldsymbol{A}\boldsymbol{B}) = \mathrm{tr}(\mathrm{sy}(\boldsymbol{A})\,\mathrm{sy}(\boldsymbol{B})) + \mathrm{tr}(\mathrm{as}(\boldsymbol{A})\,\mathrm{as}(\boldsymbol{B})) \tag{4.21}$$

with $\mathrm{sy}(\boldsymbol{A}) = \frac{1}{2}(\boldsymbol{A} + \boldsymbol{A}^T)$ *and* $\mathrm{as}(\boldsymbol{A}) = \frac{1}{2}(\boldsymbol{A} - \boldsymbol{A}^T)$ *denoting the symmetric and asymmetric part of a matrix. If $\boldsymbol{A} = \boldsymbol{A}^T$ is symmetric and $\boldsymbol{B} = -\boldsymbol{B}^T$ skew-symmetric, this simplifies to* $\mathrm{tr}(\boldsymbol{A}\boldsymbol{B}) = 0$.

As shown by Zhang and Fasse (2000), the potential function (4.18) can also be formulated using the trace of the rotation matrix $\boldsymbol{R}_{DA}$

$$\begin{aligned} V^{K,\mathrm{rot}} &= -\,\mathrm{tr}\left(\boldsymbol{R}_{DA}^T\,(\boldsymbol{R}_D^T\,\boldsymbol{R}_{\mathrm{ref}}\,\tilde{\boldsymbol{K}}\,\boldsymbol{R}_{\mathrm{ref}}^T\,\boldsymbol{R}_D)\right) + \mathrm{tr}(\tilde{\boldsymbol{K}}) \\ &= -\,\mathrm{tr}\left(\boldsymbol{R}_A^T\,(\boldsymbol{R}_{\mathrm{ref}}\,\tilde{\boldsymbol{K}}\,\boldsymbol{R}_{\mathrm{ref}}^T)\,\boldsymbol{R}_D\right) + \mathrm{tr}(\tilde{\boldsymbol{K}}) \end{aligned} \tag{4.22}$$

considering (4.2). The proof that (4.18) and (4.22) are identical is given in Appendix A. The term $\tilde{\boldsymbol{K}} = \frac{1}{2}\,\mathrm{tr}(\boldsymbol{K}_{\mathrm{rot}})\boldsymbol{I} - \boldsymbol{K}_{\mathrm{rot}}$ denotes the so-called "co-stiffness matrix", which is constant, symmetric, and expressed in the same coordinates as $\boldsymbol{K}_{\mathrm{rot}}$. Again, $\boldsymbol{R}_{\mathrm{ref}}$ is used to rotate the co-stiffness matrix $\tilde{\boldsymbol{K}}$ from the reference frame $\mathcal{F}_{\mathrm{ref}}$ into the world frame $\mathcal{W}$. The term $\mathrm{tr}(\tilde{\boldsymbol{K}})$ represents a constant offset to ensure the positive definiteness of the potential, as shown in Appendix A. According to Fasse (1997), the potential energy can be interpreted as a measure of misalignment between the base vectors of $\boldsymbol{R}_A = [\, \boldsymbol{r}_{A,1}\; \boldsymbol{r}_{A,2}\; \boldsymbol{r}_{A,3} \,]$ and $\boldsymbol{R}_D = [\, \boldsymbol{r}_{D,1}\; \boldsymbol{r}_{D,2}\; \boldsymbol{r}_{D,3} \,]$. This becomes obvious if (4.22) is reformulated as

$$V^{K,\mathrm{rot}} = -\,\mathrm{tr}\left(\boldsymbol{R}_A^T\,(\boldsymbol{R}_{\mathrm{ref}}\,\tilde{\boldsymbol{K}}\,\boldsymbol{R}_{\mathrm{ref}}^T)\,\boldsymbol{R}_D\right) + \mathrm{tr}(\tilde{\boldsymbol{K}})$$
$$= -\sum_{j=1}^{3} \boldsymbol{r}_{A,j}^T\,(\boldsymbol{R}_{\mathrm{ref}}\,\tilde{\boldsymbol{K}}\,\boldsymbol{R}_{\mathrm{ref}}^T)\,\boldsymbol{r}_{D,j} + \mathrm{tr}(\tilde{\boldsymbol{K}}) \tag{4.23}$$

using Theorem 4.2. The rotated co-stiffness matrix $(\boldsymbol{R}_{\mathrm{ref}}\,\tilde{\boldsymbol{K}}\,\boldsymbol{R}_{\mathrm{ref}}^T)$ serves as a weighting factor for penalizing the misalignment between $\boldsymbol{r}_{A,j}$ and $\boldsymbol{r}_{D,j}$.

Computing the time derivative of (4.22) using (4.1) and (4.4) yields

$$\begin{aligned}
\dot{V}^{K,\mathrm{rot}} &= -\,\mathrm{tr}\left(\dot{\boldsymbol{R}}_A^T\,(\boldsymbol{R}_{\mathrm{ref}}\,\tilde{\boldsymbol{K}}\,\boldsymbol{R}_{\mathrm{ref}}^T)\,\boldsymbol{R}_D\right) - \mathrm{tr}\left(\boldsymbol{R}_A^T\,(\dot{\boldsymbol{R}}_{\mathrm{ref}}\,\tilde{\boldsymbol{K}}\,\boldsymbol{R}_{\mathrm{ref}}^T)\,\boldsymbol{R}_D\right)\\
&\quad - \mathrm{tr}\left(\boldsymbol{R}_A^T\,(\boldsymbol{R}_{\mathrm{ref}}\,\tilde{\boldsymbol{K}}\,\dot{\boldsymbol{R}}_{\mathrm{ref}}^T)\,\boldsymbol{R}_D\right) - \mathrm{tr}\left(\boldsymbol{R}_A^T\,(\boldsymbol{R}_{\mathrm{ref}}\,\tilde{\boldsymbol{K}}\,\boldsymbol{R}_{\mathrm{ref}}^T)\,\dot{\boldsymbol{R}}_D\right)\\
&= +\,\mathrm{tr}\left(\boldsymbol{R}_A^T\,\boldsymbol{S}(\boldsymbol{\omega}_A)\,(\boldsymbol{R}_{\mathrm{ref}}\,\tilde{\boldsymbol{K}}\,\boldsymbol{R}_{\mathrm{ref}}^T)\,\boldsymbol{R}_D\right)\\
&\quad - \mathrm{tr}\left(\boldsymbol{R}_A^T\,\boldsymbol{S}(\boldsymbol{\omega}_{\mathrm{ref}})\,(\boldsymbol{R}_{\mathrm{ref}}\,\tilde{\boldsymbol{K}}\,\boldsymbol{R}_{\mathrm{ref}}^T)\,\boldsymbol{R}_D\right)\\
&\quad + \mathrm{tr}\left(\boldsymbol{R}_A^T\,(\boldsymbol{R}_{\mathrm{ref}}\,\tilde{\boldsymbol{K}}\,\boldsymbol{R}_{\mathrm{ref}}^T)\,\boldsymbol{S}(\boldsymbol{\omega}_{\mathrm{ref}})\,\boldsymbol{R}_D\right)\\
&\quad - \mathrm{tr}\left(\boldsymbol{R}_A^T\,(\boldsymbol{R}_{\mathrm{ref}}\,\tilde{\boldsymbol{K}}\,\boldsymbol{R}_{\mathrm{ref}}^T)\,\boldsymbol{S}(\boldsymbol{\omega}_D)\,\boldsymbol{R}_D\right).
\end{aligned} \tag{4.24}$$

Theorem 4.1 allows reordering of the matrices into

$$\begin{aligned}
\dot{V}^{K,\mathrm{rot}} &= +\,\mathrm{tr}\left(\boldsymbol{S}(\boldsymbol{\omega}_A)\,(\boldsymbol{R}_{\mathrm{ref}}\,\tilde{\boldsymbol{K}}\,\boldsymbol{R}_{\mathrm{ref}}^T)\,\boldsymbol{R}_D\,\boldsymbol{R}_A^T\right)\\
&\quad - \mathrm{tr}\left(\boldsymbol{S}(\boldsymbol{\omega}_{\mathrm{ref}})\,(\boldsymbol{R}_{\mathrm{ref}}\,\tilde{\boldsymbol{K}}\,\boldsymbol{R}_{\mathrm{ref}}^T)\,\boldsymbol{R}_D\,\boldsymbol{R}_A^T\right)\\
&\quad + \mathrm{tr}\left(\boldsymbol{S}(\boldsymbol{\omega}_{\mathrm{ref}})\,\boldsymbol{R}_D\,\boldsymbol{R}_A^T\,(\boldsymbol{R}_{\mathrm{ref}}\,\tilde{\boldsymbol{K}}\,\boldsymbol{R}_{\mathrm{ref}}^T)\right)\\
&\quad - \mathrm{tr}\left(\boldsymbol{S}(\boldsymbol{\omega}_D)\,\boldsymbol{R}_D\,\boldsymbol{R}_A^T\,(\boldsymbol{R}_{\mathrm{ref}}\,\tilde{\boldsymbol{K}}\,\boldsymbol{R}_{\mathrm{ref}}^T)\right).
\end{aligned} \tag{4.25}$$

Exploiting Theorem 4.3 with $\mathrm{as}(\boldsymbol{S}(\bullet)) = \boldsymbol{S}(\bullet)$ and $\mathrm{sy}(\boldsymbol{S}(\bullet)) = \boldsymbol{0}$ leads to

$$\begin{aligned}
\dot{V}^{K,\mathrm{rot}} &= +\,\mathrm{tr}\left(\boldsymbol{S}(\boldsymbol{\omega}_A)\ \mathrm{as}(\boldsymbol{R}_{\mathrm{ref}}\tilde{\boldsymbol{K}}\,\boldsymbol{R}_{\mathrm{ref}}^T\,\boldsymbol{R}_D\,\boldsymbol{R}_A^T)\right)\\
&\quad - \mathrm{tr}\left(\boldsymbol{S}(\boldsymbol{\omega}_{\mathrm{ref}})\ \mathrm{as}(\boldsymbol{R}_{\mathrm{ref}}\tilde{\boldsymbol{K}}\,\boldsymbol{R}_{\mathrm{ref}}^T\,\boldsymbol{R}_D\,\boldsymbol{R}_A^T)\right)\\
&\quad + \mathrm{tr}\left(\boldsymbol{S}(\boldsymbol{\omega}_{\mathrm{ref}})\ \mathrm{as}(\boldsymbol{R}_D\,\boldsymbol{R}_A^T\,\boldsymbol{R}_{\mathrm{ref}}\tilde{\boldsymbol{K}}\,\boldsymbol{R}_{\mathrm{ref}}^T)\right)\\
&\quad - \mathrm{tr}\left(\boldsymbol{S}(\boldsymbol{\omega}_D)\ \mathrm{as}(\boldsymbol{R}_D\,\boldsymbol{R}_A^T\,\boldsymbol{R}_{\mathrm{ref}}\tilde{\boldsymbol{K}}\,\boldsymbol{R}_{\mathrm{ref}}^T)\right).
\end{aligned} \tag{4.26}$$

It can be easily verified that (4.7) can be rewritten as

$$\dot{V}^{K,\mathrm{rot}} = \frac{1}{2}\,\mathrm{tr}\left(\boldsymbol{S}(\boldsymbol{\omega}_A)\,\boldsymbol{S}(\boldsymbol{\tau}_A^{\mathrm{cpl}})\right) + \frac{1}{2}\,\mathrm{tr}\left(\boldsymbol{S}(\boldsymbol{\omega}_D)\,\boldsymbol{S}(\boldsymbol{\tau}_D^{\mathrm{cpl}})\right), \tag{4.27}$$

which allows it to be compared with (4.26). If the reference frame is chosen to be $\mathcal{W}$ ($\boldsymbol{R}_{\text{ref}} = \boldsymbol{I}$, $\boldsymbol{\omega}_{\text{ref}} = \boldsymbol{0}$), then the torques can be identified as

$$\begin{aligned} \boldsymbol{S}(\boldsymbol{\tau}_A^{\text{cpl},K,\text{rot}}) &= +2\,\text{as}(\tilde{\boldsymbol{K}}\,\boldsymbol{R}_D\,\boldsymbol{R}_A^T) = +\tilde{\boldsymbol{K}}\,\boldsymbol{R}_D\,\boldsymbol{R}_A^T - \boldsymbol{R}_A\,\boldsymbol{R}_D^T\,\tilde{\boldsymbol{K}} \\ \boldsymbol{S}(\boldsymbol{\tau}_D^{\text{cpl},K,\text{rot}}) &= -2\,\text{as}(\boldsymbol{R}_D\,\boldsymbol{R}_A^T\,\tilde{\boldsymbol{K}}) = -\boldsymbol{R}_D\,\boldsymbol{R}_A^T\tilde{\boldsymbol{K}} + \tilde{\boldsymbol{K}}\,\boldsymbol{R}_A\,\boldsymbol{R}_D^T \\ &= -\left(\boldsymbol{R}_A\,\boldsymbol{R}_D^T\right)^T\left(\tilde{\boldsymbol{K}}\,\boldsymbol{R}_D\,\boldsymbol{R}_A^T - \boldsymbol{R}_A\,\boldsymbol{R}_D^T\,\tilde{\boldsymbol{K}}\right)\left(\boldsymbol{R}_A\,\boldsymbol{R}_D^T\right) \end{aligned} \tag{4.28}$$

by comparing (4.26) with (4.27). If the reference frame is chosen to be $\mathcal{F}_A$ ($\boldsymbol{R}_{\text{ref}} = \boldsymbol{R}_A$, $\boldsymbol{\omega}_{\text{ref}} = \boldsymbol{\omega}_A$), then the torques can be identified as

$$\begin{aligned} \boldsymbol{S}(\boldsymbol{\tau}_A^{\text{cpl},K,\text{rot}}) &= +2\,\text{as}(\boldsymbol{R}_D\,\tilde{\boldsymbol{K}}\,\boldsymbol{R}_A^T) \\ \boldsymbol{S}(\boldsymbol{\tau}_D^{\text{cpl},K,\text{rot}}) &= -2\,\text{as}(\boldsymbol{R}_D\,\tilde{\boldsymbol{K}}\,\boldsymbol{R}_A^T) \end{aligned} \tag{4.29}$$

by comparing (4.26) with (4.27). If the reference frame is chosen to be $\mathcal{F}_D$ ($\boldsymbol{R}_{\text{ref}} = \boldsymbol{R}_D$, $\boldsymbol{\omega}_{\text{ref}} = \boldsymbol{\omega}_D$), then the torques can be identified as

$$\begin{aligned} \boldsymbol{S}(\boldsymbol{\tau}_A^{\text{cpl},K,\text{rot}}) &= +2\,\text{as}(\boldsymbol{R}_D\,\tilde{\boldsymbol{K}}\,\boldsymbol{R}_A^T) \\ \boldsymbol{S}(\boldsymbol{\tau}_D^{\text{cpl},K,\text{rot}}) &= -2\,\text{as}(\boldsymbol{R}_D\,\tilde{\boldsymbol{K}}\,\boldsymbol{R}_A^T) \end{aligned} \tag{4.30}$$

by comparing (4.26) with (4.27). In conclusion, the rotational stiffness generates torques acting on the frames $\mathcal{F}_A$ and $\mathcal{F}_D$, which can be assembled from the cross product matrix $\boldsymbol{S}(\bullet)$ in (4.28), (4.29), or (4.30), respectively.

4.4 Damping

The derivation of the stiffness in Sects. 4.2 and 4.3 revealed that the wrenches $\boldsymbol{F}_A^{\text{cpl},K,\text{trans}}$ and $\boldsymbol{F}_A^{\text{cpl},K,\text{rot}}$ are connected to $\boldsymbol{F}_D^{\text{cpl},K,\text{trans}}$ and $\boldsymbol{F}_D^{\text{cpl},K,\text{rot}}$ via the mapping $\boldsymbol{\Omega}_{AD}$ specified in Table 4.4. To simplify the notation of the following chapters, the damping is designed such that it features the same mapping. In contrast to the stiffness terms, the damping is designed for translation and rotation simultaneously.

Let us define the dissipated power for deducing the translational damping as

$$\begin{aligned} P^D &= -\left(\boldsymbol{v}_A - \boldsymbol{\Omega}_{AD}\,\boldsymbol{v}_D\right)^T \begin{bmatrix} \boldsymbol{R}_{\text{ref}} & \boldsymbol{0} \\ \boldsymbol{0} & \boldsymbol{R}_{\text{ref}} \end{bmatrix} \boldsymbol{D}' \begin{bmatrix} \boldsymbol{R}_{\text{ref}} & \boldsymbol{0} \\ \boldsymbol{0} & \boldsymbol{R}_{\text{ref}} \end{bmatrix}^T \left(\boldsymbol{v}_A - \boldsymbol{\Omega}_{AD}\,\boldsymbol{v}_D\right) \\ &= -\,\boldsymbol{v}_A^T \begin{bmatrix} \boldsymbol{R}_{\text{ref}} & \boldsymbol{0} \\ \boldsymbol{0} & \boldsymbol{R}_{\text{ref}} \end{bmatrix} \boldsymbol{D}' \begin{bmatrix} \boldsymbol{R}_{\text{ref}} & \boldsymbol{0} \\ \boldsymbol{0} & \boldsymbol{R}_{\text{ref}} \end{bmatrix}^T \left(\boldsymbol{v}_A - \boldsymbol{\Omega}_{AD}\,\boldsymbol{v}_D\right) \\ &\quad + \boldsymbol{v}_D^T\,\boldsymbol{\Omega}_{AD}^T \begin{bmatrix} \boldsymbol{R}_{\text{ref}} & \boldsymbol{0} \\ \boldsymbol{0} & \boldsymbol{R}_{\text{ref}} \end{bmatrix} \boldsymbol{D}' \begin{bmatrix} \boldsymbol{R}_{\text{ref}} & \boldsymbol{0} \\ \boldsymbol{0} & \boldsymbol{R}_{\text{ref}} \end{bmatrix}^T \left(\boldsymbol{v}_A - \boldsymbol{\Omega}_{AD}\,\boldsymbol{v}_D\right) \end{aligned} \tag{4.31}$$

taking (4.3) into consideration. If the reference frame is chosen to be $\mathcal{W}$ ($\boldsymbol{R}_{\text{ref}} = \boldsymbol{I}$), then the wrenches can be identified as

$$\begin{aligned} \boldsymbol{F}_A^{\text{cpl},D} &= -\boldsymbol{D}' \left(\boldsymbol{v}_A - \boldsymbol{\Omega}_{AD}\, \boldsymbol{v}_D\right) \\ \boldsymbol{F}_D^{\text{cpl},D} &= \boldsymbol{\Omega}_{AD}^T \boldsymbol{D}' \left(\boldsymbol{v}_A - \boldsymbol{\Omega}_{AD}\, \boldsymbol{v}_D\right) \end{aligned} \tag{4.32}$$

by comparing (4.31) with (4.8). If the reference frame is chosen to be $\mathcal{F}_A$ ($\boldsymbol{R}_{\text{ref}} = \boldsymbol{R}_A$), then the wrenches can be identified as

$$\begin{aligned} \boldsymbol{F}_A^{\text{cpl},D} &= -\begin{bmatrix} \boldsymbol{R}_A & \boldsymbol{0} \\ \boldsymbol{0} & \boldsymbol{R}_A \end{bmatrix} \boldsymbol{D}' \begin{bmatrix} \boldsymbol{R}_A & \boldsymbol{0} \\ \boldsymbol{0} & \boldsymbol{R}_A \end{bmatrix}^T \left(\boldsymbol{v}_A - \boldsymbol{\Omega}_{AD}\, \boldsymbol{v}_D\right) \\ \boldsymbol{F}_D^{\text{cpl},D} &= \boldsymbol{\Omega}_{AD}^T \begin{bmatrix} \boldsymbol{R}_A & \boldsymbol{0} \\ \boldsymbol{0} & \boldsymbol{R}_A \end{bmatrix} \boldsymbol{D}' \begin{bmatrix} \boldsymbol{R}_A & \boldsymbol{0} \\ \boldsymbol{0} & \boldsymbol{R}_A \end{bmatrix}^T \left(\boldsymbol{v}_A - \boldsymbol{\Omega}_{AD}\, \boldsymbol{v}_D\right) \end{aligned} \tag{4.33}$$

by comparing (4.31) with (4.8). If the reference frame is chosen to be $\mathcal{F}_D$ ($\boldsymbol{R}_{\text{ref}} = \boldsymbol{R}_D$), then the wrenches can be identified as

$$\begin{aligned} \boldsymbol{F}_A^{\text{cpl},D} &= -\begin{bmatrix} \boldsymbol{R}_D & \boldsymbol{0} \\ \boldsymbol{0} & \boldsymbol{R}_D \end{bmatrix} \boldsymbol{D}' \begin{bmatrix} \boldsymbol{R}_D & \boldsymbol{0} \\ \boldsymbol{0} & \boldsymbol{R}_D \end{bmatrix}^T \left(\boldsymbol{v}_A - \boldsymbol{\Omega}_{AD}\, \boldsymbol{v}_D\right) \\ \boldsymbol{F}_D^{\text{cpl},D} &= \boldsymbol{\Omega}_{AD}^T \begin{bmatrix} \boldsymbol{R}_D & \boldsymbol{0} \\ \boldsymbol{0} & \boldsymbol{R}_D \end{bmatrix} \boldsymbol{D}' \begin{bmatrix} \boldsymbol{R}_D & \boldsymbol{0} \\ \boldsymbol{0} & \boldsymbol{R}_D \end{bmatrix}^T \left(\boldsymbol{v}_A - \boldsymbol{\Omega}_{AD}\, \boldsymbol{v}_D\right) \end{aligned} \tag{4.34}$$

by comparing (4.31) with (4.8).

References

Albu-Schäffer, A., Ott, C., Hirzinger, G.: A passivity based Cartesian impedance controller for flexible joint robots - part II: full state feedback, impedance design and experiments. In: IEEE International Conference on Robotics and Automation, pp. 2666–2672 (2004)

Duindam, V., Macchelli, A., Stramigioli, S., Bruyninckx, H.: Modeling and Control of Complex Physical Systems?: The Port-Hamiltonian Approach, 1st edn. Springer, Berlin, Heidelberg (2009)

Fasse, E.D.: On the spatial compliance of robotic manipulators. J. Dyn. Syst. Meas. Control **119**(4), 839–844 (1997)

Gruber, M.H.J.: Matrix Algebra for Linear Models. Wiley, New Jersey (2013)

Henze, B., Roa, M.A., Ott, C.: Passivity-based whole-body balancing for torque-controlled humanoid robots in multi-contact scenarios. Int. J. Robot. Res. **35**(12), 1522–1543 (2016)

Ott, C., Roa, M.A., Hirzinger, G.: Posture and balance control for biped robots based on contact force optimization. In: IEEE-RAS International Conference on Humanoid Robots, pp. 26–33 (2011)

Zhang, S., Fasse, E.D.: Spatial compliance modeling using a quaternion-based potential function method. Multibody Syst. Dyn. **4**(1), 75–101 (2000)

Chapter 5
Torque-Controlled Humanoid Robot TORO

The control approaches presented in this work are designed for the humanoid robot *TORO*, but can also be deployed on other humanoid robots featuring torque-controlled joints and multiple limbs. Quadrupeds are also a potential target platform, as the approaches from Chaps. 6 and 7 are able to handle an arbitrary number of contacts. In this work, the control approaches are evaluated through several experiments with the humanoid Robot *TORO* developed by the *German Aerospace Center* (*DLR*). An overview of the system architecture is given below; further details can be found in Henze et al. (2014), Englsberger et al. (2014), Ott et al. (2017).

5.1 Hardware Description

The *TOrque-controlled humanoid RObot TORO* shown in Fig. 5.1 has been developed by the *German Aerospace Center* (*DLR*) for conducting research on walking and multi-contact balancing (Henze et al. 2014; Englsberger et al. 2014). In the current version, *TORO* features 39 degrees of freedom in total, a height of 1.74 m, and a weight of 76.4 kg The 25 joints located in the legs, arms, and hip are based on the technology of the *DLR-KUKA LBR* (*Lightweight robot arm*), and can be operated both in position and torque-controlled mode (Albu-Schäffer et al. 2007). The neck comprises two DoFs, which are implemented with position-controlled servo motors by *DYNAMIXEL*. The robot has two prosthetic hands from *Touch Bionics* (*i-limb ultra*), each providing six DoFs.

In terms of sensing, *TORO* features position and torque sensors in each joint based on the *LBR* technology. Besides this, *TORO* is equipped with two 6-axis force-torque sensors (FTS), which are located between the feet and the ankles (see Fig. 5.1) in order to measure the contact wrenches acting on the feet. Furthermore, the chest contains an Inertial Measurement Unit (IMU) that provides the orientation, angular

B. Henze, *Whole-Body Control for Multi-Contact Balancing of Humanoid Robots*,
Springer Tracts in Advanced Robotics 143,
https://doi.org/10.1007/978-3-030-87212-0_5

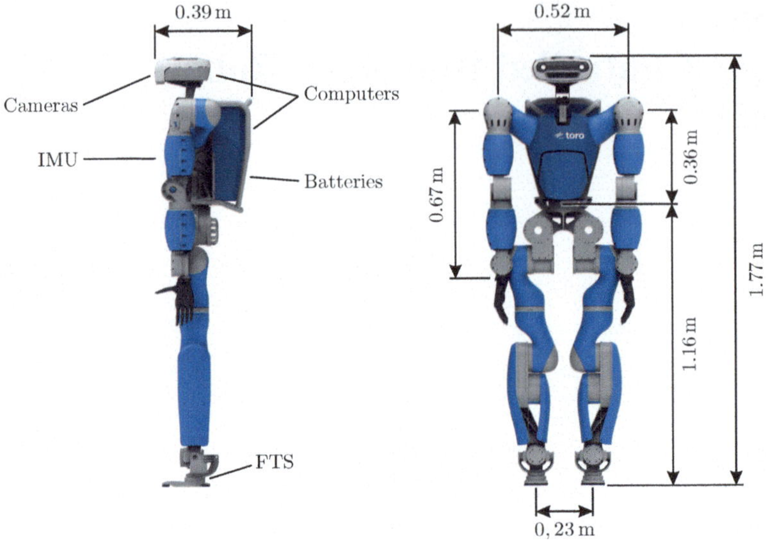

Fig. 5.1 Overview of the kinematics and the main components of *TORO*

velocity, and linear acceleration of the torso. The head contains three camera systems for localization and object detection.

To obtain an autonomous operation, all data processing and controls are done on-board: The low-level joint control is incorporated into the electronics of each *LBR* joint module (Albu-Schäffer et al. 2007). The high-level control is executed on a real-time computer in the backpack, which provides an *Intel(R) Core(TM) i7-4700EQ* CPU with four cores and utilizes the operating system *Linux* with the *PREEMPT_RT* patch. The second computer in the backpack is used for several supplementary tasks without a real-time requirement, such as the communication with the hands or the neck servo motors. The third computer is located in the head and is reserved for vision processing. In addition to the computers, the backpack also contains two batteries that can power the complete robot for roughly one hour, allowing for a completely autonomous operation of the system.

During the experiments presented in this work, all 25 *LBR* joints are operated in torque-controlled mode, which allows the robot to resolve the wrench distribution problem arising from balancing on multiple contacts (see Sect. 3.3). The only exception is the experiment presented in Sect. 8.1, where the lower body is position-controlled. Although the head is considered for computing the dynamic model (3.20), it is not actively used for controls. In fact, the two servos motors are only utilized for orienting the cameras of the head. The DoFs in the hands allow for the opening and closing of the fingers in order to perform grasping operations such as picking up an object or holding onto a handrail. Note that all experiments presented in this work assume a unilateral point contact for the hands according to Sect. 3.4.3. In case the presented control approaches are not evaluated in experiment but in simulation, the latter is conducted with *OpenHRP* by Kanehiro et al. (2002).

The proposed control approaches are implemented in *Matlab/Simulink*, using *qpOASES* (version 2.0) by Ferreau et al. (2008, 2014) for solving the constraint quadratic optimization problem (cQP) presented in Chaps. 6, 7, and 8. The controller is computed at a rate of 1 kHz, while the low-level joint controller is executed at a speed of 3 kHz (Albu-Schäffer et al. 2007). Videos of the experiments can be found on the *YouTube* channel of the *Institute of Robotics and Mechatronics* at *DLR* (RMC 2019) or can be downloaded from the publishers of the corresponding contributions.

5.2 Base Frame Estimation

The performance of a balancing controller is directly linked to the quality of the information provided on the state of the floating base of the robot. This section presents an algorithm, which we first reported in Henze et al. (2016) for estimating the state of the base frame $\mathcal{B}$ using kinematic information and data from the IMU located in *TORO*'s chest (see Fig. 5.1). As soon as the state of the base frame $\mathcal{B}$ (here the hip) is known, one can compute the state of the center of mass frame C, as detailed in Sect. 3.2.2. In contrast to Stelzer et al. (2012), Bloesch et al. (2013) who proposed Kalman filters for state estimation, our method (Henze et al. 2016) targets the direct (lag-free) estimation of the base frame and its velocity from contact information, kinematics, and IMU measurements.

Let us assume that all support contacts ($i = 1 \ldots \psi_{\text{sup}}$) are in contact with a rigid environment, which leads to $\dot{\boldsymbol{v}}_{\text{sup}} = \boldsymbol{v}_{\text{sup}} = \boldsymbol{0}$ (see Assumption 3.2). Furthermore, it is assumed that the poses of all contacts given by $\boldsymbol{x}_i$ and $\boldsymbol{R}_i$ are well-known with respect to the world frame $\mathcal{W}$. Then, each contact $(\boldsymbol{x}_i, \boldsymbol{R}_i)$ can be used to compute exactly one estimate for the base frame $\mathcal{B}$ by using the kinematic information of the robot $(\boldsymbol{q}, \dot{\boldsymbol{q}})$. Each estimation provides a value for $\boldsymbol{x}_b$, $\dot{\boldsymbol{x}}_b$, $\boldsymbol{R}_b$ and $\boldsymbol{\omega}_b$, representing the whole state of the base frame. Another estimate for $\boldsymbol{R}_b$ and $\boldsymbol{\omega}_b$ is provided by an on-board IMU attached to the base of the robot. In order to reduce the influence of uncertainties and measurement noise, the information provided by the different data sources (contacts and IMU) is averaged according to the scheme sketched in Fig. 5.2. The method consists of four steps that sequentially compute $\boldsymbol{R}_b$, $\boldsymbol{\omega}_b$, $\boldsymbol{x}_b$ and $\dot{\boldsymbol{x}}_b$, in this order.

The computation of $\boldsymbol{R}_b$ comprises two steps: First, the estimates provided by the contacts are averaged based on Appendix B. Here, $\alpha_i \in [0, 1]$ with $\sum_{i=1}^{\psi} \alpha_i = 1$ represent the weights that determine the influence of each contact on the resulting mean orientation. The result is then combined with the base orientation estimated by the on-board IMU. As the IMU is operated in a rather noisy environment in terms of electromagnetic interference, it is not possible to evaluate the earth's magnetic field properly. Thus, the orientation measured by the IMU can show a drift about the vertical axis of the world frame. In order to remove the drift, the estimation

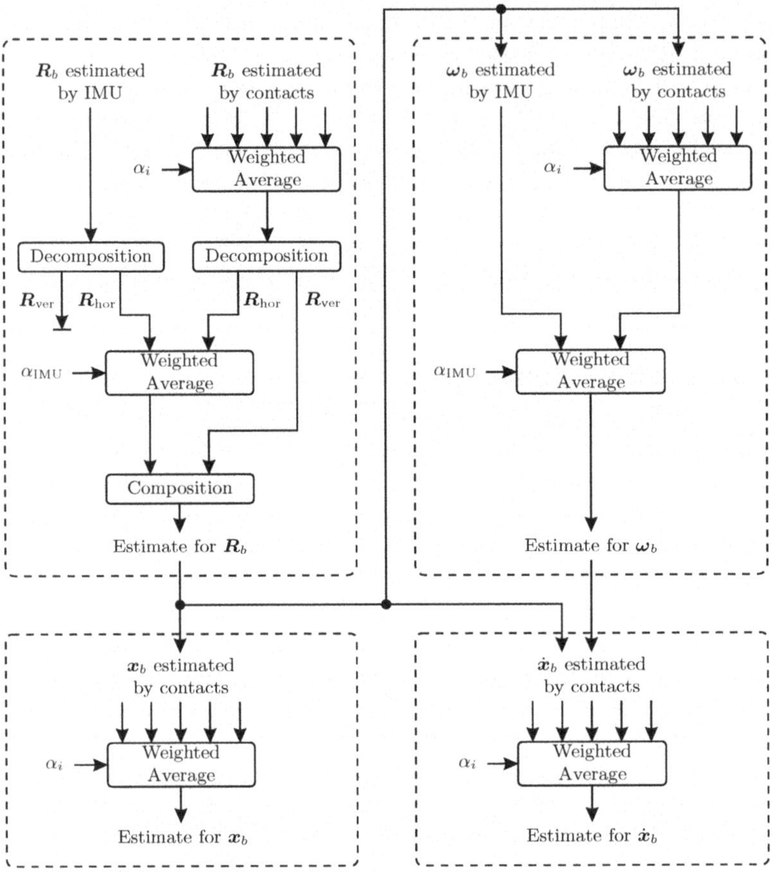

Fig. 5.2 Algorithm used for estimating the state of the base frame (Henze et al. 2016)

provided by the IMU and the average provided by the contacts are both decomposed into two consecutive rotations.[1] The first one, $\boldsymbol{R}_{\text{ver}}$, is a rotation about the vertical axis of the world frame $\mathcal{W}$. The second one, $\boldsymbol{R}_{\text{hor}}$, is a rotation about an axis lying in the horizontal plane of $\mathcal{W}$. The horizontal rotations are averaged based on the weight $\alpha_{\text{IMU}} \in [0, 1]$ (see Appendix B) that determines the trust given to the IMU data. Then, the result is composed with the vertical estimate provided by the kinematics in order to provide an estimate for the overall orientation $\boldsymbol{R}_b$ of the base frame. The vertical rotation provided by the IMU is not processed any further because it is compromised by the aforementioned drift.

The computation of $\boldsymbol{\omega}_b$ consists of two analogous steps: First, the estimates provided by the contacts are averaged based on the weights α_i, which can be used to

[1] This decomposition was previously used in Fritsch et al. (2013) for attitude control in quadrotor helicopters.

specify the confidence in each contact. Afterwards, the result is averaged with the estimation of $\boldsymbol{\omega}_b$ provided by the IMU using the weight α_{IMU}. The computation of $\boldsymbol{x}_b$ and $\dot{\boldsymbol{x}}_b$ is much simpler, as the only relevant source of information is the kinematics. For both computations, the estimates are averaged based on the weights α_i.

In order to increase accuracy, the algorithm shown in Fig. 5.2 computes $\boldsymbol{R}_b$, $\boldsymbol{\omega}_b$, $\boldsymbol{x}_b$, and $\dot{\boldsymbol{x}}_b$ in this exact order. For instance, computing the estimate for $\boldsymbol{\omega}_b$ from the contact information requires the base orientation $\boldsymbol{R}_b$. To minimize a propagation of errors, the computation of $\boldsymbol{\omega}_b$ utilizes the averaged base orientation instead of the individual estimates provided by kinematics or the IMU. Analogously, the averaged angular velocity $\boldsymbol{\omega}_b$ is used for computing $\boldsymbol{x}_b$ and $\dot{\boldsymbol{x}}_b$.

By adapting the weights α_i and α_{IMU}, the algorithm can consider the reliability of each type of sensor information. If the parameter α_{IMU} is set to $\alpha_{\text{IMU}} = 1$, then the algorithm only uses the information about the orientation provided by the IMU. This corresponds to the assumption of the balancing contacts being point contacts. The opposite scenario is $\alpha_{\text{IMU}} = 0$, which causes the algorithm to rely exclusively on the kinematic information. This corresponds to the assumption of non-rotating contact surfaces. Besides this, if one contact surface lifts off from the environment, then the corresponding weight must be set to $\alpha_i = 0$.

For instance, the experiments presented in Chaps. 6 and 7 only evaluate the information on the orientation provided by the IMU ($\alpha_{\text{IMU}} = 1$) in order to obtain a higher robustness against uncertainties of the floor (see e.g. the experiment in Fig. 6.5). The contacts are only used to obtain information on the translation of the base. The experiments do not evaluate the hand contacts ($\alpha_{\text{HandR}} = \alpha_{\text{HandL}} = 0$) because they are not very reliable due to the uneven surface of the palm (see Fig. 3.8a). Instead, the algorithm exploits the contacts at the feet or the knees, as shown in Table 5.1. However, the contacts at the feet are preferred due to a higher robustness.

Note that the algorithm presented above requires $\boldsymbol{x}_i$ and $\boldsymbol{R}_i$ to be known. During the initialization of the controller, one can choose the world frame arbitrarily but fixed to the environment, thus leading to an initial $\boldsymbol{x}_i$ and $\boldsymbol{R}_i$. In case of repositioning an end-effector or contact, $\boldsymbol{x}_i$ and $\boldsymbol{R}_i$ must be updated before the balancing contact is used again for estimating the base frame. To achieve this, the pose of the contacts relative to the world frame $\mathcal{W}$ can be measured by using the estimation of the base

Table 5.1 Parameterization of the base frame estimation used for the experiments presented in Chaps. 6 and 7

Available contacts				Parameters			
FootR	FootL	KneeR	KneeL	α_{FootR}	α_{FootL}	α_{KneeR}	α_{KneeL}
Yes	Yes	Optional	Optional	0.5	0.5	0	0
Yes	No	Optional	Optional	1	0	0	0
No	Yes	Optional	Optional	0	1	0	0
No	No	Yes	Yes	0	0	0.5	0.5
No	No	Yes	No	0	0	1	0
No	No	No	Yes	0	0	0	1

frame $\mathcal{B}$ provided by the other contacts and the IMU. Of course, the corresponding weight must be set to $\alpha_i = 0$ until the update is completed.

References

Albu-Schäffer, A., Haddadin, S., Ott, C., Stemmer, A., Wimböck, T., Hirzinger, G.: The DLR lightweight robot - design and control concepts in human environments. Ind. Robot: Int. J. **34**(5), 376–385 (2007)

Bloesch, M., Gehring, C., Fankhauser, P., Hutter, M., Hoepflinger, M.A., Siegwart, R.: State estimation for legged robots on unstable and slippery terrain. In: IEEE/RSJ International Conference on Intelligent Robots and Systems, pp. 6058–6064 (2013)

Englsberger, J., Werner, A., Ott, C., Henze, B., Roa, M.A., Garofalo, G., Burger, R., Beyer, A., Eiberger, O., Schmid, K., Albu-Schäffer, A.: Overview of the torque-controlled humanoid robot TORO. In: IEEE-RAS International Conference on Humanoid Robots, pp. 916–923 (2014)

Ferreau, H.J., Bock, H.G., Diehl, M.: An online active set strategy to overcome the limitations of explicit MPC. Int. J. Robust Nonlinear Control **18**(8), 816–830 (2008)

Ferreau, H.J., Kirches, C., Potschka, A., Bock, H.G., Diehl, M.: qpOASES: a parametric active-set algorithm for quadratic programming. Math. Program. Comput. **6**(4), 327–363 (2014)

Fritsch, O., Henze, B., Lohmann, B.: Fast and saturating thrust direction control for a quadrotor helicopter. Automatisierungstechnik **61**(3), 172–182 (2013)

Henze, B., Werner, A., Roa, M.A., Garofalo, G., Englsberger, J., Ott, C.: Control applications of TORO - a torque controlled humanoid robot. In: IEEE-RAS International Conference on Humanoid Robots, pp. 841–841 (2014)

Henze, B., Roa, M.A., Ott, C.: Passivity-based whole-body balancing for torque-controlled humanoid robots in multi-contact scenarios. Int. J. Robot. Res. **35**(12), 1522–1543 (2016)

Kanehiro, F., Fujiwara, K., Kajita, S., Yokoi, K., Kaneko, K., Hirukawa, H., Nakamura, Y., Yamane, K.: OpenHRP: open architecture humanoid robotics platform. In: IEEE International Conference on Robotics and Automation, pp. 24–30 (2002)

Ott, C., Roa, M.A., Schmidt, F., Friedl, W., Englsberger, J., Burger, R., Werner, A., Dietrich, A., Leidner, D., Henze, B., Eiberger, O., Beyer, A., Bäuml, B., Borst, C., Albu-Schäffer, A.: Mechanisms and design of DLR humanoid robots. Humanoid Robotics: a Reference, pp. 1–26. Springer, Netherlands (2017)

RMC, YouTube channel of the Institute of Robotics and Mechatronics at the German Aerospace Center (DLR) (2019). www.youtube.com/user/DLRRMC. Accessed 13 Sept 2019

Stelzer, A., Hirschmüller, H., Görner, M.: Stereo-vision-based navigation of a six-legged walking robot in unknown rough terrain. Int. J. Robot. Res. **31**(4), 381–402 (2012)

Chapter 6
Whole-Body Control for Multi-contact Balancing

The presented control framework for passivity-based multi-contact balancing originates from the works of Ott et al. (2011), who developed a balancing controller for torque-controlled, legged robots. The controller by Ott et al. (2011) features a Cartesian compliance at the center of mass for stabilizing the robot and utilizes the legs for generating the required support. We extended the concept in Henze et al. (2016b) such that the formulation does not only comprise the legs but also the arms of the robot, which yields to the presented framework for Whole-Body Control (WBC) of humanoid robots. The framework allows the robot to either use the end-effectors to generate the required support or to interact with the environment (see support and interaction end-effectors as introduced in Sect. 3.2.1).

Besides the generalization of the framework, we proposed several extensions to the whole-body controller in Henze et al. (2016b, 2018), Abi-Farraj et al. (2019), which are compiled into this chapter. The extension to hierarchical whole-body balancing reported in Henze et al. (2016a, 2018) is presented separately in Chap. 7.

For instance, Sect. 6.1 presents the basic concept of the whole-body controller for the regulation case (MCB), which is extended to the tracking case in Sect. 6.3 (MCB+). For both, the passivity and stability properties are discussed assuming a stationary ground floor. The basic controller is extended in Sect. 6.4 (MCB-PC) in order to guarantee passivity even on movable and deformable surfaces using the methods from the field of Passivity Observer-Passivity Control (PO-PC). A method for handling interaction tasks with high forces is presented in Sect. 6.5 (MCB-ACT) utilizing a suitable motion of the CoM as compensation. The properties and performance of the control framework are discussed for each variant in the corresponding sections with the help of experiments conducted with the humanoid robot *TORO* (see Chap. 5).

B. Henze, *Whole-Body Control for Multi-Contact Balancing of Humanoid Robots*,
Springer Tracts in Advanced Robotics 143,
https://doi.org/10.1007/978-3-030-87212-0_6

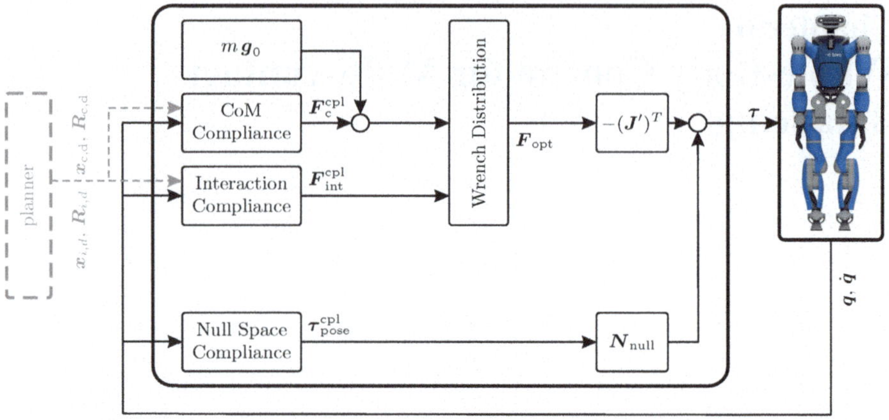

Fig. 6.1 Overview of the architecture of the MCB control framework

6.1 Basic Concept (MCB)

The main goal of the presented control framework is to enable a humanoid robot to maintain balance while performing a manipulation or interaction task. For this purpose, it is assumed that all task frames are exclusively located at the end-effectors. Consequently, the ψ end-effectors can be divided into the group of supporting end-effectors ($1 \leq i \leq \psi_{sup}$) and interaction end-effectors ($\psi_{sup} + 1 \leq i \leq \psi$). A method that can also handle task frames located all over the robot's body is presented in Chap. 7. As the presented framework is able to balance on multiple contacts, it will be called Multi-Contact Balancing controller (MCB) for the remainder of this manuscript. The controller consists of two different components, as shown in Fig. 6.1: The upper branch shows the multi-contact balancing controller for stabilizing the center of mass location, the hip orientation, and the Cartesian pose of each interaction end-effector. In order to achieve this, several Cartesian compliances generate a suitable set of wrenches, which are distributed to the end-effectors and then mapped to the joint torques. The second component is a compliant null space controller that stabilizes the redundant kinematics of the robot by generating torques that are projected onto the null space. The torques of both components are added and applied to the robot.

6.1.1 Controller Derivation

In order to maintain balance, the controller features a Cartesian compliance that stabilizes the CoM frame C at a predefined location $\boldsymbol{x}_{c,d} \in \mathbb{R}^3$ and orientation $\boldsymbol{R}_{c,d} \in SO(3)$, which e.g. can be provided by an external planning algorithm, as discussed in Sect. 6.6. As already mentioned in Sect. 3.2.2, the CoM frame C has the same orientation $\boldsymbol{R}_\mathrm{c} = \boldsymbol{R}_b$ as the hip, which ultimately means that the CoM compliance stabilizes the CoM location and hip orientation. The wrench[1] generated by the CoM compliance is given by

$$\boldsymbol{F}_\mathrm{c}^\mathrm{cpl} = -(\nabla V_\mathrm{c}^K)^T - \boldsymbol{D}_\mathrm{c}\,\Delta\boldsymbol{v}_\mathrm{c} \tag{6.1}$$

based on definition (4.10). The associated potential and damping matrix are denoted by $V_\mathrm{c}^K \geq 0$ and $\boldsymbol{D}_\mathrm{c} \geq 0$, respectively. The velocity error is compressed into

$$\Delta\boldsymbol{v}_\mathrm{c} = \boldsymbol{v}_\mathrm{c} - \boldsymbol{\Omega}_\mathrm{c}\,\boldsymbol{v}_\mathrm{c,d} \tag{6.2}$$

with $\boldsymbol{v}_\mathrm{c,d} \in \mathbb{R}^6$ denoting the desired Cartesian velocities of frame C. As the basic version of the whole-body balancer is designed for the regulation case, $\boldsymbol{v}_\mathrm{c,d}$ equals zero. Analogously, each interaction end-effector ($\psi_\mathrm{sup} + 1 \leq i \leq \psi$) is stabilized by a Cartesian compliance based on a given desired position $\boldsymbol{x}_{i,d} \in \mathbb{R}^3$ and orientation $\boldsymbol{R}_{i,d} \in SO(3)$. The corresponding desired velocities are denoted by $\boldsymbol{v}_{i,d} \in \mathbb{R}^6$, which can be assumed to be zero for the regulation case. Based on definition (4.10), the generated wrench is given by

$$\boldsymbol{F}_i^\mathrm{cpl} = -(\nabla V_i^K)^T - \boldsymbol{D}_i\,\Delta\boldsymbol{v}_i \tag{6.3}$$

with $\Delta\boldsymbol{v}_i = \boldsymbol{v}_i - \boldsymbol{\Omega}_i\,\boldsymbol{v}_{i,d}$ for all i satisfying $(\psi_\mathrm{sup} + 1) \leq i \leq \psi$. The associated potential and damping matrix are denoted by $V_i^K \geq 0$ and $\boldsymbol{D}_i \geq 0$, respectively. For simplicity of notation, the wrenches of the interaction compliances are stacked analogously to (3.17) into

$$\boldsymbol{F}_\mathrm{int}^\mathrm{cpl} = \begin{pmatrix} \boldsymbol{F}_{(\psi_\mathrm{sup}+1)}^\mathrm{cpl} \\ \vdots \\ \boldsymbol{F}_\psi^\mathrm{cpl} \end{pmatrix} = -\underbrace{\begin{pmatrix} (\nabla V_{(\psi_\mathrm{sup}+1)}^K)^T \\ \vdots \\ (\nabla V_\psi^K)^T \end{pmatrix}}_{(\nabla V_\mathrm{int}^K)^T} - \underbrace{\begin{bmatrix} \boldsymbol{D}_{\psi_\mathrm{sup}+1} & \cdots & \boldsymbol{0} \\ \vdots & \ddots & \vdots \\ \boldsymbol{0} & \cdots & \boldsymbol{D}_\psi \end{bmatrix}}_{\boldsymbol{D}_\mathrm{int}} \underbrace{\begin{pmatrix} \Delta\boldsymbol{v}_{(\psi_\mathrm{sup}+1)} \\ \vdots \\ \Delta\boldsymbol{v}_\psi \end{pmatrix}}_{\Delta\boldsymbol{v}_\mathrm{int}} \tag{6.4}$$

with $V_\mathrm{int}^K = V_{(\psi_\mathrm{sup}+1)}^K + \cdots + V_\psi$. The velocity errors are combined into

$$\underbrace{\begin{pmatrix} \Delta\boldsymbol{v}_{(\psi_\mathrm{sup}+1)} \\ \vdots \\ \Delta\boldsymbol{v}_\psi \end{pmatrix}}_{\Delta\boldsymbol{v}_\mathrm{int}} = \underbrace{\begin{pmatrix} \boldsymbol{v}_{(\psi_\mathrm{sup}+1)} \\ \vdots \\ \boldsymbol{v}_\psi \end{pmatrix}}_{\boldsymbol{v}_\mathrm{int}} - \underbrace{\begin{bmatrix} \boldsymbol{\Omega}_{(\psi_\mathrm{sup}+1)} & \cdots & \boldsymbol{0} \\ \vdots & \ddots & \vdots \\ \boldsymbol{0} & \cdots & \boldsymbol{\Omega}_\psi \end{bmatrix}}_{\boldsymbol{\Omega}_\mathrm{int}} \underbrace{\begin{pmatrix} \boldsymbol{v}_{(\psi_\mathrm{sup}+1),d} \\ \vdots \\ \boldsymbol{v}_{\psi,d} \end{pmatrix}}_{\boldsymbol{v}_\mathrm{int}} . \tag{6.5}$$

The basic idea of the balancing controller is to use the support end-effectors to generate suitable contact wrenches $\boldsymbol{F}_\mathrm{sup}$ that counteract the influence of gravity and

---

[1] Note that the definition of a Cartesian compliance according to Chap. 4 features an opposite sign compared to the definition used in Henze et al. (2016b).

the wrenches of the CoM and interaction compliances. The first line of the system dynamics (3.20) and of the mapping (3.24) suggests formulating this relationship via

$$\boldsymbol{Ad}_{\text{sup}}^T \boldsymbol{F}_{\text{sup}} = m\boldsymbol{g}_0 + \boldsymbol{F}_{\text{c}}^{\text{cpl}} + \boldsymbol{Ad}_{\text{int}}^T \boldsymbol{F}_{\text{int}}^{\text{cpl}}. \tag{6.6}$$

The right-hand side describes the wrench at the CoM frame C required for gravity compensation and the Cartesian compliances. The left-hand side represents the wrench at the CoM resulting from the support contacts. This equation is a direct consequence of the fact that the DoFs of the center of mass are not actuated due to the floating base. Instead, the support contacts must be used to generate the required wrench at the CoM. Thus, condition (6.6) will be referred to as the "underactuation constraint".

The balancing controller can be derived by formulating a desired dynamics for the closed-loop system:

$$\boldsymbol{M}\begin{pmatrix}\dot{\boldsymbol{v}}_{\text{c}}\\ \ddot{\boldsymbol{q}}\end{pmatrix} + \boldsymbol{C}\begin{pmatrix}\boldsymbol{v}_{\text{c}}\\ \dot{\boldsymbol{q}}\end{pmatrix} = \begin{pmatrix}\boldsymbol{F}_{\text{c}}^{\text{cpl}}\\ \boldsymbol{0}\end{pmatrix} + \begin{bmatrix}\boldsymbol{Ad}_{\text{sup}}^T\\ (\boldsymbol{J}'_{\text{sup}})^T\end{bmatrix}(-\boldsymbol{F}_{\text{sup}}) + \begin{bmatrix}\boldsymbol{Ad}_{\text{int}}^T\\ (\boldsymbol{J}'_{\text{int}})^T\end{bmatrix}\boldsymbol{F}_{\text{int}}^{\text{cpl}} + \boldsymbol{\tau}_{\text{ext}}. \tag{6.7}$$

The Cartesian compliance stabilizing the CoM frame C is directly applied at the CoM coordinates. The desired or commanded end-effector wrenches are mapped via the transpose of their respective Jacobian matrix to the configuration space, as suggested by (3.24). For instance, the interaction compliances $\boldsymbol{F}_{\text{int}}^{\text{cpl}}$ are mapped via the transpose of $\boldsymbol{J}_{\text{int}} = [\, \boldsymbol{Ad}_{\text{int}} \; \boldsymbol{J}'_{\text{int}} \,]$. The negative support wrenches $(-\boldsymbol{F}_{\text{sup}})$ are mapped analogously via the transpose of $\boldsymbol{J}_{\text{sup}} = [\, \boldsymbol{Ad}_{\text{sup}} \; \boldsymbol{J}'_{\text{sup}} \,]$. The negative sign in front of $\boldsymbol{F}_{\text{sup}}$ might seem unusual at first glance, but can be explained by comparing (6.7) with the system dynamics (3.20), which yields the control law

$$\begin{pmatrix}m\boldsymbol{g}_0 + \boldsymbol{F}_{\text{c}}^{\text{cpl}}\\ -\boldsymbol{\tau}\end{pmatrix} = \begin{bmatrix}\boldsymbol{Ad}_{\text{sup}}^T & \boldsymbol{Ad}_{\text{int}}^T\\ (\boldsymbol{J}'_{\text{sup}})^T & (\boldsymbol{J}'_{\text{int}})^T\end{bmatrix}\begin{pmatrix}\boldsymbol{F}_{\text{sup}}\\ -\boldsymbol{F}_{\text{int}}^{\text{cpl}}\end{pmatrix}. \tag{6.8}$$

It can be easily verified, that the negative sign is necessary for the first line of the control law to satisfy the underactuation condition (6.6). Another explanation is offered in Fig. 6.2: It is customary in compliance control to define the commanded end-effector wrench as the wrench that the robot exerts on the last body of the kinematic chain (Ott 2008). Consequently, the external and the commanded wrench are in equilibrium if and only if $\boldsymbol{F}_{\text{int}}^{\text{cpl}} = -\boldsymbol{F}_{\text{int}}^{\text{ext}}$ holds. However in balance control, the commanded contact wrench is usually perceived as a wrench that the contact exerts on the robot, which is also known as the Ground Reaction Force (GRF), as mentioned e.g. in Wensing et al. (2013), Hyon et al. (2007). Therefore, the commanded and the external contact wrench are in equilibrium if and only if $\boldsymbol{F}_{\text{sup}} = \boldsymbol{F}_{\text{sup}}^{\text{ext}}$ holds.

As already mentioned in Sect. 3.3, the Contact Map $\boldsymbol{Ad}_{\text{sup}}^T \in \mathbb{R}^{6\times 6\psi_{\text{sup}}}$ is subjective but not injective for more than one contact ($6\psi_{\text{sup}} > 6$). Therefore, the controller can select $\boldsymbol{F}_{\text{sup}}$ from a set of feasible solutions to generate the required overall wrench at the center of mass, which is also known as the wrench distribution problem. This

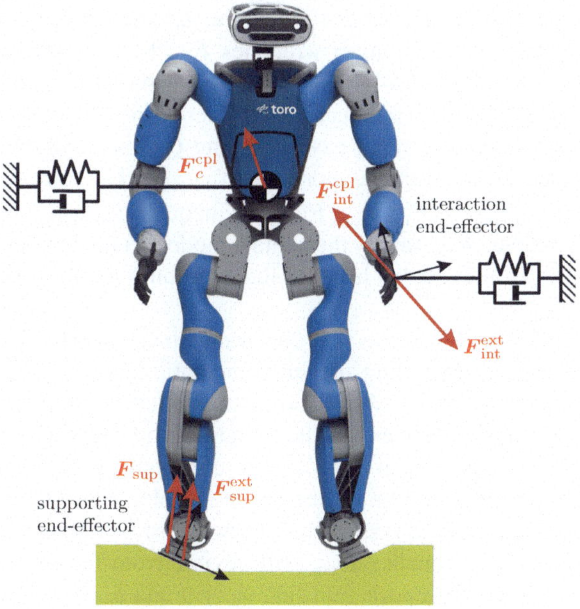

Fig. 6.2 Definition of the commanded support wrenches in comparison to the commanded interaction wrenches. The end-effector wrenches are shown in equilibrium

redundancy in the space of the contact wrenches $\boldsymbol{F}_{\text{sup}}$ can be resolved by computing the following Constrained Quadratic Optimization Problem (cQP):

$$\boldsymbol{F}^{\text{opt}} = \underset{\boldsymbol{F}}{\operatorname{argmin}} \left(\frac{1}{2}\boldsymbol{\delta}_{\text{c}}^T \boldsymbol{Q}_{\text{c}}\boldsymbol{\delta}_{\text{c}} + \frac{1}{2}\boldsymbol{\delta}_{\text{sup}}^T \boldsymbol{Q}_{\text{sup}}\boldsymbol{\delta}_{\text{sup}} + \frac{1}{2}\boldsymbol{\delta}_{\text{int}}^T \boldsymbol{Q}_{\text{int}}\boldsymbol{\delta}_{\text{int}} \right) \tag{6.9}$$

minimizing the residua

$$\boldsymbol{\delta}_{\text{c}} = \boldsymbol{Ad}^T \boldsymbol{F} - m\boldsymbol{g}_0 - \boldsymbol{F}_{\text{c}}^{\text{cpl}} \tag{6.10}$$

$$\boldsymbol{\delta}_{\text{sup}} = \boldsymbol{F}_{\text{sup}} - \boldsymbol{F}_{\text{sup}}^{\text{def}} \tag{6.11}$$

$$\boldsymbol{\delta}_{\text{int}} = \boldsymbol{F}_{\text{int}} + \boldsymbol{F}_{\text{int}}^{\text{cpl}} \tag{6.12}$$

with respect to the inequality constraints

$$\boldsymbol{A}_i \boldsymbol{F}_i \leq \boldsymbol{b}_i \qquad \forall\, i = 1 \ldots \psi_{\text{sup}} \tag{6.13}$$

$$\boldsymbol{B}_\tau \boldsymbol{F} \leq \boldsymbol{\tau}^{\max}. \tag{6.14}$$

Here, the commanded end-effector wrenches are stacked into $\boldsymbol{F} = (\,\boldsymbol{F}_{\text{sup}}^T\; \boldsymbol{F}_{\text{int}}^T\,)^T$. The residua within the cost function (6.9) are weighted with the symmetric and positive semi-definite matrices $\boldsymbol{Q}_{\text{c}} \in \mathbb{R}^{6\times 6}$, $\boldsymbol{Q}_{\text{sup}} \in \mathbb{R}^{6\psi_{\text{sup}}\times 6\psi_{\text{sup}}}$, and $\boldsymbol{Q}_{\text{int}} \in \mathbb{R}^{6\psi_{\text{int}}\times 6\psi_{\text{int}}}$.

The first residuum (6.10) implements the underactuation condition (6.6). The second one (6.11) regularizes the contact wrenches $\boldsymbol{F}_{\text{sup}}$ to a default wrench distribution $\boldsymbol{F}_{\text{sup}}^{\text{def}} \in \mathbb{R}^{6\psi_{\text{sup}}}$, which is ideally provided by an external planning algorithm as part of the reference trajectory (see Sect. 6.6). If there is no trajectory available, then $\boldsymbol{F}_{\text{sup}}^{\text{def}}$ can also be set manually. In this case, $\boldsymbol{F}_{\text{sup}}^{\text{def}}$ and $\boldsymbol{Q}_{\text{sup}}$ serve as tuning parameters for shifting the load between the support end-effectors. The third residuum (6.12) substitutes $\boldsymbol{F}_{\text{int}}^{\text{cpl}}$ in (6.6) with $\boldsymbol{F}_{\text{int}}^{\text{cpl}} = -\boldsymbol{F}_{\text{int}}$. The inequality constraint (6.13) ensures that all contact wrenches $\boldsymbol{F}_{\text{sup}}$ satisfy the contact model detailed in Sect. 3.4. A model of the actuators is given by (6.14) limiting the maximum joint torque to $\boldsymbol{\tau}^{\max} \in \mathbb{R}^n$ using $\boldsymbol{B}_\tau \in \mathbb{R}^{n\times\psi}$.

Formulating (6.12) and (6.10) as soft constraints instead of hard constraints features several advantages regarding the implementation, as discussed in Sect. 6.7. Besides the practical aspects, the above formulation allows for a prioritization of the residua or optimization objectives via the weighting matrices $\boldsymbol{Q}_{\text{c}}$, $\boldsymbol{Q}_{\text{sup}}$, and $\boldsymbol{Q}_{\text{int}}$ (see Table 6.1). The contact model (6.13) can be considered as the objective with the highest priority, as it will always be satisfied by $\boldsymbol{F}$. By choosing $\boldsymbol{Q}_{\text{c}} \gg \boldsymbol{Q}_{\text{int}} \gg \boldsymbol{Q}_{\text{sup}}$, the residuum $\boldsymbol{\delta}_{\text{sup}}$ becomes the goal with the lowest priority level. This means that the commanded contact wrenches $\boldsymbol{F}_{\text{sup}}$ will deviate from the default wrench distribution $\boldsymbol{F}_{\text{sup}}^{\text{def}}$ if it is necessary for fulfilling $\boldsymbol{\delta}_{\text{c}} \approx \boldsymbol{0}$ and $\boldsymbol{\delta}_{\text{int}} \approx \boldsymbol{0}$. By adjusting the weights inside $\boldsymbol{Q}_{\text{sup}}$, one can specify how much each end-effector is allowed to deviate from $\boldsymbol{F}_{\text{sup}}^{\text{def}}$. If the optimization cannot find a solution satisfying the contact model without violating the objectives $\boldsymbol{\delta}_{\text{c}} \approx \boldsymbol{0}$ and $\boldsymbol{\delta}_{\text{int}} \approx \boldsymbol{0}$, one of them will be sacrificed to render the optimization problem feasible again. Sacrificing $\boldsymbol{\delta}_{\text{c}} \approx \boldsymbol{0}$ or $\boldsymbol{\delta}_{\text{int}} \approx \boldsymbol{0}$ corresponds to a modification of the required overall wrench at the center of mass (right-hand side of (6.6)) such that the counteracting support wrench (left-hand side of (6.6)) becomes feasible with respect to the contact model. It is debatable whether $\boldsymbol{\delta}_{\text{c}} \approx \boldsymbol{0}$ or $\boldsymbol{\delta}_{\text{int}} \approx \boldsymbol{0}$ should be sacrificed first, but as the CoM is crucial for balancing, it is probably recommendable to assign the higher priority to the residuum $\boldsymbol{\delta}_{\text{c}}$ (see Table 6.1).

Note that using the weighting matrices $\boldsymbol{Q}_{\text{c}}$, $\boldsymbol{Q}_{\text{sup}}$, and $\boldsymbol{Q}_{\text{int}}$ does not result in a strict prioritization of the optimization objectives in contrast to methods from the field of hierarchical optimization. Here, the cQP is decomposed into a series of optimizations, each representing one of the objectives. In order to achieve a prioritization, each objective is projected onto the null space of all other objectives with a higher priority. The advantage of this method is that the series of optimizations can be solved sequentially, which is used for example in Escande et al. (2014), Herzog et al. (2016)

Table 6.1 Prioritization of the optimization objectives within the cQP

Priority level	Objective
1 (highest)	$\boldsymbol{A}_i \boldsymbol{F}_i \leq \boldsymbol{b}_i$ and $\boldsymbol{B}_\tau \boldsymbol{F} \leq \boldsymbol{\tau}^{\max}$
2	$\boldsymbol{\delta}_{\text{c}} \approx \boldsymbol{0}$
3	$\boldsymbol{\delta}_{\text{int}} \approx \boldsymbol{0}$
4 (lowest)	$\boldsymbol{\delta}_{\text{sup}} \approx \boldsymbol{0}$

for planning and control of humanoid robots. But as the task assignment of the end-effectors changes (e.g. during locomotion), the size of the individual optimizations must be adjusted online. For this particular reason, this work utilizes a non-strict hierarchy implemented via weighting matrices because it allows for a formulation of the cQP (see Sect. 6.7.2), featuring a constant size, which is independent of the task assignment. Furthermore, the weights can be exploited to implement a smooth contact transition, as shown in Sect. 6.7.3.

For the remainder of this work, we make the following assumption to simplify the discussion of the controller properties:

Assumption 6.1 (Fulfillment of $\boldsymbol{\delta}_{\mathrm{c}} \approx \mathbf{0}$ and $\boldsymbol{\delta}_{\mathrm{int}} \approx \mathbf{0}$) The residua $\boldsymbol{\delta}_{\mathrm{c}}$ and $\boldsymbol{\delta}_{\mathrm{int}}$ of the constrained quadratic optimization problem (cQP) are always sufficiently fulfilled, such that $\boldsymbol{\delta}_{\mathrm{c}} \approx \mathbf{0}$ and $\boldsymbol{\delta}_{\mathrm{int}} \approx \mathbf{0}$ hold.

This assumption implies that the commanded support wrenches $\boldsymbol{F}_{\mathrm{sup}}$ always comply with their respective contact model (6.13) and actuator limitations (6.14). Thus, there is no need for the optimization to violate the soft constraints (6.10) or (6.12). As a consequence, it is always possible to generate the required overall support wrench $\boldsymbol{F}_{\mathrm{sup}}^{\mathrm{res}}$, as demanded by Assumption 3.1.

Remark 6.1 Another interesting property is revealed if the wrench distribution problem is solved without considering the contact model or the actuator limitations. If we assume that the residua (6.10) and (6.12) are entirely fulfilled as in Assumption 6.1, then the optimization (6.9) simplifies to

$$\boldsymbol{F}_{\mathrm{sup}}^{\mathrm{opt}} = \underset{\boldsymbol{F}_{\mathrm{sup}}}{\mathrm{argmin}}\ \frac{1}{2}\left(\boldsymbol{F}_{\mathrm{sup}} - \boldsymbol{F}_{\mathrm{sup}}^{\mathrm{def}}\right)^T \boldsymbol{Q}_{\mathrm{sup}} \left(\boldsymbol{F}_{\mathrm{sup}} - \boldsymbol{F}_{\mathrm{sup}}^{\mathrm{def}}\right) \tag{6.15}$$

with respect to the underactuation constraint (6.6). The analytical solution of (6.15) is given by

$$\boldsymbol{F}_{\mathrm{sup}}^{\mathrm{opt}} = (\boldsymbol{Ad}_{\mathrm{sup}})^{\boldsymbol{Q}_{\mathrm{sup}}+,T} \left(m\boldsymbol{g}_0 + \boldsymbol{F}_{\mathrm{c}}^{\mathrm{cpl}} + \boldsymbol{Ad}_{\mathrm{int}}^T \boldsymbol{F}_{\mathrm{int}}^{\mathrm{cpl}}\right) + \boldsymbol{N}_{\mathrm{sup}} \boldsymbol{F}_{\mathrm{sup}}^{\mathrm{def}} \tag{6.16}$$

with $(\boldsymbol{Ad}_{\mathrm{sup}})^{\boldsymbol{Q}_{\mathrm{sup}}+} = \left(\boldsymbol{Ad}_{\mathrm{sup}}^T \boldsymbol{Q}_{\mathrm{sup}}^{-1} \boldsymbol{Ad}_{\mathrm{sup}}\right)^{-1} \boldsymbol{Ad}_{\mathrm{sup}}^T \boldsymbol{Q}_{\mathrm{sup}}^{-1}$ denoting the weighted pseudoinverse of the contact map $\boldsymbol{Ad}_{\mathrm{sup}}^T$. The associated null space projector is represented by $\boldsymbol{N}_{\mathrm{sup}} = \boldsymbol{I} - (\boldsymbol{Ad}_{\mathrm{sup}})^{\boldsymbol{Q}_{\mathrm{sup}}+,T} \boldsymbol{Ad}_{\mathrm{sup}}^T$. The pseudoinverse distributes the desired overall wrench at the CoM to the contacts without creating internal stress (see Sect. 3.3). Therefore, the distribution via the pseudoinverse can be considered a "minimal solution". The default wrench distribution $\boldsymbol{F}_{\mathrm{sup}}^{\mathrm{def}}$ acts in the null space of the Contact Map $\boldsymbol{Ad}_{\mathrm{sup}}^T$ and can be used to (a) shift load from one end-effector to another and (b) artificially generate internal stress.

After the end-effector wrenches $\boldsymbol{F}^{\mathrm{opt}}$ are computed via the constrained quadratic optimization using the upper part of the control law (6.8), the lower part is exploited to compute the control torques via

$$\boldsymbol{\tau} = -(\boldsymbol{J}')^T \boldsymbol{F}^{\mathrm{opt}}. \tag{6.17}$$

Note that this basic version of the passivity-based whole-body controller does not require an inversion of the Jacobian matrix, which makes it robust against singularities and redundant kinematics. A motion in the kinematic null space can be regulated by adding a null space controller, as shown in Sect. 6.2 (see also Fig. 6.1).

6.1.2 Properties of the Controller

6.1.2.1 End-Effector Wrenches

The purpose of the contact model presented in Sect. 3.4 is to prevent the supporting end-effectors from unintentionally loosing contact with the environment. Therefore, the basic version of the whole-body controller (MCB) accounts for the contact model via (6.13). However, the controller invokes the contact model on the commanded contact wrenches $\boldsymbol{F}_{\mathrm{sup}}$ although the stability of the contacts depends on the external or actual support wrenches $\boldsymbol{F}^{\mathrm{ext}}_{\mathrm{sup}}$. Thus, a more detailed analysis of the end-effector wrenches and the static equilibrium of the controller appears to be in order.

If the soft constraints (6.10) and (6.12) are exactly satisfied by the optimization (see Assumption 6.1), the closed-loop dynamics of the system is given by (6.7). If we furthermore assume that all external forces and torques act solely at the end-effectors as in (3.24), then (6.7) becomes

$$\boldsymbol{M}\begin{pmatrix}\dot{\boldsymbol{v}}_{\mathrm{c}}\\ \ddot{\boldsymbol{q}}\end{pmatrix} + \boldsymbol{C}\begin{pmatrix}\boldsymbol{v}_{\mathrm{c}}\\ \dot{\boldsymbol{q}}\end{pmatrix} = \begin{pmatrix}\boldsymbol{F}^{\mathrm{cpl}}_{\mathrm{c}}\\ \boldsymbol{0}\end{pmatrix} + \begin{bmatrix}\boldsymbol{Ad}^T\\ (\boldsymbol{J}')^T\end{bmatrix}\begin{pmatrix}-\boldsymbol{F}_{\mathrm{sup}} + \boldsymbol{F}^{\mathrm{ext}}_{\mathrm{sup}}\\ \boldsymbol{F}^{\mathrm{cpl}}_{\mathrm{int}} + \boldsymbol{F}^{\mathrm{ext}}_{\mathrm{int}}\end{pmatrix}. \tag{6.18}$$

Let us first consider the static case of (6.18) with $\dot{\boldsymbol{v}}_{\mathrm{c}} = \boldsymbol{v}_{\mathrm{c}} = \boldsymbol{0}$ and $\ddot{\boldsymbol{q}} = \dot{\boldsymbol{q}} = \boldsymbol{0}$:

$$\boldsymbol{0} = \begin{pmatrix}\boldsymbol{F}^{\mathrm{cpl}}_{\mathrm{c}}\\ \boldsymbol{0}\end{pmatrix} + \begin{bmatrix}\boldsymbol{Ad}^T\\ (\boldsymbol{J}')^T\end{bmatrix}\begin{pmatrix}-\boldsymbol{F}_{\mathrm{sup}} + \boldsymbol{F}^{\mathrm{ext}}_{\mathrm{sup}}\\ \boldsymbol{F}^{\mathrm{cpl}}_{\mathrm{int}} + \boldsymbol{F}^{\mathrm{ext}}_{\mathrm{int}}\end{pmatrix}. \tag{6.19}$$

If the Sub-Jacobian matrix $\boldsymbol{J}'$ is of full row-rank excluding singular configurations, one can conclude that

$$\boldsymbol{F}_{\mathrm{sup}} = \boldsymbol{F}^{\mathrm{ext}}_{\mathrm{sup}} \tag{6.20}$$

and

$$\boldsymbol{F}^{\mathrm{cpl}}_{\mathrm{int}} = -\boldsymbol{F}^{\mathrm{ext}}_{\mathrm{int}} \tag{6.21}$$

must hold. Reinserting (6.20) and (6.21) into (6.19) leads to

$$\boldsymbol{F}^{\mathrm{cpl}}_{\mathrm{c}} = \boldsymbol{0} \tag{6.22}$$

considering that the stacked adjoint matrix $\boldsymbol{Ad}$ is always of rank six. The relation (6.21) implies that in steady state, the external interaction wrenches $\boldsymbol{F}_{\text{int}}^{\text{ext}}$ are balanced by the interaction compliances $\boldsymbol{F}_{\text{int}}^{\text{cpl}}$. Therefore, the interaction end-effectors can show a steady state error according to (6.4) with $\boldsymbol{v}_i = \boldsymbol{v}_{i,d} = \boldsymbol{0}$, which is caused by $\boldsymbol{F}_{\text{int}}^{\text{ext}}$. Besides this, one can conclude from (6.22) that the center of mass frame C coincides with its desired configuration even in the presence of $\boldsymbol{F}_{\text{int}}^{\text{ext}}$ due to (6.1) with $\boldsymbol{v}_{\text{c}} = \boldsymbol{v}_{\text{c,d}} = \boldsymbol{0}$. Relation (6.20) is a rather important property because it states that the external contact wrenches $\boldsymbol{F}_{\text{sup}}^{\text{ext}}$ match the commanded ones $\boldsymbol{F}_{\text{sup}}$ in steady-state. Thus, the contact model successfully prevents the tilting, sliding, or lift-off of the contacts in the case of static or quasi-static balancing, although it is invoked on the commanded wrenches $\boldsymbol{F}_{\text{sup}}$ and not on external (actual) ones $\boldsymbol{F}_{\text{sup}}^{\text{ext}}$. If the robot is performing a more dynamic motion, (6.18) implies that $\boldsymbol{F}_{\text{sup}}^{\text{ext}}$ can deviate from $\boldsymbol{F}_{\text{sup}}$ and cause the support end-effectors to lose contact even despite $\boldsymbol{F}_{\text{sup}}$ satisfying the contact model.

If the dynamic motion is caused by an external disturbance, then this problem can only be remedied by applying the contact model to the actual contact wrenches. The latter requires an explicit measurement of $\boldsymbol{F}_{\text{sup}}^{\text{ext}}$, although this can cause secondary problems regarding causality or additional measurement noise, which is introduced into the controller. For this reason, this work applies the contact model only to the commanded and not to the actual contact wrenches. But if a dynamic trajectory is commanded to the robot, which means that the robot deliberately performs a dynamic motion, the discrepancy between $\boldsymbol{F}_{\text{sup}}$ and $\boldsymbol{F}_{\text{sup}}^{\text{ext}}$ can be avoided by equipping the controller with additional feedforward terms, as shown in Sect. 6.3 (MCB+ controller).

6.1.2.2 Passivity on Stationary Ground

The passivity of the closed-loop system (6.7) can be shown following (Ott 2008; Khalil 2014) by considering the kinetic energy of the robot plus the potential energy of the compliances

$$V = \frac{1}{2}\boldsymbol{\nu}^T \boldsymbol{M}\boldsymbol{\nu} + V_{\text{c}}^K + V_{\text{int}}^K \tag{6.23}$$

as positive definite storage function. Computing the time derivative and inserting the closed-loop dynamics (6.7) yields

$$\begin{aligned} \dot{V} &= \boldsymbol{\nu}^T \boldsymbol{M}\dot{\boldsymbol{\nu}} + \frac{1}{2}\boldsymbol{\nu}^T \dot{\boldsymbol{M}}\boldsymbol{\nu} + \dot{V}_{\text{c}}^K + \dot{V}_{\text{int}}^K \\ &= \boldsymbol{\nu}^T \left\{ -\boldsymbol{C}\boldsymbol{\nu} + \begin{pmatrix} \boldsymbol{F}_{\text{c}}^{\text{cpl}} \\ \boldsymbol{0} \end{pmatrix} + \begin{bmatrix} \boldsymbol{Ad}^T \\ (\boldsymbol{J}')^T \end{bmatrix} \begin{pmatrix} -\boldsymbol{F}_{\text{sup}} \\ \boldsymbol{F}_{\text{int}}^{\text{cpl}} \end{pmatrix} + \boldsymbol{\tau}_{\text{ext}} \right\} \\ &+ \frac{1}{2}\boldsymbol{\nu}^T \dot{\boldsymbol{M}}\boldsymbol{\nu} + \dot{V}_{\text{c}}^K + \dot{V}_{\text{int}}^K. \end{aligned} \tag{6.24}$$

Exploiting the fact that $\dot{M} - 2C$ is skew-symmetric (Ott 2008) allows for a simplification of $\dot{V}$ into

$$\begin{aligned}\dot{V} &= \nu^T \left\{ \begin{pmatrix} F_c^{cpl} \\ 0 \end{pmatrix} + \begin{bmatrix} Ad^T \\ (J')^T \end{bmatrix} \begin{pmatrix} -F_{sup} \\ F_{int}^{cpl} \end{pmatrix} + \tau_{ext} \right\} + \dot{V}_c^K + \dot{V}_{int}^K \\ &= v_c^T F_c^{cpl} - v_{sup}^T F_{sup} + v_{int}^T F_{int}^{cpl} + \nu^T \tau_{ext} + \dot{V}_c^K + \dot{V}_{int}^K\end{aligned} \tag{6.25}$$

considering (3.21) and (3.22). Inserting (6.1) and (6.4) with $v_{c,d} = 0$ and $v_{int,d} = 0$ for the regulation case yields

$$\begin{aligned}\dot{V} = &- \nabla V_c v_c - v_c^T D_c v_c + \dot{V}_c^K \\ &- \nabla V_{int} v_{int} - v_{int}^T D_{int} v_{int} + \dot{V}_{int}^K \\ &- v_{sup}^T F_{sup} + \nu^T \tau_{ext}.\end{aligned} \tag{6.26}$$

The latter can be simplified to

$$\dot{V} = \underbrace{-v_c^T D_c v_c - v_{int}^T D_{int} v_{int}}_{\leq 0} - v_{sup}^T F_{sup} + \underbrace{\nu^T \tau_{ext}}_{\text{port}} \tag{6.27}$$

considering $\dot{V}_c = \nabla V_c v_c$ and $\dot{V}_{int} = \nabla V_{int} v_{int}$. If we assume that the robot is balancing on a non-movable and non-deformable floor, then the term $v_{sup}^T F_{sup}$ vanishes due to $v_{sup} = 0$ (see Assumption 3.2). Thus,

$$\dot{V} \leq \nu^T \tau_{ext} \tag{6.28}$$

holds, which proves the passivity of the closed-loop system (6.7) with respect to the power port $(v_c^T \; \dot{q}^T) \, \tau_{ext}$ based on the definition given by Khalil (2014). The port represents the energy transfer between the robot and the environment as, for instance, a human disturbing the robot.

Note that the above analysis applies to the desired closed-loop dynamics (6.7), which implies that the soft constraints (6.10) and (6.12) are exactly fulfilled, as demanded by Assumption 6.1. A passivity and stability analysis of the case where the soft constraints are not exactly fulfilled would be considerably more cumbersome. In this case, the constraints will affect the quality of the task fulfillment according to their corresponding priorities (see Table 6.1), which would lead to a different closed-loop dynamics.

6.1.2.3 Stability on Stationary Ground

Asymptotic stability of the closed-loop dynamics (6.7) can be shown following the Invariant Set Theorem presented in Slotine and Lee (1991), Khalil (2014), which requires $\dot{V} \leq 0$ for the autonomous closed-loop system.

Theorem 6.1 (**Local Invariant Set Theorem** Slotine and Lee 1991) *Consider the autonomous system $\dot{\boldsymbol{x}} = \boldsymbol{f}(\boldsymbol{x})$ with $\boldsymbol{f}(\boldsymbol{x})$ continuous, and let $V(\boldsymbol{x})$ be a scalar function with continuous first partial derivatives. Assume that*

- *for some $l \geq 0$, the region $\mathcal{X}_l$ defined by $V(\boldsymbol{x}) \leq l$ is bounded*
- *and $\dot{V}(\boldsymbol{x}) \leq 0$ for all $\boldsymbol{x}$ in $\mathcal{X}_l$.*

Let $\mathcal{R}$ be the set of all points within $\mathcal{X}_l$ where $\dot{V}(\boldsymbol{x}) = \boldsymbol{0}$, and $\mathcal{M}$ be the largest invariant set in $\mathcal{R}$. Then, every solution $\boldsymbol{x}(t)$ originating in $\mathcal{X}_l$ tends to $\mathcal{M}$ as $t \to \infty$.

For simplicity, let us exclude singular configurations and redundant kinematics from our considerations. Furthermore, let us assume that the residua $\boldsymbol{\delta}_{\text{c}}$ and $\boldsymbol{\delta}_{\text{int}}$ are sufficiently fulfilled (see Assumption 6.1). The closed-loop dynamics (6.7) becomes autonomous if there are no external loads acting on the robots except for the external contact wrenches $\boldsymbol{F}_{\text{sup}}^{\text{ext}}$. Then (6.27) becomes

$$\begin{aligned} \dot{V} &= -\boldsymbol{v}_{\text{c}}^T \boldsymbol{D}_{\text{c}} \boldsymbol{v}_{\text{c}} - \boldsymbol{v}_{\text{int}}^T \boldsymbol{D}_{\text{int}} \boldsymbol{v}_{\text{int}} + \boldsymbol{\nu}^T \boldsymbol{\tau}_{\text{ext}} \\ &= -\boldsymbol{v}_{\text{c}}^T \boldsymbol{D}_{\text{c}} \boldsymbol{v}_{\text{c}} - \boldsymbol{v}_{\text{int}}^T \boldsymbol{D}_{\text{int}} \boldsymbol{v}_{\text{int}} + \boldsymbol{\nu}^T \boldsymbol{J}_{\text{sup}}^T \boldsymbol{F}_{\text{sup}}^{\text{ext}} \\ &= -\boldsymbol{v}_{\text{c}}^T \boldsymbol{D}_{\text{c}} \boldsymbol{v}_{\text{c}} - \boldsymbol{v}_{\text{int}}^T \boldsymbol{D}_{\text{int}} \boldsymbol{v}_{\text{int}} + \boldsymbol{v}_{\text{sup}}^T \boldsymbol{F}_{\text{sup}}^{\text{ext}} \\ &= -\boldsymbol{v}_{\text{c}}^T \boldsymbol{D}_{\text{c}} \boldsymbol{v}_{\text{c}} - \boldsymbol{v}_{\text{int}}^T \boldsymbol{D}_{\text{int}} \boldsymbol{v}_{\text{int}} \leq 0. \end{aligned} \tag{6.29}$$

considering (3.21), a stationary floor ($\boldsymbol{v}_{\text{sup}} = \boldsymbol{0}$), and Assumption 3.2. The time derivative $\dot{V}$ only vanishes if $\boldsymbol{\nu} = \left(\boldsymbol{v}_{\text{c}}^T \ \boldsymbol{v}_{\text{int}}^T\right)^T = \boldsymbol{0}$ and if singular configurations and redundant kinematics are excluded from the considerations. Thus, the set $\mathcal{R}$ contains all points in the states space of the system (6.7) with $\boldsymbol{\nu} = \boldsymbol{0}$. Inserting this into (6.7) reveals

$$\boldsymbol{M}\dot{\boldsymbol{\nu}} = \begin{pmatrix} \boldsymbol{F}_{\text{c}}^{\text{cpl}} \\ \boldsymbol{0} \end{pmatrix} + \begin{bmatrix} \boldsymbol{Ad}^T \\ (\boldsymbol{J}')^T \end{bmatrix} \begin{pmatrix} \boldsymbol{F}_{\text{sup}}^{\text{ext}} - \boldsymbol{F}_{\text{sup}} \\ \boldsymbol{F}_{\text{int}}^{\text{cpl}} \end{pmatrix}. \tag{6.30}$$

The largest invariant set $\mathcal{M}$ within $\mathcal{R}$ is given by $\boldsymbol{\nu} = \boldsymbol{0}$ and $\dot{\boldsymbol{\nu}} = \boldsymbol{0}$. In order for $\dot{\boldsymbol{\nu}}$ to vanish, $\boldsymbol{F}_{\text{sup}} = \boldsymbol{F}_{\text{sup}}^{\text{ext}}$ and $\boldsymbol{F}_{\text{int}}^{\text{cpl}} = \boldsymbol{0}$ must hold because of the Sub-Jacobian matrix $\boldsymbol{J}'$ being of full rank (no singularities and redundancies). Furthermore, one can conclude from the upper part of (6.30) that $\boldsymbol{F}_{\text{c}}^{\text{cpl}} = \boldsymbol{0}$ must hold as well. It can be concluded from (6.1) and (6.4) that $\boldsymbol{F}_{\text{c}}^{\text{cpl}}$ and $\boldsymbol{F}_{\text{int}}^{\text{cpl}}$ only vanish if the CoM frame $\mathcal{C}$ and all interaction frames $\mathcal{T}_i$ coincide with their desired configurations or if they show a rotation error of 180°. If we also exclude configurations with a rotation error of 180°, then the largest invariant set $\mathcal{M}$ within $\mathcal{R}$ is given by the static desired configuration of the robot. Consequently, the closed-loop system tends to the desired configuration as $t \to \infty$.

6.1.3 Experimental Evaluation

The performance of the MCB control framework was evaluated in Henze et al. (2016b) using the torque-controlled humanoid robot *TORO* developed at *DLR*. A brief description of the robot hardware is given in Chap. 5.

6.1.3.1 Redundancy in the Wrench Distribution

The first experiment demonstrates the influence of the constrained quadratic optimization problem (6.9) to (6.14) on the wrench distribution (Henze et al. 2016b). The setup of the experiment is shown in Fig. 6.3a, where the robot is using the right hand and both feet as supporting end-effectors. The left hand is assigned to the interaction task. The robot is commanded into two different configurations by shifting the CoM 17 cm to the right or 10 cm to the left. The recorded telemetry data in Fig. 6.3b reveals that the CoM position shows a static control error of less than 0.03 m due to joint friction. However, this could be counteracted by adding an integral term to the CoM compliance (6.1). The combined center of pressure of both feet $\boldsymbol{p}_{\text{Feet}}$ also shows a good congruence of the commanded and measured values, which is an indicator that the external contact wrenches $\boldsymbol{F}^{\text{ext}}_{\text{sup}}$ are sufficiently close to the commanded ones $\boldsymbol{F}_{\text{sup}}$.

The most interesting aspect of this experiment is the influence of the CoM position on the wrench distribution. If the CoM is shifted to the left, the right hand cannot contribute to supporting the robot because of the unilateral contact (the hand is not closed). Instead, the optimization commands the minimal allowed contact force $f^{\text{min}}_{\text{HandR},y} = 20$ N according to Sect. 3.4. As a result, the CoP of the feet is shifted 10 cm to the left to support the robot's weight. Note that the CoP is not located directly under the CoM because of the torque that the right hand introduces about the x-axis of the world frame $\mathcal{W}$.

If the robot moves to the right, the optimization can exploit the additional hand contact by generating a force of $f_{\text{HandR},y} = 65$ N. Due to the high lever arm of the hand (height of 1.42 m above floor), the hand can completely counteract the torque, which the CoM displacement induces about the x-axis of $\mathcal{W}$. As a result, the CoP of the feet stays centered at $p_{\text{Feet},y} = 0$ cm. This observation is remarkable, as the CoM is shifted further to the right (17 cm) than to the left (10 cm).

In summary, the controller can exploit the redundancy in the wrench distribution problem to obtain an optimal distribution of contact wrenches according to the cost function (6.9). In doing so, the optimization must respect the contact model (6.13) limiting the contribution of each contact to the overall support wrench $\boldsymbol{F}^{\text{res}}_{\text{sup}}$. To achieve the required $\boldsymbol{F}^{\text{res}}_{\text{sup}}$ nevertheless, the controller can exploit the redundancy by redistributing the wrenches between the contacts.

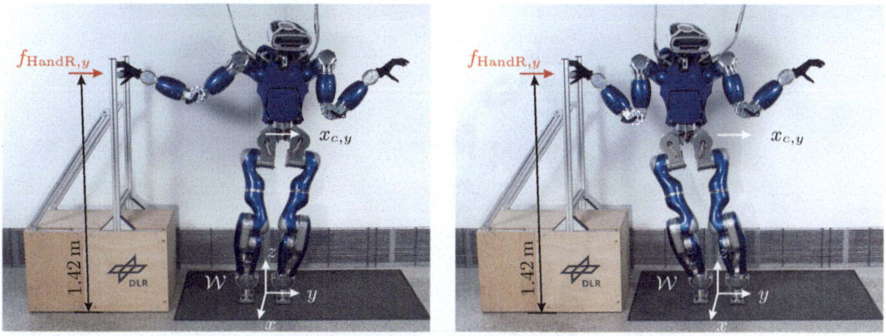

(a) CoM shifted to the left and to the right.

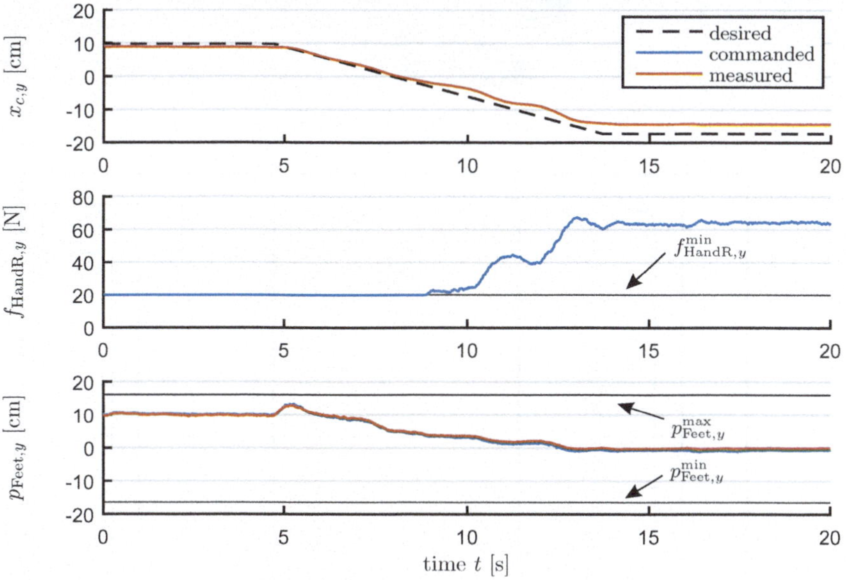

(b) CoM position and contact wrenches.

Fig. 6.3 Exploiting the redundancy in the wrench distribution: the right hand contributes to the required overall support wrench $\boldsymbol{F}^{\text{res}}_{\text{sup}}$ depending on the CoM position

6.1.3.2 Balancing While Performing an Interaction Task

This experiment shows the capability of the whole-body balancer to maintain balance while performing an interaction task (Henze et al. 2016b). As shown in Fig. 6.4a, the robot *TORO* stands in front of a table, from which it lifts up a box of bottles. The weight of the box is 12.2 kg, which corresponds to 16% of the robot mass (76.4 kg). In this particular setup, the hands are assigned to the interaction task and thereby governed by the Cartesian compliance (6.4). In order to lift the box, the desired

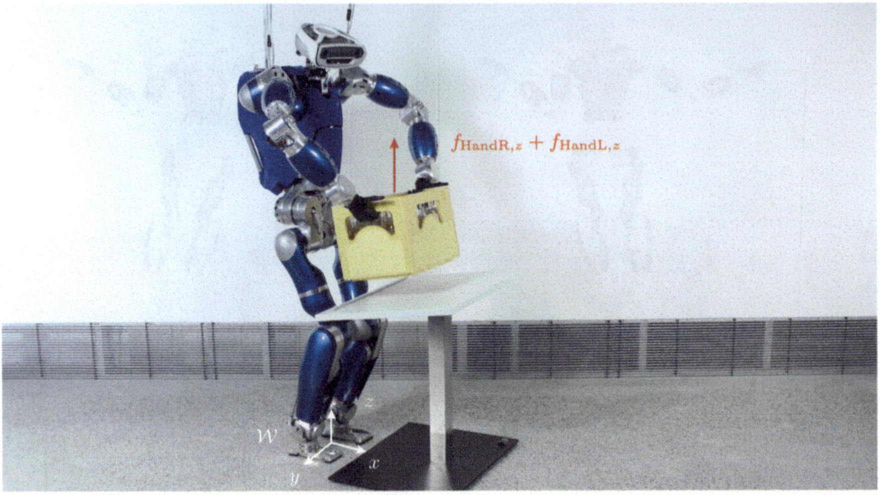

(a) Setup of the experiment.

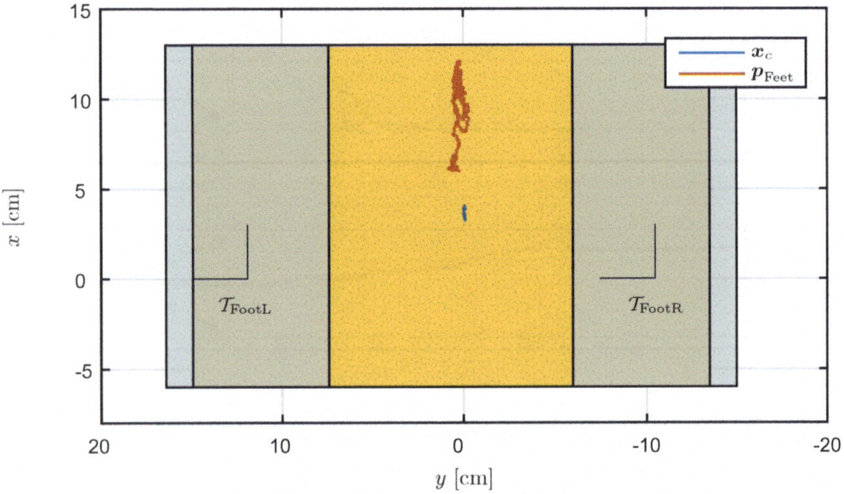

(b) Trajectory of the CoM and the CoP relative to the support polygon $\mathcal{S}_{\mathrm{SP}}$ (yellow). The contact areas of the feet $\mathcal{S}_{\mathrm{FootR}}$ and $\mathcal{S}_{\mathrm{FootL}}$ are given in gray.

Fig. 6.4 Lifting a 12.2 kg box from a table: the robot maintains the CoM position by compensating for the additional load with its feet

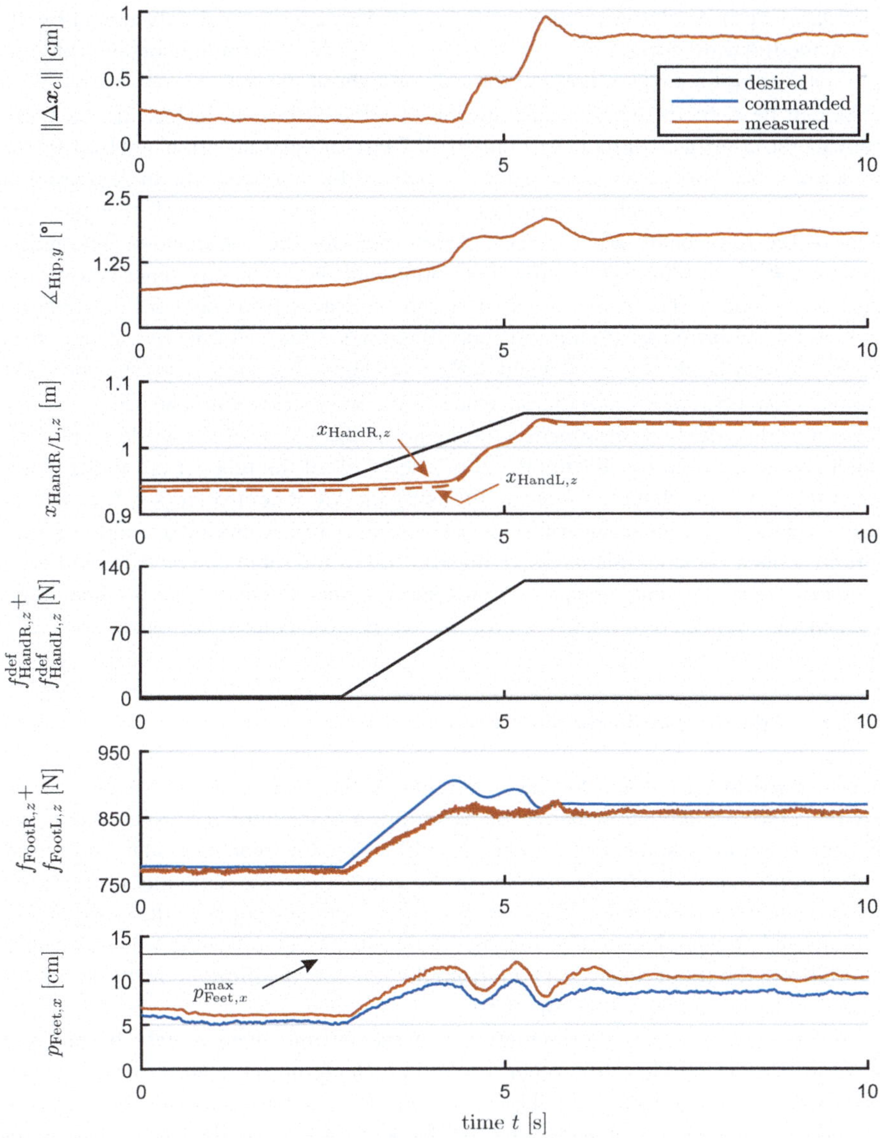

(c) CoM position, hip orientation, and end-effector wrenches.

Fig. 6.4 (continued)

position of both hand compliances ($\boldsymbol{x}_{\text{HandR,d}}$ and $\boldsymbol{x}_{\text{HandL,d}}$) are slowly raised by 10 cm. A feedforward force ($f^{\text{def}}_{\text{HandR},z} = f^{\text{def}}_{\text{HandL},z} = 62$ N) is simultaneously added to each hand compliance to compensate for the weight of the box.

As can be seen in Fig. 6.4c, lifting the box has almost no effect on the CoM frame C: the control error increases only by 0.7 cm for the translation and by 1.0° for the orientation. This observation can be explained by analyzing the underactuation condition (6.6): The contact wrenches $\boldsymbol{F}^{\text{opt}}_{\text{sup}}$ are always chosen such that they are equal to the right-hand side of (6.6), which includes the commanded interaction wrenches $\boldsymbol{F}^{\text{opt}}_{\text{int}}$. In other words, the feet counteract the wrenches generated by the hand compliances. The same conclusion can be drawn from Sect. 6.1.2, studying the static equilibrium of the closed-loop dynamics: One can conclude from (6.1) that (6.22) only holds if $\Delta\boldsymbol{x}_{\text{c}} = \boldsymbol{0}$ and $\Delta\boldsymbol{R}_{\text{c}} = \boldsymbol{I}$. Thus, the control error for the CoM frame C must be zero even in the presence of the interaction wrenches $\boldsymbol{F}^{\text{cpl}}_{\text{int}}$.

Besides this, the box introduces a torque about the y-axis of the world frame $\mathcal{W}$, which is counteracted by shifting the combined CoP of the feet 4.1 cm to the front. Note that the weight that the robot can lift using the MCB controller (see Sect. 6.1) is limited by the CoP constraints of the feet (see Fig. 6.4b). Although a weight of 12.2 kg represents a considerable payload, the abilities of the robot can be extended even further by using the control approach presented in Sect. 6.5 (MCB-ACT). The same experiment is repeated in Sect. 6.5.3 for the MCB-ACT control approach.

6.1.3.3 Balancing on Compliant Support

As discussed in Sect. 6.1.2, the basic version of the presented control framework (MCB) is passive and stable under the assumption of a stationary ground floor. We presented an experiment in Henze et al. (2016b) testing the robustness of the proposed control approach against uncertainties in the contact surface deliberately violating the above-named assumption. To generate a soft, compliant support, the robot *TORO* is positioned on three layers of gym mats,[2] as shown in Fig. 6.5a. After the controller is initialized, the robot receives a kick from the left side resulting in the perturbation of the CoM position shown in Fig. 6.5b.

The robot counteracts the disturbance in the frontal plane mainly by shifting its weight from one leg to the other, instead of changing the center of pressure of each foot ($p_{\text{FootR},y}$ and $p_{\text{FootL},y}$). The observed effect is caused by the choice of the weighting matrix $\boldsymbol{Q}_{\text{sup}}$ in Henze et al. (2016b), which penalizes the foot torques more than the foot forces. For balancing in the sagittal plane, the controller does not have any other choice than using the foot torques. Thus, the CoPs of both feet ($p_{\text{FootR},x}$ and $p_{\text{FootL},x}$) exhibit a high excitation covering the whole length of the feet. The strength of the kick brings the robot to the limit of feasible balancing: At 0.5 s, the vertical force of the left foot ($f_{\text{FootL},z}$) almost reaches the lower limit of $f^{\text{min}}_{\text{FootL},z} = 50$ N. The corresponding commanded CoP $\boldsymbol{p}_{\text{FootL}}$ moves to the front-right corner of the foot

[2] The gym mats have a size of 2 m $\times$2 m $\times$ 8 cm, and are made of a compound foam with a density of 20 $^{\text{kg}}/_{\text{m}^3}$.

indicated by $p_{\text{FootL},x} = p^{\max}_{\text{FootL},x} = +13$ cm and $p_{\text{FootL},y} = p^{\min}_{\text{FootL},y} = -4.5$ cm. The latter causes the heel of the left foot to lift off such that only the front part of the left foot is in contact with the mattresses (see Fig. 6.5a).

The difficulty in balancing on a compliant support lies in the unmodeled dynamics of the ground floor: The material of the support must be deformed first before the desired contact wrench can be generated. Therefore, the vertical contact forces show a temporary deviation between the commanded and the measured values of up to 83 N for the right and 111 N for the left leg (see Fig. 6.5c). Also, the CoP positions exhibit a strong deviation between command and measurement. According to the sensor readings, the CoP of the right foot even leaves the contact area at $t = 1.5$ s, which appears to be impossible according to Definition 3.3. However, the force-torque sensor for measuring the contact wrench is not located at the sole of the foot but at the ankle (see Fig. 5.1). Although the computation of the CoP accounts for the additional lever arm, dynamic effects can still impair the measurement. The recorded video (see *YouTube*-channel RMC 2019 of the *Institute of Robotics and Mechatronics*) shows that the feet are constantly moving during the whole experiment. In particular, the right foot has a high acceleration at $t = 1.5$ s, which might cause the inertia of the right foot to influence the measurement.

The algorithm presented in Sect. 5.2 for estimating the state of the base/hip frame $\mathcal{B}$, is one of the key elements for achieving a robust balancing behavior. Due to the compliant support, the feet are constantly moving in order to adapt the support surface. By using only the IMU signal ($\alpha_{\text{IMU}} = 1$ in Sect. 5.2), the estimated orientation of the base $\boldsymbol{R}_b$ becomes independent of the foot motion. The influence on the base position $\boldsymbol{x}_b$ is minimized by averaging the translational information provided by the kinematics of the right and left leg ($\alpha_{\text{FootR}} = \alpha_{\text{FootL}} = 1$ and $\alpha_{\text{HandR}} = \alpha_{\text{HandL}} = 0$ in Sect. 5.2). Note that Fig. 6.5 shows the position of the CoM that was computed online by the robot using the estimation algorithm. The estimated CoM location is also used for control, which demonstrates the robustness of the presented approach with respect to unmodeled contact dynamics.

6.2 Null Space Controller

As the basic version of the multi-contact balancing framework (MCB) presented in Sect. 6.1 regulates the configuration of the robot in Cartesian or task coordinates, there might exist a null space within the configuration space of the robot. To prevent a drift within the null space, a conventional null space controller is added by Henze et al. (2016b) to the control scheme, as shown in Fig. 6.1. Instead of commanding the joint torque $\boldsymbol{\tau}$ in (6.17) directly to the robot, it is superimposed with a joint compliance $\boldsymbol{\tau}^{\text{cpl}}_{\text{pose}} \in \mathbb{R}^n$ according to

$$\boldsymbol{\tau}' = \boldsymbol{\tau} + \boldsymbol{N}_{\text{null}}\, \boldsymbol{\tau}^{\text{cpl}}_{\text{pose}}. \tag{6.31}$$

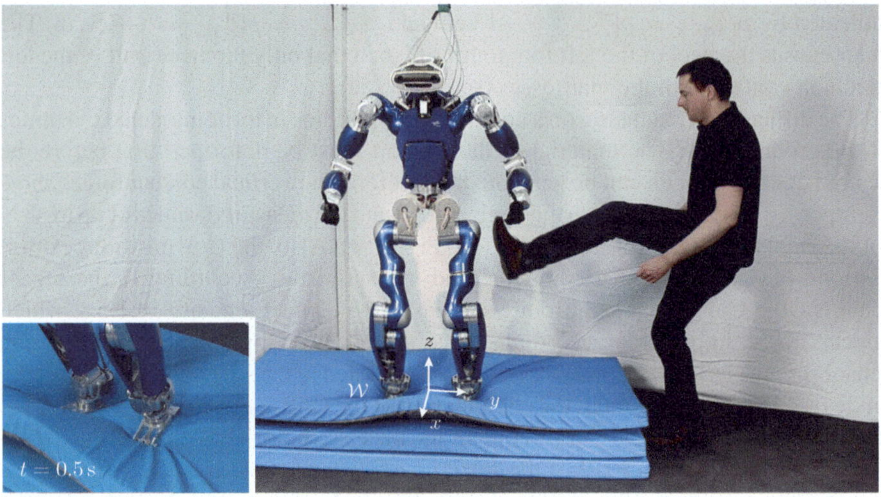

(a) Setup of the experiment.

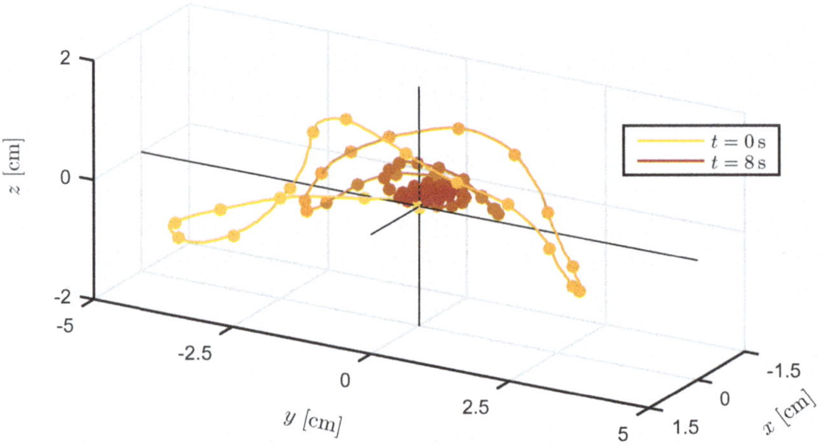

(b) Trajectory of the CoM in three-dimensional space. Each dot represents a measurement every $0.1\,\mathrm{s}$.

Fig. 6.5 Balancing on compliant support: the perturbation of the robot is caused by a kick from the left

The null space projector $\boldsymbol{N}_{\text{null}} = \boldsymbol{I} - (\boldsymbol{J}')^T (\boldsymbol{J}')^{\boldsymbol{M}_{\text{qq}}+,T}$ is computed from the dynamically consistent pseudoinverse

$$(\boldsymbol{J}')^{\boldsymbol{M}_{\text{qq}}+} = \boldsymbol{M}_{\text{qq}}^{-1} (\boldsymbol{J}')^T \left((\boldsymbol{J}') \boldsymbol{M}_{\text{qq}} (\boldsymbol{J}')^T \right)^{-1}, \tag{6.32}$$

as given in Dietrich (2016). The weighting matrix consists of $\boldsymbol{M}_{\text{qq}}$ in (3.20), which is the block of the inertia matrix $\boldsymbol{M}$ correlating with a motion in joint space. Note that

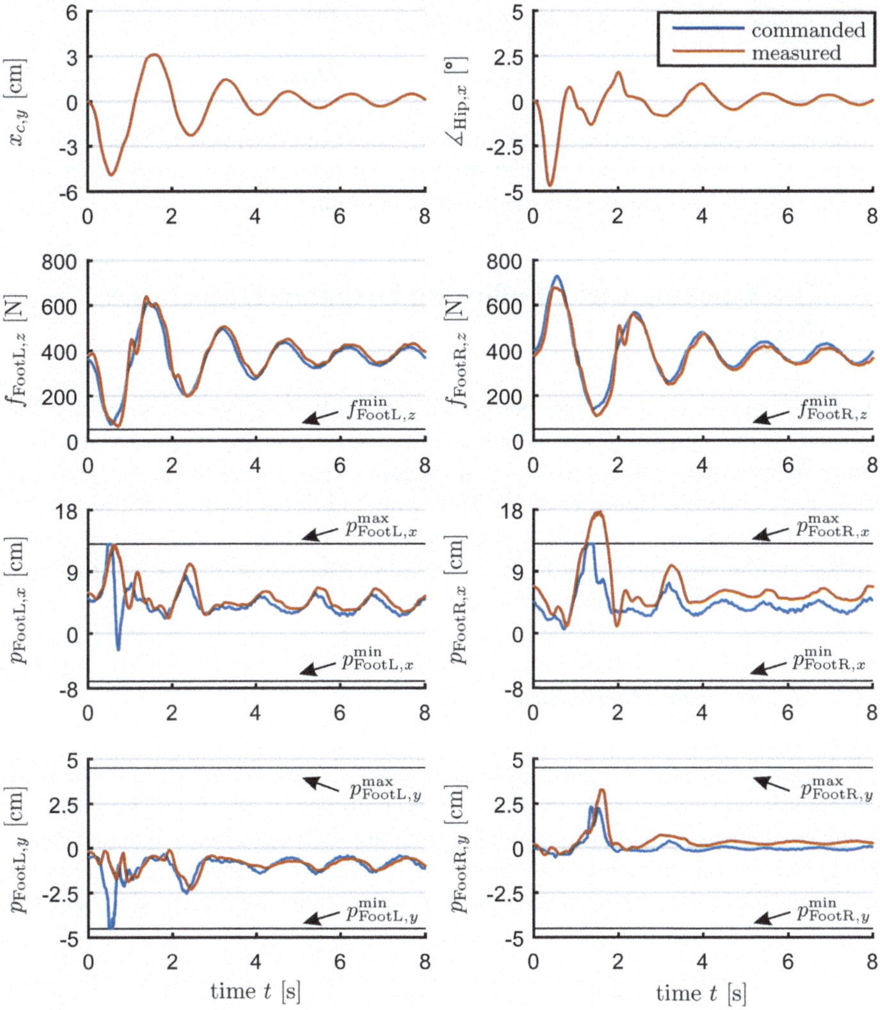

(c) CoM position, hip orientation, and end-effector wrenches.

Fig. 6.5 (continued)

the null space projector $\boldsymbol{N}_{\mathrm{null}}$ is computed from the Sub-Jacobian matrix $\boldsymbol{J}'$ instead of the complete Jacobian $\boldsymbol{J} = [\,\boldsymbol{Ad}\ \boldsymbol{J}'\,]$ (see (3.23)). Although this is a rather heuristic choice, several experiments have revealed that it leads to a very robust behavior of the robot even in challenging situations (Henze et al. 2019; Werner et al. 2016). A theoretically correct approach is presented in Chap. 7 in the form of a framework for hierarchical whole-body control. See also the discussion in Sect. 7.2.2 on the relation between the hierarchical framework and the MCB controller.

The torque $\boldsymbol{\tau}^{\text{cpl}}_{\text{pose}} \in \mathbb{R}^n$ in (6.31) is provided by a PD-controller

$$\boldsymbol{\tau}^{\text{cpl}}_{\text{pose}} = -\boldsymbol{K}_{\text{pose}} (\boldsymbol{q} - \boldsymbol{q}_d) - \boldsymbol{D}_{\text{pose}}\, \dot{\boldsymbol{q}} \,. \tag{6.33}$$

stabilizing the desired joint configuration $\boldsymbol{q}_d$, which can be specified by an external planning algorithm, for instance. The stiffness and damping matrices $\boldsymbol{K}_{\text{pose}} \in \mathbb{R}^{n\times n}$ and $\boldsymbol{D}_{\text{pose}} \in \mathbb{R}^{n\times n}$ are symmetric and positive definite.

6.3 The Tracking Case: Following Dynamic Trajectories (MCB+)

The basic version of the whole-body controller (MCB) presented in Sect. 6.1 is designed for the regulation case, which enables static balancing or quasi-static locomotion on multiple contacts. To account for more dynamic motions, the basic version was extended in Henze et al. (2016b) with additional feedforward terms (see Fig. 6.6) for tracking a dynamic trajectory. As the resulting closed-loop dynamics shows a similar structure as in PD+ control, the extended approach will be called MCB+ controller for the remainder of this work.

6.3.1 Controller Derivation

The extended framework consists of three branches, as shown in Fig. 6.6: The feedback control loop is identical to the basic MCB controller (cf. Fig. 6.1). The additional feedforward terms (shown in green in Fig. 6.6) are used to follow a dynamic trajectory. In order to handle robots with redundant kinematics or singular configurations, the framework is also equipped with the null space controller presented in Sect. 6.2.

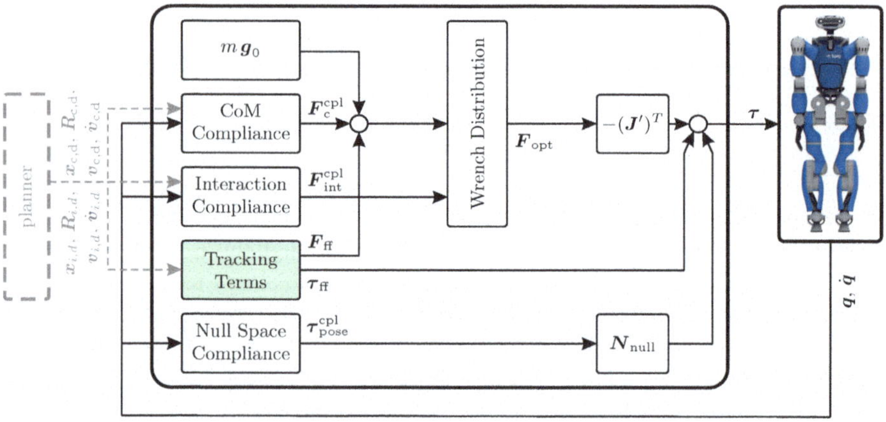

Fig. 6.6 Overview of the architecture of the MCB+ control framework extended to the tracking case

In order to deduce the tracking terms as proposed in Henze et al. (2016b), the dynamic model (3.20) is transfered into task coordinates, which requires the Sub-Jacobian matrix $\boldsymbol{J}'$ to be square and invertible. However, this assumption is only required for the feedforward terms and not for the feedback loop. If the Sub-Jacobian matrix $\boldsymbol{J}'$ is singular or not square, then the task space can be augmented with additional coordinates, as shown e.g. in Ott (2008), such that the augmented Jacobian matrix is invertible again. But for the sake of simplicity, we will restrict our theoretical considerations to $\boldsymbol{J}'$ being square and invertible. To use the feedforward terms for redundant robots and singular configurations nevertheless, we propose the use of a damped pseudoinverse, as discussed in (6.53).

Based on the assumption that Sub-Jacobian matrix $\boldsymbol{J}'$ is square and invertible, the joint velocities $\dot{\boldsymbol{q}}$ are replaced in the dynamic model (3.20) with the Cartesian velocities $\boldsymbol{v}$ of the task or end-effector frames. The corresponding transformation $\boldsymbol{T}_{\mathrm{TS}}$ to task space is defined by

$$\begin{pmatrix} \boldsymbol{v}_{\mathrm{c}} \\ \boldsymbol{v} \end{pmatrix} = \underbrace{\begin{bmatrix} \boldsymbol{I} & \boldsymbol{0} \\ \boldsymbol{Ad} & \boldsymbol{J}' \end{bmatrix}}_{\boldsymbol{T}_{\mathrm{TS}}} \begin{pmatrix} \boldsymbol{v}_{\mathrm{c}} \\ \dot{\boldsymbol{q}} \end{pmatrix}, \tag{6.34}$$

which allows for the formulation of the inverse transformation $\boldsymbol{T}_{\mathrm{TS}}^{-1}$ as

$$\begin{pmatrix} \boldsymbol{v}_{\mathrm{c}} \\ \dot{\boldsymbol{q}} \end{pmatrix} = \underbrace{\begin{bmatrix} \boldsymbol{I} & \boldsymbol{0} \\ -(\boldsymbol{J}')^{-1}\boldsymbol{Ad} & (\boldsymbol{J}')^{-1} \end{bmatrix}}_{\boldsymbol{T}_{\mathrm{TS}}^{-1}} \begin{pmatrix} \boldsymbol{v}_{\mathrm{c}} \\ \boldsymbol{v} \end{pmatrix}. \tag{6.35}$$

Applying the inverse transformation $\boldsymbol{T}_{\mathrm{TS}}^{-1}$ to the dynamic model (3.20) leads to a formulation of the robot dynamics in task coordinates

$$\boldsymbol{\Lambda} \begin{pmatrix} \dot{\boldsymbol{v}}_{\mathrm{c}} \\ \dot{\boldsymbol{v}} \end{pmatrix} + \boldsymbol{\mu} \begin{pmatrix} \boldsymbol{v}_{\mathrm{c}} \\ \boldsymbol{v} \end{pmatrix} + \underbrace{\begin{pmatrix} m\boldsymbol{g}_0 \\ \boldsymbol{0} \end{pmatrix}}_{\boldsymbol{g}} = \begin{bmatrix} -\boldsymbol{Ad}^T \\ \boldsymbol{I} \end{bmatrix} (\boldsymbol{J}')^{-T}\boldsymbol{\tau} + \boldsymbol{T}_{\mathrm{TS}}^{-T}\boldsymbol{\tau}_{\mathrm{ext}} \tag{6.36}$$

with $\boldsymbol{\Lambda} = \boldsymbol{T}_{\mathrm{TS}}^{-T}\boldsymbol{M}\boldsymbol{T}_{\mathrm{TS}}^{-1}$ and $\boldsymbol{\mu} = \boldsymbol{T}_{\mathrm{TS}}^{-T}\boldsymbol{C}\boldsymbol{T}_{\mathrm{TS}}^{-1} + \boldsymbol{T}_{\mathrm{TS}}^{-T}\boldsymbol{M}\frac{d}{dt}(\boldsymbol{T}_{\mathrm{TS}}^{-1})$ denoting the Cartesian inertia and Coriolis matrix. It can be easily verified that the gravity vector $\boldsymbol{g} = \boldsymbol{T}_{\mathrm{TS}}^{-T}\boldsymbol{g}$ is invariant under the transformation $\boldsymbol{T}_{\mathrm{TS}}^{-1}$.

As in Henze et al. (2016b), the desired closed-loop dynamics can be defined as

$$\boldsymbol{\Lambda} \begin{pmatrix} \Delta\dot{\boldsymbol{v}}_{\mathrm{c}} \\ \Delta\dot{\boldsymbol{v}}_{\mathrm{sup}} \\ \Delta\dot{\boldsymbol{v}}_{\mathrm{int}} \end{pmatrix} + \boldsymbol{\mu} \begin{pmatrix} \Delta\boldsymbol{v}_{\mathrm{c}} \\ \Delta\boldsymbol{v}_{\mathrm{sup}} \\ \Delta\boldsymbol{v}_{\mathrm{int}} \end{pmatrix} = \begin{pmatrix} \boldsymbol{F}_{\mathrm{c}}^{\mathrm{cpl}} \\ -\boldsymbol{F}_{\mathrm{sup}} \\ \boldsymbol{F}_{\mathrm{int}}^{\mathrm{cpl}} \end{pmatrix} + \boldsymbol{T}_{\mathrm{TS}}^{-T}\boldsymbol{\tau}_{\mathrm{ext}} \tag{6.37}$$

with $\Delta\boldsymbol{v}_{\mathrm{c}}$, $\Delta\boldsymbol{v}_{\mathrm{sup}}$, and $\Delta\boldsymbol{v}_{\mathrm{int}}$ denoting the deviation from the desired trajectory. The velocity errors of the CoM $\Delta\boldsymbol{v}_{\mathrm{c}}$ and the interaction end-effectors $\Delta\boldsymbol{v}_{\mathrm{int}}$ are already defined in (6.2) and (6.5), respectively. Let us analogously define the velocity error for the support task as

$$\underbrace{\begin{pmatrix} \Delta \boldsymbol{v}_1 \\ \vdots \\ \Delta \boldsymbol{v}_{\psi_{\text{sup}}} \end{pmatrix}}_{\Delta \boldsymbol{v}_{\text{sup}}} = \underbrace{\begin{pmatrix} \boldsymbol{v}_1 \\ \vdots \\ \boldsymbol{v}_{\psi_{\text{sup}}} \end{pmatrix}}_{\boldsymbol{v}_{\text{sup}}} - \underbrace{\begin{bmatrix} \boldsymbol{\Omega}_1 & \dots & \mathbf{0} \\ \vdots & \ddots & \vdots \\ \mathbf{0} & \dots & \boldsymbol{\Omega}_{\psi_{\text{sup}}} \end{bmatrix}}_{\boldsymbol{\Omega}_{\text{sup}}} \underbrace{\begin{pmatrix} \boldsymbol{v}_{1,d} \\ \vdots \\ \boldsymbol{v}_{\psi_{\text{sup}},d} \end{pmatrix}}_{\boldsymbol{v}_{\text{sup}}} . \tag{6.38}$$

Note that the actual velocities of the support contacts $\boldsymbol{v}_{\text{sup}}$ as well as the desired ones $\boldsymbol{v}_{\text{sup,d}}$ are zero in the case of a stationary and rigid ground surface leading to $\Delta \boldsymbol{v}_{\text{sup}} = \mathbf{0}$. The compliances $\boldsymbol{F}_{\text{c}}^{\text{cpl}}$ and $\boldsymbol{F}_{\text{int}}^{\text{cpl}}$ are defined according to (6.1) and (6.4) based on the desired configurations and velocities given by the trajectory. Analogously to (6.7), the right-hand side of (6.37) also contains the negative support wrenches $(-\boldsymbol{F}_{\text{sup}})$. Note that the structure of (6.37) is inspired by a PD+ control of an ordinary fixed-base robotic manipulator, which is well-known in control theory (Paden and Panja 1988; Whitcomb et al. 1993).

Comparing the desired closed-loop behavior (6.37) with the robot dynamics (6.36) yields

$$\boldsymbol{\Lambda}\,\boldsymbol{\Omega} \begin{pmatrix} \dot{\boldsymbol{v}}_{\text{c,d}} \\ \dot{\boldsymbol{v}}_{\text{sup,d}} \\ \dot{\boldsymbol{v}}_{\text{int,d}} \end{pmatrix} + (\boldsymbol{\mu}\,\boldsymbol{\Omega} + \boldsymbol{\Lambda}\,\dot{\boldsymbol{\Omega}}) \begin{pmatrix} \boldsymbol{v}_{\text{c,d}} \\ \boldsymbol{v}_{\text{sup,d}} \\ \boldsymbol{v}_{\text{int,d}} \end{pmatrix} + \begin{pmatrix} m\boldsymbol{g}_0 \\ \mathbf{0} \\ \mathbf{0} \end{pmatrix} =$$
$$\begin{bmatrix} -\boldsymbol{Ad}_{\text{sup}}^T & -\boldsymbol{Ad}_{\text{int}}^T \\ \boldsymbol{I} & \mathbf{0} \\ \mathbf{0} & \boldsymbol{I} \end{bmatrix} (\boldsymbol{J}')^{-T}\boldsymbol{\tau} + \begin{pmatrix} -\boldsymbol{F}_{\text{c}}^{\text{cpl}} \\ \boldsymbol{F}_{\text{sup}} \\ -\boldsymbol{F}_{\text{int}}^{\text{cpl}} \end{pmatrix} \tag{6.39}$$

representing the control law. The matrices $\boldsymbol{\Omega}_{\text{c}}$, $\boldsymbol{\Omega}_{\text{sup}}$, and $\boldsymbol{\Omega}_{\text{int}}$ are combined into the block diagonal matrix $\boldsymbol{\Omega} = \text{blkdiag}(\boldsymbol{\Omega}_{\text{c}}, \boldsymbol{\Omega}_{\text{sup}}, \boldsymbol{\Omega}_{\text{int}})$. Note that $\boldsymbol{F}_{\text{sup}}$ and $\boldsymbol{\tau}$ are the remaining free variables in (6.39), which can be determined by exploiting the structure of (6.39): Analogously to Sect. 6.1, an optimization is used first to compute the end-effector wrenches representing the external load condition of the robot. Then the control torques $\boldsymbol{\tau}$ are computed characterizing the internal load inside the robot structure.

To obtain $\boldsymbol{F}_{\text{sup}}$, the control law (6.39) is partitioned into

$$\boldsymbol{\Lambda}_u\,\boldsymbol{\Omega} \begin{pmatrix} \dot{\boldsymbol{v}}_{\text{c,d}} \\ \dot{\boldsymbol{v}}_{\text{sup,d}} \\ \dot{\boldsymbol{v}}_{\text{int,d}} \end{pmatrix} + (\boldsymbol{\mu}_u\,\boldsymbol{\Omega} + \boldsymbol{\Lambda}_u\,\dot{\boldsymbol{\Omega}}) \begin{pmatrix} \boldsymbol{v}_{\text{c,d}} \\ \boldsymbol{v}_{\text{sup,d}} \\ \boldsymbol{v}_{\text{int,d}} \end{pmatrix} + m\boldsymbol{g}_0 = -\boldsymbol{Ad}^T(\boldsymbol{J}')^{-T}\boldsymbol{\tau} - \boldsymbol{F}_{\text{c}}^{\text{cpl}} \tag{6.40}$$

$$\boldsymbol{\Lambda}_l\,\boldsymbol{\Omega} \begin{pmatrix} \dot{\boldsymbol{v}}_{\text{c,d}} \\ \dot{\boldsymbol{v}}_{\text{sup,d}} \\ \dot{\boldsymbol{v}}_{\text{int,d}} \end{pmatrix} + (\boldsymbol{\mu}_l\,\boldsymbol{\Omega} + \boldsymbol{\Lambda}_l\,\dot{\boldsymbol{\Omega}}) \begin{pmatrix} \boldsymbol{v}_{\text{c,d}} \\ \boldsymbol{v}_{\text{sup,d}} \\ \boldsymbol{v}_{\text{int,d}} \end{pmatrix} = (\boldsymbol{J}')^{-T}\boldsymbol{\tau} + \begin{pmatrix} \boldsymbol{F}_{\text{sup}} \\ -\boldsymbol{F}_{\text{int}}^{\text{cpl}} \end{pmatrix} \tag{6.41}$$

with $\boldsymbol{\Lambda}_u \in \mathbb{R}^{6\times 6+6\psi}$, $\boldsymbol{\Lambda}_l \in \mathbb{R}^{6\psi\times 6+6\psi}$, $\boldsymbol{\mu}_u \in \mathbb{R}^{6\times 6+6\psi}$, and $\boldsymbol{\mu}_l \in \mathbb{R}^{6\psi\times 6+6\psi}$ originating from $\boldsymbol{\Lambda} = [\,\boldsymbol{\Lambda}_u^T\ \boldsymbol{\Lambda}_u^T\,]^T$ and $\boldsymbol{\mu} = [\,\boldsymbol{\mu}_u^T\ \boldsymbol{\mu}_u^T\,]^T$. Eliminating the common variable $(\boldsymbol{J}')^T\boldsymbol{\tau}$ yields

$$\boldsymbol{F}_{\mathrm{ff}} + m\boldsymbol{g}_0 + \boldsymbol{F}_{\mathrm{c}}^{\mathrm{cpl}} = \boldsymbol{Ad}^T \begin{pmatrix} \boldsymbol{F}_{\mathrm{sup}} \\ -\boldsymbol{F}_{\mathrm{int}}^{\mathrm{cpl}} \end{pmatrix}. \tag{6.42}$$

Note that the above equation corresponds to the underactuation condition (6.6), which has been extended with a feedforward wrench

$$\boldsymbol{F}_{\mathrm{ff}} = \left(\boldsymbol{\Lambda}_u + \boldsymbol{Ad}^T\,\boldsymbol{\Lambda}_l\right)\boldsymbol{\Omega}\begin{pmatrix}\dot{\boldsymbol{v}}_{\mathrm{c,d}}\\ \dot{\boldsymbol{v}}_{\mathrm{sup,d}}\\ \dot{\boldsymbol{v}}_{\mathrm{int,d}}\end{pmatrix} + \left(\boldsymbol{\mu}_u\,\boldsymbol{\Omega} + \boldsymbol{\Lambda}_u\,\dot{\boldsymbol{\Omega}} + \boldsymbol{Ad}^T\,\boldsymbol{\mu}_l\,\boldsymbol{\Omega} + \boldsymbol{Ad}^T\,\boldsymbol{\Lambda}_l\,\dot{\boldsymbol{\Omega}}\right)\begin{pmatrix}\boldsymbol{v}_{\mathrm{c,d}}\\ \boldsymbol{v}_{\mathrm{sup,d}}\\ \boldsymbol{v}_{\mathrm{int,d}}\end{pmatrix} \tag{6.43}$$

accounting for the dynamic effects of the desired trajectory on the CoM. Consequently, the end-effector wrenches $\boldsymbol{F}^{\mathrm{opt}}$ can be obtained by extending the constrained quadratic optimization problem in Sect. 6.1 with $\boldsymbol{F}_{\mathrm{ff}}$:

$$\boldsymbol{F}^{\mathrm{opt}} = \underset{\boldsymbol{F}}{\operatorname{argmin}}\left(\frac{1}{2}\boldsymbol{\delta}_{\mathrm{c}}^T\boldsymbol{Q}_{\mathrm{c}}\boldsymbol{\delta}_{\mathrm{c}} + \frac{1}{2}\boldsymbol{\delta}_{\mathrm{sup}}^T\boldsymbol{Q}_{\mathrm{sup}}\boldsymbol{\delta}_{\mathrm{sup}} + \frac{1}{2}\boldsymbol{\delta}_{\mathrm{int}}^T\boldsymbol{Q}_{\mathrm{int}}\boldsymbol{\delta}_{\mathrm{int}}\right) \tag{6.44}$$

minimizing the residua

$$\boldsymbol{\delta}_{\mathrm{c}} = \boldsymbol{Ad}^T\,\boldsymbol{F} - m\boldsymbol{g}_0 - \boldsymbol{F}_{\mathrm{c}}^{\mathrm{cpl}} - \boldsymbol{F}_{\mathrm{ff}} \tag{6.45}$$

$$\boldsymbol{\delta}_{\mathrm{sup}} = \boldsymbol{F}_{\mathrm{sup}} - \boldsymbol{F}_{\mathrm{sup}}^{\mathrm{def}} \tag{6.46}$$

$$\boldsymbol{\delta}_{\mathrm{int}} = \boldsymbol{F}_{\mathrm{int}} + \boldsymbol{F}_{\mathrm{int}}^{\mathrm{cpl}} \tag{6.47}$$

with respect to the inequality constraints

$$\boldsymbol{A}_i\boldsymbol{F}_i \leq \boldsymbol{b}_i \qquad \forall i = 1\ldots\psi_{\mathrm{sup}} \tag{6.48}$$

$$\boldsymbol{B}_\tau\boldsymbol{F} \leq \boldsymbol{\tau}^{\mathrm{max}}. \tag{6.49}$$

After computing the end-effector wrenches $\boldsymbol{F}^{\mathrm{opt}}$, the corresponding joint torques

$$\boldsymbol{\tau} = -(\boldsymbol{J}')^T\boldsymbol{F}^{\mathrm{opt}} + \boldsymbol{\tau}_{\mathrm{ff}} \tag{6.50}$$

can be obtained by multiplying (6.41) with $(\boldsymbol{J}')^T$. Here,

$$\boldsymbol{\tau}_{\mathrm{ff}} = (\boldsymbol{J}')^T\boldsymbol{\Lambda}_l\,\boldsymbol{\Omega}\begin{pmatrix}\dot{\boldsymbol{v}}_{\mathrm{c,d}}\\ \dot{\boldsymbol{v}}_{\mathrm{sup,d}}\\ \dot{\boldsymbol{v}}_{\mathrm{int,d}}\end{pmatrix} + (\boldsymbol{J}')^T\left(\boldsymbol{\mu}_l\,\boldsymbol{\Omega} + \boldsymbol{\Lambda}_l\,\dot{\boldsymbol{\Omega}}\right)\begin{pmatrix}\boldsymbol{v}_{\mathrm{c,d}}\\ \boldsymbol{v}_{\mathrm{sup,d}}\\ \boldsymbol{v}_{\mathrm{int,d}}\end{pmatrix} \tag{6.51}$$

represents another feedforward term accounting for the dynamics of the desired trajectory in joint space. Note that the MCB controller, which is presented in Sect. 6.1 for the regulation case, can be considered a special case of the MCB+ controller addressing the tracking case. The only difference is given by the additional tracking terms $\boldsymbol{F}_{\mathrm{ff}}$ and $\boldsymbol{\tau}_{\mathrm{ff}}$, as shown in Fig. 6.6. Note that only the feedforward terms require the inverse of the Sub-Jacobian $\boldsymbol{J}'$ to compute the Cartesian inertia $\boldsymbol{\Lambda}$ and Coriolis matrix $\boldsymbol{\mu}$. In contrast, the feedback loop, which corresponds to the MCB controller for the regulation case (see Sect. 6.1), is robust against singularities and redundant kinematics. To achieve a similar robustness for the feedforward terms, the inverse Sub-Jacobian $(\boldsymbol{J}')^{-1}$ in (6.35) can be replaced with a generalized damped pseudoinverse

$$(\boldsymbol{J}')^{+\rho} = \left((\boldsymbol{J}')^T(\boldsymbol{J}') + \rho^2\boldsymbol{I}\right)^{-1}(\boldsymbol{J}')^T \tag{6.52}$$

with $\rho \in \mathbb{R}$ denoting the damping factor. A motion in the kinematic null space can be regulated by adding a null space controller, as shown in Sect. 6.2 (see also Fig. 6.6). The choice of the damping ρ in $(\boldsymbol{J}')^{+\rho}$ is a compromise between robustness against singularities and the accuracy of the solution. As already mentioned, one could also use the task space augmentation method (Ott 2008) to deal with $\boldsymbol{J}'$ not being invertible. But for the sake of simplicity, we will use the more heuristic approach in the form of the damped pseudoinverse (6.52). It can be seen from (6.42) and (6.50) that this approximation only affects the feedforward control action ($\boldsymbol{F}_{\mathrm{ff}}$ and $\boldsymbol{\tau}_{\mathrm{ff}}$) and not the feedback control. Thus, the tracking behavior of the overall controller might deteriorate if the robot is close to a singularity, but the feedback loop remains unaffected.

6.3.2 *Properties of the Controller*

6.3.2.1 End-Effector Wrenches

A problem of the MCB controller presented in Sect. 6.1 is that the commanded contact wrenches $\boldsymbol{F}_{\mathrm{sup}}^{\mathrm{ext}}$ can deviate from the actual ones $\boldsymbol{F}_{\mathrm{sup}}$ during dynamic motions (see Sect. 6.1.2), which can be reduced by adding the tracking terms $\boldsymbol{F}_{\mathrm{ff}}$ and $\boldsymbol{\tau}_{\mathrm{ff}}$. If Assumption 6.1 holds and singular configurations and redundant robots are excluded from the configurations, then the closed-loop dynamics is given by (6.37). Furthermore, if we assume that all external forces and torques act solely at the end-effectors as in (3.24), then the closed-loop behavior (6.37) becomes

$$\boldsymbol{\Lambda}\begin{pmatrix}\Delta\dot{\boldsymbol{v}}_{\mathrm{c}}\\ \Delta\dot{\boldsymbol{v}}_{\mathrm{sup}}\\ \Delta\dot{\boldsymbol{v}}_{\mathrm{int}}\end{pmatrix} + \boldsymbol{\mu}\begin{pmatrix}\Delta\boldsymbol{v}_{\mathrm{c}}\\ \Delta\boldsymbol{v}_{\mathrm{sup}}\\ \Delta\boldsymbol{v}_{\mathrm{int}}\end{pmatrix} = \begin{pmatrix}\boldsymbol{F}_{\mathrm{c}}^{\mathrm{cpl}}\\ -\boldsymbol{F}_{\mathrm{sup}} + \boldsymbol{F}_{\mathrm{sup}}^{\mathrm{ext}}\\ \boldsymbol{F}_{\mathrm{int}}^{\mathrm{cpl}} + \boldsymbol{F}_{\mathrm{int}}^{\mathrm{ext}}\end{pmatrix} \tag{6.53}$$

due to $\boldsymbol{T}_{\mathrm{TS}}^{-T}\left[\,Ad\,(\boldsymbol{J}')\,\right]^T = \left[\begin{smallmatrix}\boldsymbol{0}\\ \boldsymbol{I}\end{smallmatrix}\right]$. Considering the equilibrium given by $\Delta\dot{\boldsymbol{v}}_{\mathrm{c}} = \Delta\boldsymbol{v}_{\mathrm{c}} = \boldsymbol{0}$, $\Delta\dot{\boldsymbol{v}}_{\mathrm{sup}} = \Delta\boldsymbol{v}_{\mathrm{sup}} = \boldsymbol{0}$, and $\Delta\dot{\boldsymbol{v}}_{\mathrm{int}} = \Delta\boldsymbol{v}_{\mathrm{int}} = \boldsymbol{0}$ leads to the conclusion that

$$\boldsymbol{F}_{\mathrm{c}}^{\mathrm{cpl}} = \mathbf{0} \tag{6.54}$$

$$\boldsymbol{F}_{\mathrm{sup}} = \boldsymbol{F}_{\mathrm{sup}}^{\mathrm{ext}} \tag{6.55}$$

$$\boldsymbol{F}_{\mathrm{int}}^{\mathrm{cpl}} = \boldsymbol{F}_{\mathrm{int}}^{\mathrm{ext}} \tag{6.56}$$

must hold. Again, the commanded interaction wrenches are balanced by the external ones leading to a static control error according to (6.4) with $\Delta \boldsymbol{v}_i = \mathbf{0}$. The center of mass coincides with the desired CoM frame according to (6.1) with $\Delta \boldsymbol{v}_{\mathrm{c}} = \mathbf{0}$ even in the presence of $\boldsymbol{F}_{\mathrm{int}}^{\mathrm{ext}}$. Furthermore, the actual contact wrenches $\boldsymbol{F}_{\mathrm{sup}}^{\mathrm{ext}}$ match the commanded ones $\boldsymbol{F}_{\mathrm{sup}}$, which legitimatizes invoking the contact model (6.48) on $\boldsymbol{F}_{\mathrm{sup}}$ instead of $\boldsymbol{F}_{\mathrm{sup}}^{\mathrm{ext}}$. However, this only holds in the case of an equilibrium, which means that the robot is exactly following the desired trajectory. If the robot deviates from the desired trajectory, (6.53) implies that $\boldsymbol{F}_{\mathrm{sup}}^{\mathrm{ext}}$ can differ from $\boldsymbol{F}_{\mathrm{sup}}$ and cause the support end-effectors to lose contact. In conclusion, there is still the danger that the supporting end-effectors might unintentionally lose contact if the robot is performing a dynamic motion due to an unknown external disturbance. But if the robot is performing a deliberate dynamic motion by following a desired trajectory, the feedforward terms in Sect. 6.3.1 ensure that the difference between $\boldsymbol{F}_{\mathrm{sup}}$ and $\boldsymbol{F}_{\mathrm{sup}}^{\mathrm{ext}}$ is sufficiently small.

6.3.2.2 Passivity and Stability on Stationary Ground

As the MCB+ controller presented in Sect. 6.3.1 is designed for the tracking case, the desired trajectory can inject and remove an arbitrary amount of energy into the controller and the robot. Thus, it is not possible to reuse the same storage function V as in Sect. 6.1.2. Instead, the passivity of the closed-loop system (6.37) can be shown by using the positive definite storage function

$$V = \frac{1}{2} \begin{pmatrix} \Delta \boldsymbol{v}_{\mathrm{c}} \\ \Delta \boldsymbol{v}_{\mathrm{sup}} \\ \Delta \boldsymbol{v}_{\mathrm{int}} \end{pmatrix}^T \boldsymbol{M} \begin{pmatrix} \Delta \boldsymbol{v}_{\mathrm{c}} \\ \Delta \boldsymbol{v}_{\mathrm{sup}} \\ \Delta \boldsymbol{v}_{\mathrm{int}} \end{pmatrix} + V_{\mathrm{c}}^K + V_{\mathrm{int}}^K \tag{6.57}$$

in Paden and Panja (1988), which serves as a metric for the deviation of the robot from the desired trajectory. Computing the time derivative, inserting the closed-loop dynamics (6.37), and exploiting the property that $\dot{\boldsymbol{\Lambda}} - 2\boldsymbol{\mu}$ is skew-symmetric (Ott 2008) yields

$$
\begin{aligned}
\dot{V} &= \begin{pmatrix} \Delta \boldsymbol{v}_{\mathrm{c}} \\ \Delta \boldsymbol{v}_{\mathrm{sup}} \\ \Delta \boldsymbol{v}_{\mathrm{int}} \end{pmatrix}^{T} \boldsymbol{\Lambda} \begin{pmatrix} \Delta \dot{\boldsymbol{v}}_{\mathrm{c}} \\ \Delta \dot{\boldsymbol{v}}_{\mathrm{sup}} \\ \Delta \dot{\boldsymbol{v}}_{\mathrm{int}} \end{pmatrix} + \frac{1}{2} \begin{pmatrix} \Delta \boldsymbol{v}_{\mathrm{c}} \\ \Delta \boldsymbol{v}_{\mathrm{sup}} \\ \Delta \boldsymbol{v}_{\mathrm{int}} \end{pmatrix}^{T} \dot{\boldsymbol{\Lambda}} \begin{pmatrix} \Delta \boldsymbol{v}_{\mathrm{c}} \\ \Delta \boldsymbol{v}_{\mathrm{sup}} \\ \Delta \boldsymbol{v}_{\mathrm{int}} \end{pmatrix} + \dot{V}_{\mathrm{c}}^{K} + \dot{V}_{\mathrm{int}}^{K} \\
&= \begin{pmatrix} \Delta \boldsymbol{v}_{\mathrm{c}} \\ \Delta \boldsymbol{v}_{\mathrm{sup}} \\ \Delta \boldsymbol{v}_{\mathrm{int}} \end{pmatrix}^{T} \left\{ -\boldsymbol{\mu} \begin{pmatrix} \Delta \boldsymbol{v}_{\mathrm{c}} \\ \Delta \boldsymbol{v}_{\mathrm{sup}} \\ \Delta \boldsymbol{v}_{\mathrm{int}} \end{pmatrix} + \begin{pmatrix} \boldsymbol{F}_{\mathrm{c}}^{\mathrm{cpl}} \\ -\boldsymbol{F}_{\mathrm{sup}} \\ \boldsymbol{F}_{\mathrm{int}}^{\mathrm{cpl}} \end{pmatrix} + \boldsymbol{T}_{\mathrm{TS}}^{-T} \boldsymbol{\tau}_{\mathrm{ext}} \right\} \\
&\quad + \frac{1}{2} \begin{pmatrix} \Delta \boldsymbol{v}_{\mathrm{c}} \\ \Delta \boldsymbol{v}_{\mathrm{sup}} \\ \Delta \boldsymbol{v}_{\mathrm{int}} \end{pmatrix}^{T} \dot{\boldsymbol{\Lambda}} \begin{pmatrix} \Delta \boldsymbol{v}_{\mathrm{c}} \\ \Delta \boldsymbol{v}_{\mathrm{sup}} \\ \Delta \boldsymbol{v}_{\mathrm{int}} \end{pmatrix} + \dot{V}_{\mathrm{c}}^{K} + \dot{V}_{\mathrm{int}}^{K} \\
&= \begin{pmatrix} \Delta \boldsymbol{v}_{\mathrm{c}} \\ \Delta \boldsymbol{v}_{\mathrm{sup}} \\ \Delta \boldsymbol{v}_{\mathrm{int}} \end{pmatrix}^{T} \begin{pmatrix} \boldsymbol{F}_{\mathrm{c}}^{\mathrm{cpl}} \\ -\boldsymbol{F}_{\mathrm{sup}} \\ \boldsymbol{F}_{\mathrm{int}}^{\mathrm{cpl}} \end{pmatrix} + \begin{pmatrix} \Delta \boldsymbol{v}_{\mathrm{c}} \\ \Delta \boldsymbol{v}_{\mathrm{sup}} \\ \Delta \boldsymbol{v}_{\mathrm{int}} \end{pmatrix}^{T} \boldsymbol{T}_{\mathrm{TS}}^{-T} \boldsymbol{\tau}_{\mathrm{ext}} + \dot{V}_{\mathrm{c}}^{K} + \dot{V}_{\mathrm{int}}^{K}.
\end{aligned}
\tag{6.58}
$$

Inserting the Cartesian compliances (6.1) and (6.4) leads to

$$
\begin{aligned}
\dot{V} = &-\nabla V_{\mathrm{c}} \Delta \boldsymbol{v}_{\mathrm{c}} - \Delta \boldsymbol{v}_{\mathrm{c}}^{T} \boldsymbol{D}_{\mathrm{c}} \Delta \boldsymbol{v}_{\mathrm{c}} + \dot{V}_{\mathrm{c}}^{K} \\
&- \nabla V_{\mathrm{int}} \Delta \boldsymbol{v}_{\mathrm{int}} - \Delta \boldsymbol{v}_{\mathrm{int}}^{T} \boldsymbol{D}_{\mathrm{int}} \Delta \boldsymbol{v}_{\mathrm{int}} + \dot{V}_{\mathrm{int}}^{K} \\
&- \Delta \boldsymbol{v}_{\mathrm{sup}}^{T} \boldsymbol{F}_{\mathrm{sup}} + \begin{pmatrix} \Delta \boldsymbol{v}_{\mathrm{c}} \\ \Delta \boldsymbol{v}_{\mathrm{sup}} \\ \Delta \boldsymbol{v}_{\mathrm{int}} \end{pmatrix}^{T} \boldsymbol{T}_{\mathrm{TS}}^{-T} \boldsymbol{\tau}_{\mathrm{ext}},
\end{aligned}
\tag{6.59}
$$

which can be simplified to

$$
\dot{V} = \underbrace{-\Delta \boldsymbol{v}_{\mathrm{c}}^{T} \boldsymbol{D}_{\mathrm{c}} \Delta \boldsymbol{v}_{\mathrm{c}} - \Delta \boldsymbol{v}_{\mathrm{int}} \boldsymbol{D}_{\mathrm{int}} \Delta \boldsymbol{v}_{\mathrm{int}}}_{\leq 0} - \Delta \boldsymbol{v}_{\mathrm{sup}}^{T} \boldsymbol{F}_{\mathrm{sup}} + \underbrace{\begin{pmatrix} \Delta \boldsymbol{v}_{\mathrm{c}} \\ \Delta \boldsymbol{v}_{\mathrm{sup}} \\ \Delta \boldsymbol{v}_{\mathrm{int}} \end{pmatrix}^{T} \boldsymbol{T}_{\mathrm{TS}}^{-T} \boldsymbol{\tau}_{\mathrm{ext}}}_{\mathrm{port}}
\tag{6.60}
$$

considering $\dot{V}_{\mathrm{c}} = \nabla V_{\mathrm{c}} \Delta \boldsymbol{v}_{\mathrm{c}}$ and $\dot{V}_{\mathrm{int}} = \nabla V_{\mathrm{int}} \Delta \boldsymbol{v}_{\mathrm{int}}$. If we assume that the robot is balancing on a non-movable and non-deformable floor, then the term $\Delta \boldsymbol{v}_{\mathrm{sup}}^{T} \boldsymbol{F}_{\mathrm{sup}}$ vanishes due to $\Delta \boldsymbol{v}_{\mathrm{sup}} = \mathbf{0}$ (see Assumption 3.2). Thus,

$$
\dot{V} \leq \begin{pmatrix} \Delta \boldsymbol{v}_{\mathrm{c}} \\ \Delta \boldsymbol{v}_{\mathrm{sup}} \\ \Delta \boldsymbol{v}_{\mathrm{int}} \end{pmatrix}^{T} \boldsymbol{T}_{\mathrm{TS}}^{-T} \boldsymbol{\tau}_{\mathrm{ext}}
\tag{6.61}
$$

holds, which proves the passivity of the closed-loop system (6.37) with respect to the power port $\left(\Delta \boldsymbol{v}_{\mathrm{c}}^{T} \ \Delta \boldsymbol{v}_{\mathrm{sup}}^{T} \ \Delta \boldsymbol{v}_{\mathrm{int}}^{T} \right) (\boldsymbol{T}_{\mathrm{TS}}^{-T} \boldsymbol{\tau}_{\mathrm{ext}})$ based on the definition given by Khalil (2014). The port represents the energy transfer between the robot and the environment, such as when a human disturbs the robot.

The stability of the controller can be shown by following (Paden and Panja 1988). Note that the matrices $\boldsymbol{\Lambda}$ and $\boldsymbol{\mu}$ in (6.36) can be written as a function of the control error ($\Delta \boldsymbol{x}_{\mathrm{c}}$, $\Delta \boldsymbol{x}_{i}$, $\Delta \boldsymbol{R}_{\mathrm{c}}$, $\Delta \boldsymbol{R}_{i}$, $\Delta \boldsymbol{v}_{\mathrm{c}}$, $\Delta \boldsymbol{v}_{i}$) and of the desired trajectory ($\boldsymbol{x}_{\mathrm{c,d}}$, $\boldsymbol{x}_{i,d}$,

$\boldsymbol{R}_{\text{c,d}}$, $\boldsymbol{R}_{i,d}$, $\boldsymbol{v}_{\text{c,d}}$, $\boldsymbol{v}_{i,d}$). Therefore, standard invariance principles are not applicable because the closed-loop dynamics (6.37) is time-variant. However, Paden and Panja (1988) showed asymptotic stability for a PD+ controller based on the Matrosov-Theorem. Moreover, Whitcomb et al. (1993) used strict Lyapunov functions to show asymptotic stability. If we assume that the floor is rigid and stationary, $\Delta\boldsymbol{v}_{\text{sup}} = \boldsymbol{0}$ can be exploited to bring (6.37) into the required form. Applying the transformation

$$\begin{pmatrix} \Delta\boldsymbol{v}_{\text{c}} \\ \Delta\boldsymbol{v}_{\text{sup}} \\ \Delta\boldsymbol{v}_{\text{int}} \end{pmatrix} = \underbrace{\begin{bmatrix} \boldsymbol{I} & \boldsymbol{0} \\ \boldsymbol{0} & \boldsymbol{0} \\ \boldsymbol{0} & \boldsymbol{I} \end{bmatrix}}_{\boldsymbol{T}_{\text{red}}} \begin{pmatrix} \Delta\boldsymbol{v}_{\text{c}} \\ \Delta\boldsymbol{v}_{\text{int}} \end{pmatrix} \tag{6.62}$$

to the closed-loop system (6.37) results in

$$\boldsymbol{\Lambda}_{\text{red}} \begin{pmatrix} \Delta\dot{\boldsymbol{v}}_{\text{c}} \\ \Delta\dot{\boldsymbol{v}}_{\text{int}} \end{pmatrix} + \boldsymbol{\mu}_{\text{red}} \begin{pmatrix} \Delta\boldsymbol{v}_{\text{c}} \\ \Delta\boldsymbol{v}_{\text{int}} \end{pmatrix} = \begin{pmatrix} \boldsymbol{F}_{\text{c}}^{\text{cpl}} \\ \boldsymbol{F}_{\text{int}}^{\text{cpl}} \end{pmatrix} + \boldsymbol{T}_{\text{red}}^{T} \boldsymbol{T}_{\text{TS}}^{-T} \boldsymbol{\tau}_{\text{ext}}. \tag{6.63}$$

In order to follow the proof in Paden and Panja (1988), the autonomous system with $\boldsymbol{T}_{\text{red}}^{T} \boldsymbol{T}_{\text{TS}}^{-T} \boldsymbol{\tau}_{\text{ext}} = \boldsymbol{0}$ needs to be considered. Note that in general $\boldsymbol{\tau}_{\text{ext}} \neq \boldsymbol{0}$ due to the external contact wrenches $\boldsymbol{F}_{\text{sup}}^{\text{ext}}$. But the influence of $\boldsymbol{F}_{\text{sup}}^{\text{ext}}$ is removed by the transformation $\boldsymbol{T}_{\text{red}}$ discarding the Cartesian coordinates of the support end-effectors. Note that (6.57) can be used as a Lyapunov function leading to

$$\dot{V} \leq \begin{pmatrix} \Delta\boldsymbol{v}_{\text{c}} \\ \Delta\boldsymbol{v}_{\text{int}} \end{pmatrix}^{T} \boldsymbol{T}_{\text{red}}^{T} \boldsymbol{T}_{\text{TS}}^{-T} \boldsymbol{\tau}_{\text{ext}} = 0 \tag{6.64}$$

by applying $\boldsymbol{T}_{\text{red}}$ to (6.61).

The above analysis applies to the desired closed-loop dynamics (6.37), which implies that the soft constraints (6.45) and (6.47) are exactly fulfilled, as demanded by Assumption 6.1. A passivity and stability analysis of the case where the soft constraints are not exactly fulfilled would be considerably more cumbersome. In this case, the constraints will affect the quality of the task fulfillment according to their corresponding priorities (see Table 6.1), which would lead to a different closed-loop dynamics. Furthermore, the analysis only holds for the case of the Sub-Jacobian matrix $\boldsymbol{J}'$ being square and invertible excluding redundant robots and singular configurations. An analysis of the case that $\boldsymbol{J}'$ is not invertible using, for instance, the task space augmentation method (Ott 2008) would be considerably more cumbersome.

6.3.3 Experimental Evaluation

The performance of the MCB+ control framework was evaluated in Henze et al. (2016b) using the torque-controlled humanoid robot *TORO* developed at *DLR*. A brief description of the robot hardware is given in Chap. 5.

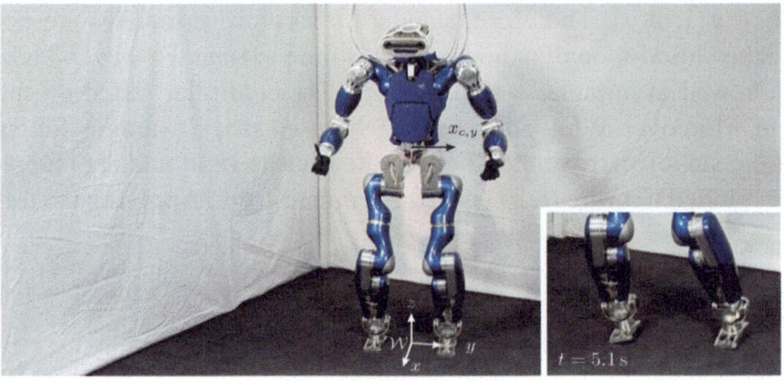

(a) Setup of the experiment.

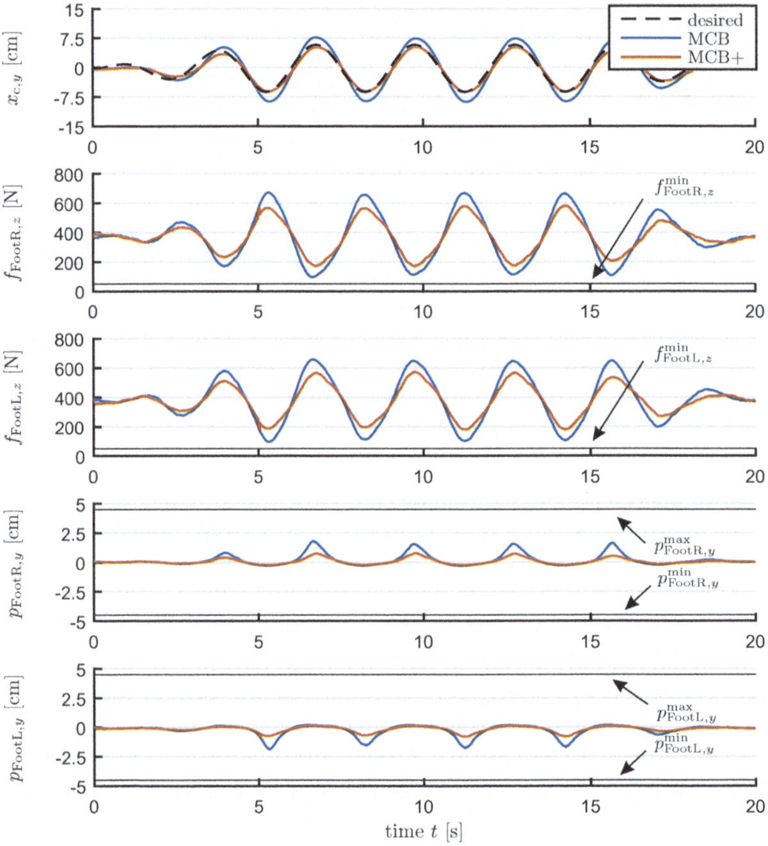

(b) Behavior for a maximum amplitude of 6 cm.

Fig. 6.7 Comparison of the tracking behavior using the MCB and the MCB+ controller: the trajectory consists of a sinusoidal signal with a frequency of 1/3 Hz

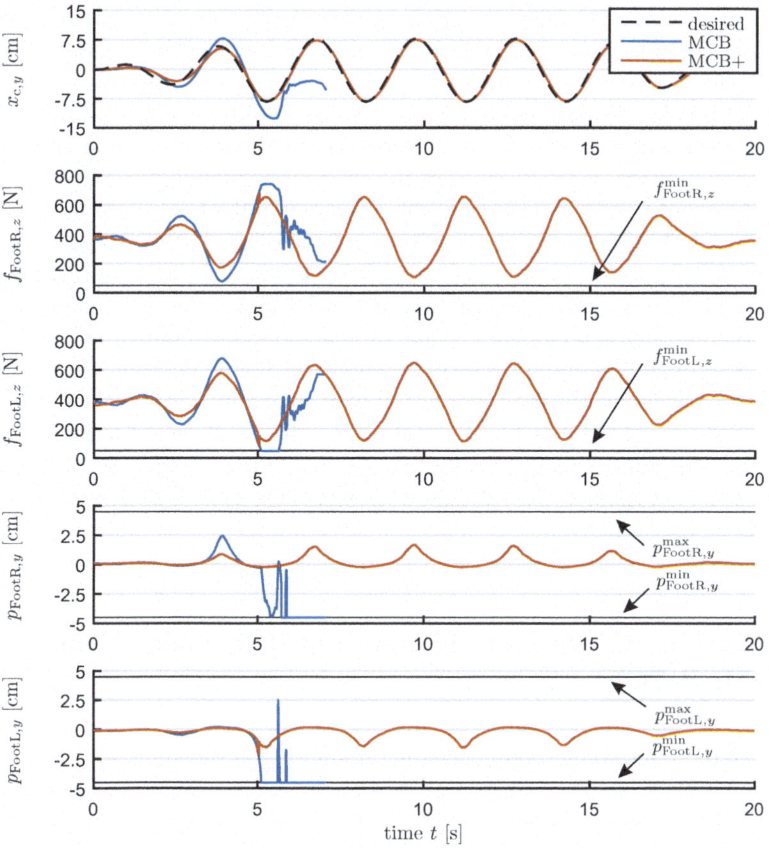

(c) Behavior for a maximum amplitude of 8 cm.

Fig. 6.7 (continued)

As shown in Fig. 6.7a, the humanoid robot *TORO* balances with both feet on the ground, while the arms are operated in interaction mode. The robot is supposed to follow a dynamic trajectory, which moves the CoM along the y-axis of the world frame $\mathcal{W}$. The trajectory consists of a sinusoidal signal for the desired CoM position with a frequency of 1/3 Hz. Within the first 5 s, the amplitude of the signal is linearly increased from 0 to 6 cm, held constant for 10 s, and then decreased to 0 cm again within the last 5 s of the experiment.

The results of the experiment can be found in Fig. 6.7b: If the motion is performed using the MCB controller (see Sect. 6.1), that means without feedforward terms, then the horizontal CoM position shows an overshoot of up to 2.5 cm. Using the MCB+ control approach from Sect. 6.3, which in contrast features feedforward terms, reduces the tracking error significantly to less than 0.7 cm. Besides this, the feedforward terms also reduce the contact wrenches by 14 and 12% for the right and left leg, respectively. However, the effect becomes even more apparent by analyzing the center of pressure of each foot. Here, the excitation of the CoPs is reduced by 50% for the right and by 59% for the left foot. In any case, both controllers are able to maintain balance following the trajectory with an amplitude of 6 cm.

The experiment was repeated in a second run with an increased amplitude of 8 cm, as shown in Fig. 6.7c. Because of the stronger excitation, the MCB controller (no feedforward terms) is no longer able to maintain the balance of the robot. As can be seen in Fig. 6.7c, the vertical contact force of the left foot reaches its minimum of $f^{\min}_{\text{FootL},z} = 50$ N at $t = 5.1$ s. In addition, the CoPs of both feet reach the boundary of the respective contact area at $t = 5.1$ s and $t = 5.7$ s. Consequently, the MCB controller is no longer capable of generating the required overall support wrench $\boldsymbol{F}^{\text{res}}_{\text{sup}}$ to follow the trajectory. As shown in Fig. 6.7a, both feet start to tilt to the right, which causes the robot to fall over. In contrast to the MCB controller (no feedforward terms), the MCB+ controller (with feedforward terms) is able to follow the trajectory without loosing balance even for an amplitude of 8 cm. The MCB+ controller again shows a decent tracking performance with an error of 0.9 cm for the CoM position.

6.4 Passivity Control on Movable and Deformable Ground (MCB-PC)

The passivity of the whole-body balancing framework is shown in Sects. 6.1 (MCB) and 6.3 (MCB+) for the regulation as well as for the tracking case. Both analyses were conducted under the assumption of a static ground surface. As a by-product, the balancing capabilities on a deformable ground surface were also demonstrated in Henze et al. (2016b) or Sect. 6.1.3, respectively, although the passivity of the controller is not specifically formalized for this particular scenario. Also, many other works assume that the ground floor is static or that the dynamics of the floor is known (e.g. Sentis 2010; Mistry et al. 2010; Righetti et al. 2013; Herzog et al. 2016). However, to make the control approach robust against a large variety of support surfaces, the concept of passivity can be exploited to make as few assumptions on the floor as possible.

The basic version of the balancing framework (MCB, Sect. 6.1) was extended in Henze et al. (2018) by modifying the controller such that the combination of floor, robot, and controller remains passive even if the floor surface is movable and deformable. The modifications require the assumptions that

- the ground floor is passive itself
- and stiff enough to carry the weight of the robot.

Note that Assumption 3.2 still holds, which postulates that the balancing end-effectors never detach unintentionally from the ground floor although the latter is movable and deformable.

Stramigioli (2015) recently showed that for every active controller, an environment can be found that destabilizes the closed-loop system even if the environment is passive. Consequently, there is no guarantee that the basic version of our control framework (see Sect. 6.1) will remain stable on a movable and deformable support surface. In Henze et al. (2018), we employed the concept of Passivity Observer-Passivity Control (PO-PC) as presented by Hannaford and Ryu (2002) to passivate the closed-loop system. Thus, the presented framework will be called Multi-Contact Balancer with Passivity Control (MCB-PC) for the remainder of the manuscript. Passivation is made less conservative by exploiting the concept of energy tanks (Franken et al. 2011) in Sect. 6.4.4.

6.4.1 Problem Description

The problem of ensuring passivity on movable and deformable ground surfaces already becomes apparent in (6.27): The term $\boldsymbol{\nu}^T \boldsymbol{\tau}_{\mathrm{ext}}$ represents the power port with which the robot can exchange energy with the environment, such as a human disturbing the robot, or, in this particular case, a movable or deformable ground floor. However, the problem does not originate from the power port, but rather from the term $\boldsymbol{v}_{\mathrm{sup}}^T \boldsymbol{F}_{\mathrm{sup}}$ in (6.27). The latter can be conceived as an active element inside the controller, which can dissipate ($\boldsymbol{v}_{\mathrm{sup}}^T \boldsymbol{F}_{\mathrm{sup}} > 0$) but also inject energy into the controller ($\boldsymbol{v}_{\mathrm{sup}}^T \boldsymbol{F}_{\mathrm{sup}} < 0$). In the case of a rigid and stationary ground floor, as assumed in (6.27), the active element vanishes due to $\boldsymbol{v}_{\mathrm{sup}} = \boldsymbol{0}$. However, if the floor is movable or deformable, then the active element can compromise passivity.

The basic idea of the control strategy can be motivated with the simplified example shown in Fig. 6.8, in which a humanoid robot is standing with both feet on a moving platform. Using the compliant whole-body controller from Sect. 6.1 (MCB), the center of mass frame C as well as the interaction end-effectors (the hands in this case) are stabilized by their associated compliances (6.1) and (6.4) with respect to the stationary world frame $\mathcal{W}$. The feet are used as supporting end-effectors to carry the robot by generating suitable contact wrenches $\boldsymbol{F}_{\mathrm{sup}}$.

Let us first consider the situation where the platform is moving upward: In order to maintain the position of the CoM and of the hands relative to the world frame $\mathcal{W}$, the robot must retract the legs. The term $\boldsymbol{v}_{\mathrm{sup}}^T \boldsymbol{F}_{\mathrm{sup}}$ in (6.27) becomes positive, which means that the active element dissipates power. Therefore, the system remains passive.

If the ground floor is moving downwards, the robot must extend the legs to maintain the CoM and hand positions. Now, the term $\boldsymbol{v}_{\mathrm{sup}}^T \boldsymbol{F}_{\mathrm{sup}}$ in (6.27) becomes negative, which means that the active element injects power. Thus, the negative semi-definiteness of $\dot{V} \leq 0$ and therefore passivity can no longer be guaranteed. In order to ensure passivity of the closed-loop system (floor + robot + controller) nevertheless, the following control strategy is proposed in Henze et al. (2018):

If the controller is about to become active ($\boldsymbol{v}_{\mathrm{sup}}^T \boldsymbol{F}_{\mathrm{sup}} < 0$), the robot shall move the setpoints/desired configurations of the CoM and the interaction compliances to dissipate the same amount of power as injected by $\boldsymbol{v}_{\mathrm{sup}}^T \boldsymbol{F}_{\mathrm{sup}}$. A logical choice for

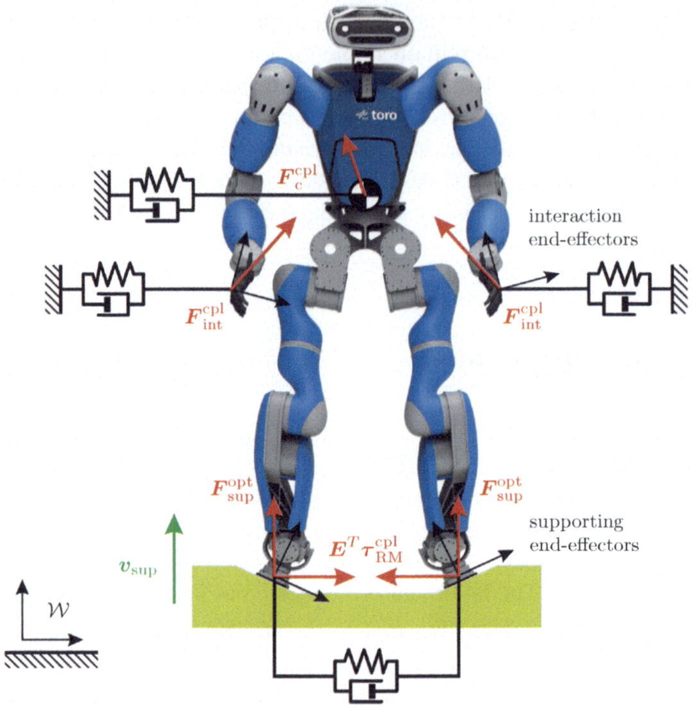

Fig. 6.8 Example of a humanoid robot balancing on a moving platform

the motion of the setpoints is to let the CoM and the hands follow the motion of the ground floor, similar to a person riding an elevator. If the term $v_{\text{sup}}^T F_{\text{sup}}$ in (6.27) is positive, the power can either be dissipated by doing nothing or used to move the setpoints back into their initial configuration. The latter is again inspired by the elevator example.

6.4.2 Controller Derivation

The solution for movable and deformable floors (MCB-PC) reuses the basic version of the whole-body controller (MCB) from Sect. 6.1 with minor modifications. On top of the MCB controller, there are two additional modules, as shown in Fig. 6.9, for ensuring passivity: The first one implements an "admittance type" passivity control by moving the setpoints of the CoM and the interaction compliances. It also realizes an "impedance type" passivity control by modifying the gravity compensation of the controller. The second module facilitates an "impedance type" passivity control by adjusting a variable damper. The latter is part of a relative compliance that connects the support end-effectors with each other. As detailed in this section, the

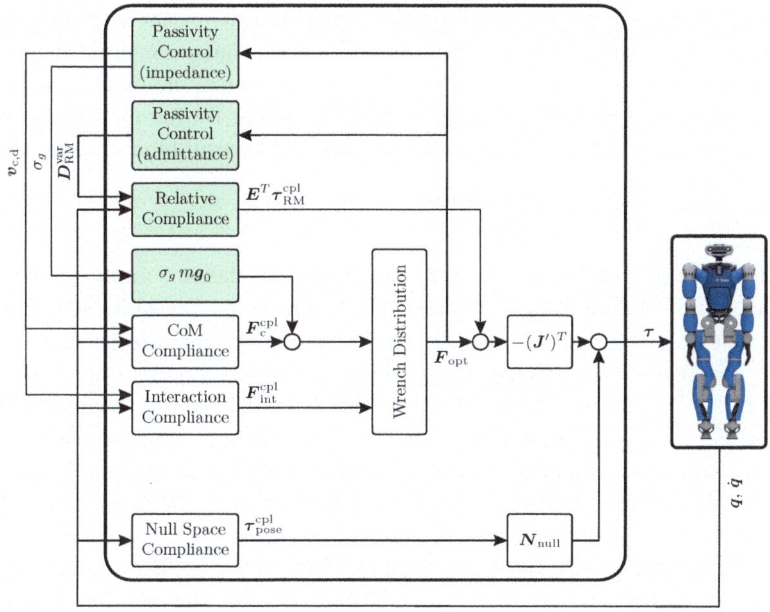

Fig. 6.9 Overview of the architecture of the MCB-PC control framework extended to movable and deformable support surfaces

admittance type is used to passivate the group motion of the support contacts, while the impedance type is applied to the relative motion. According to Fig. 6.9, both modules violate causality by modifying and evaluating the commanded end-effector wrenches $\boldsymbol{F}_{\text{opt}}$ at the same time. The work by Hannaford and Ryu (2002) on PO-PC also includes a time-discrete implementation that resolves the causality problem. But for reasons of simplicity, in Henze et al. (2018) we inserted a unit delay (not shown in Fig. 6.9) to approximate the solution by Hannaford and Ryu (2002). The use of energy tanks for rendering the method less conservative is detailed in Sect. 6.4.4. In order to handle robots with a redundant kinematics or in singular configurations, the framework is also equipped with the null space controller presented in Sect. 6.2.

The controller reuses the compliances for the CoM and the interaction end-effectors according to (6.1) and (6.4), respectively. But in contrast to Sect. 6.1, the desired velocities $\boldsymbol{v}_{\text{c,d}}$ and $\boldsymbol{v}_{\text{int,d}}$ can be different from zero to facilitate a motion of the setpoints (see Fig. 6.9). Furthermore, the gravity compensation is altered such that it can be disabled via a binary parameter $\sigma_g = \{0, 1\}$. The latter can be used to accelerate the center of mass during a downward motion of the support contacts. This results in a modified version

$$\boldsymbol{Ad}_{\text{sup}}^T\, \boldsymbol{F}_{\text{sup}} = \sigma_g\, m \boldsymbol{g}_0 + \boldsymbol{F}_{\text{c}}^{\text{cpl}} + \boldsymbol{Ad}_{\text{int}}^T\, \boldsymbol{F}_{\text{int}}^{\text{cpl}}. \tag{6.65}$$

of the underactuation constraint (6.6). The compliance wrenches $\boldsymbol{F}_{\mathrm{c}}^{\mathrm{cpl}}$ and $\boldsymbol{F}_{\mathrm{int}}^{\mathrm{cpl}}$ are distributed to the supporting contacts by solving the same constrained quadratic optimization problem (6.9) to (6.14) as before. Of course, the residuum (6.10) is replaced with $\boldsymbol{\delta}_{\mathrm{c}} = \boldsymbol{Ad}^T\, \boldsymbol{F} - \sigma_g\, m \boldsymbol{g}_0 - \boldsymbol{F}_{\mathrm{c}}^{\mathrm{cpl}}$ to account for the modified underactuation condition (6.65). In contrast to Sect. 6.1, the control torques are computed via

$$\boldsymbol{\tau} = -(\boldsymbol{J}')^T \boldsymbol{F}^{\mathrm{opt}} + (\boldsymbol{J}'_{\mathrm{sup}})^T \boldsymbol{E}^T \boldsymbol{\tau}_{\mathrm{RM}}^{\mathrm{cpl}} \tag{6.66}$$

rather than via (6.17). The second term establishes a relative compliance based on the concept of virtual linkage connecting the support contacts with each other (see Sect. 3.5 and Williams and Khatib (1993)). The vector of generalized torques $\boldsymbol{\tau}_{\mathrm{RM}}^{\mathrm{cpl}} \in \mathbb{R}^{6(\psi_{\mathrm{sup}}-1)}$ issued by the relative compliance is mapped via the transpose of the matrix $\boldsymbol{E} \in \mathbb{R}^{6\psi_{\mathrm{sup}} \times 6(\psi_{\mathrm{sup}}-1)}$ from the space of the virtual linkage to contact forces[3] and then via $(\boldsymbol{J}'_{\mathrm{sup}})^T$ to joint space. The basic idea behind $\boldsymbol{\tau}_{\mathrm{RM}}^{\mathrm{cpl}}$ is to stabilize the relative motion of the support end-effectors. As we have learned from the introductory example shown in Fig. 6.8, a motion of the setpoints can be used to dissipate the power that is injected by a group motion of the support end-effectors. The latter is generally conceived as a rigid body motion of the end-effectors maintaining their relative posture. If the relative posture of the end-effectors changes, then this is considered a relative motion. As a relative motion of the support contacts can also inject energy, the relative compliance

$$\boldsymbol{\tau}_{\mathrm{RM}}^{\mathrm{cpl}} = -(\nabla V_{\mathrm{RM}}^{K})^T - (\boldsymbol{D}_{\mathrm{RM}} + \boldsymbol{D}_{\mathrm{RM}}^{\mathrm{var}})\, \boldsymbol{v}_{\mathrm{RM}} \tag{6.67}$$

is used to dissipate the injected power. The relative stiffness is represented by the positive definite potential $V_{\mathrm{RM}}^{K} \in \mathbb{R}$. The default damping is given the constant and positive definite matrix $\boldsymbol{D}_{\mathrm{RM}} \in \mathbb{R}^{6(\psi_{\mathrm{sup}}-1)\times 6(\psi_{\mathrm{sup}}-1)}$. The semi-positive definite matrix $\boldsymbol{D}_{\mathrm{RM}}^{\mathrm{var}} \in \mathbb{R}^{6(\psi_{\mathrm{sup}}-1)\times 6(\psi_{\mathrm{sup}}-1)}$ represents an additional variable damper, which can be used to damp out the energy that is injected by the active element $\boldsymbol{v}_{\mathrm{sup}}^T \boldsymbol{F}_{\mathrm{sup}}^{\mathrm{opt}}$ into the relative motion. The velocities of the virtual linkage are given by $\boldsymbol{v}_{\mathrm{RM}} = \boldsymbol{E}\, \boldsymbol{J}'_{\mathrm{sup}} \dot{\boldsymbol{q}}$ with $\boldsymbol{v}_{\mathrm{RM}} \in \mathbb{R}^{6(\psi_{\mathrm{sup}}-1)}$. The advantage of using the concept of virtual linkage is that $\boldsymbol{\tau}_{\mathrm{RM}}^{\mathrm{cpl}}$ does not alter the resulting overall support wrench (see Sect. 3.5). Therefore, it does not need to be considered by the residuum $\boldsymbol{\delta}_{\mathrm{c}}$ in (6.10). However, $\boldsymbol{\tau}_{\mathrm{RM}}^{\mathrm{cpl}}$ affects the support wrenches $\boldsymbol{F}_{\mathrm{sup}}^{\mathrm{opt}}$, which can lead to a violation of the contact model. A solution would be to consider $\boldsymbol{\tau}_{\mathrm{RM}}^{\mathrm{cpl}}$ inside the constrained quadratic optimization (6.9) to (6.14) to ensure that the contact model is respected. But then the contact model could modify $\boldsymbol{\tau}_{\mathrm{RM}}^{\mathrm{cpl}}$, which would be problematic regarding the passivation of the system.

In the previous sections, the passivity analysis was conducted by formulating the dynamics of the closed-loop system comprising the controller and the robot.

[3] Note Williams and Khatib (1993) define the matrix $\boldsymbol{E}$ such that it maps only internal forces to the contacts, while the mapping including internal torques is more complicated. For simplicity of notation, this work will use $\boldsymbol{E}$ for describing the mapping of both internal forces and torques.

As the floor was assumed to be stationary, it was not necessary to incorporate its dynamics into the closed-loop system. The MCB-PC control approach, presented in this section, explicitly extends the passivity considerations from stationary to movable and deformable ground surfaces. But, the intention of the approach is to make as little assumptions on the floor as possible. Therefore, the dynamic equation of the floor is considered unknown. Instead, the passivity consideration first focuses on the subsystem containing only the controller and the robot. Afterwards, the passivity analysis will be extended to the complete closed-loop system combining controller, robot and the ground floor. The dynamics of the subsystem, which contains the dynamics of the controller and of the robot,

$$\boldsymbol{M}\dot{\boldsymbol{v}} + \boldsymbol{C}\boldsymbol{v} + \begin{pmatrix} (1-\sigma_g)\, m\boldsymbol{g}_0 \\ \boldsymbol{0} \end{pmatrix} = \begin{pmatrix} \boldsymbol{F}_{\mathrm{c}}^{\mathrm{cpl}} \\ (\boldsymbol{J}'_{\mathrm{sup}})^T \boldsymbol{E}^T \boldsymbol{\tau}_{\mathrm{RM}}^{\mathrm{cpl}} \end{pmatrix} + \boldsymbol{J}^T \begin{pmatrix} -\boldsymbol{F}_{\mathrm{sup}}^{\mathrm{opt}} \\ \boldsymbol{F}_{\mathrm{int}}^{\mathrm{cpl}} \end{pmatrix} + \boldsymbol{\tau}_{\mathrm{ext}} \tag{6.68}$$

can be obtained by inserting (6.66) and (6.65) into (3.20) under the assumption that the soft constraints (6.10) and (6.12) are exactly fulfilled (see Assumption 6.1).

The rules for moving the setpoints via $\boldsymbol{v}_{\mathrm{c,d}}$ and $\boldsymbol{v}_{\mathrm{int,d}}$, disabling gravity compensation via σ_g, and for modulating the variable damper $\boldsymbol{D}_{\mathrm{c}}^{\mathrm{var}}$ can be derived as follows: Let us consider

$$V = \frac{1}{2}\boldsymbol{v}^T \boldsymbol{M}\boldsymbol{v} + \underbrace{m\boldsymbol{g}_0^T \boldsymbol{x}_{\mathrm{c}}}_{V_g(\boldsymbol{x}_{\mathrm{c}})} + V_{\mathrm{c}}^K + V_{\mathrm{int}}^K + V_{\mathrm{RM}}^K \tag{6.69}$$

as a candidate for a storage function. The first two terms represent the physical energy storages of the robot, i.e. inertia and gravity. The remaining terms represent the energy storages of the compliances inside the controller: The potentials V_{c}^K, V_{int}^K, and V_{RM}^K are introduced by compliances for the CoM, the interaction end-effectors, and the relative feet motion, respectively. Note that the potential (6.69) has no lower bound due to $V_g(\boldsymbol{x}_{\mathrm{c}})$, which is addressed at the end of this section.

Computing the time derivative, inserting the dynamics of the subsystem (6.68) and accounting for the skew-symmetry of $\dot{\boldsymbol{M}} - 2\boldsymbol{C}$ (Ott 2008) yields

$$\begin{aligned} \dot{V} &= \boldsymbol{v}^T \boldsymbol{M}\dot{\boldsymbol{v}} + \frac{1}{2}\boldsymbol{v}^T \dot{\boldsymbol{M}}\boldsymbol{v} + \dot{V}_{\mathrm{c}}^K + \dot{V}_{\mathrm{int}}^K + \dot{V}_{\mathrm{RM}}^K + m\boldsymbol{g}_0^T \boldsymbol{v}_{\mathrm{c}} \\ &= \boldsymbol{v}_{\mathrm{c}}^T \boldsymbol{F}_{\mathrm{c}}^{\mathrm{cpl}} + \dot{\boldsymbol{q}}^T (\boldsymbol{J}'_{\mathrm{sup}})^T \boldsymbol{E}^T \boldsymbol{\tau}_{\mathrm{RM}}^{\mathrm{cpl}} - \boldsymbol{v}^T \boldsymbol{J}_{\mathrm{sup}}^T \boldsymbol{F}_{\mathrm{sup}}^{\mathrm{opt}} + \boldsymbol{v}^T \boldsymbol{J}_{\mathrm{int}}^T \boldsymbol{F}_{\mathrm{int}}^{\mathrm{cpl}} \\ &+ \dot{V}_{\mathrm{c}}^K + \dot{V}_{\mathrm{int}}^K + \dot{V}_{\mathrm{RM}}^K + \sigma_g\, m\boldsymbol{g}_0^T \boldsymbol{v}_{\mathrm{c}} + \boldsymbol{v}^T \boldsymbol{\tau}_{\mathrm{ext}}. \end{aligned} \tag{6.70}$$

Considering (3.21), (3.22), and $\boldsymbol{v}_{\mathrm{RM}} = \boldsymbol{E}\,\boldsymbol{J}'_{\mathrm{sup}}\dot{\boldsymbol{q}}$ simplifies $\dot{V}$ to

$$\begin{aligned} \dot{V} &= \boldsymbol{v}_{\mathrm{c}}^T \boldsymbol{F}_{\mathrm{c}}^{\mathrm{cpl}} + \boldsymbol{v}_{\mathrm{int}}^T \boldsymbol{F}_{\mathrm{int}}^{\mathrm{cpl}} + \boldsymbol{v}_{\mathrm{RM}}^T \boldsymbol{\tau}_{\mathrm{RM}}^{\mathrm{cpl}} \\ &+ \dot{V}_{\mathrm{c}}^K + \dot{V}_{\mathrm{int}}^K + \dot{V}_{\mathrm{RM}}^K \\ &+ \sigma_g\, m\boldsymbol{g}_0^T \boldsymbol{v}_{\mathrm{c}} - \boldsymbol{v}_{\mathrm{sup}}^T \boldsymbol{F}_{\mathrm{sup}}^{\mathrm{opt}} + \boldsymbol{v}^T \boldsymbol{\tau}_{\mathrm{ext}}. \end{aligned} \tag{6.71}$$

Considering $\dot{V}_{\mathrm{c}}^{K} = \Delta\boldsymbol{v}_{\mathrm{c}}^{T}(\nabla V_{\mathrm{c}}^{K})^{T}$, $\dot{V}_{\mathrm{int}}^{K} = \Delta\boldsymbol{v}_{\mathrm{int}}^{T}(\nabla V_{\mathrm{int}}^{K})^{T}$, $\dot{V}_{\mathrm{RM}}^{K} = \Delta\boldsymbol{v}_{\mathrm{RM}}^{T}(\nabla V_{\mathrm{RM}}^{K})^{T}$ and inserting (6.1), (6.4), (6.67) yields

$$
\begin{aligned}
\dot{V} &= \boldsymbol{v}_{\mathrm{c}}^{T}\boldsymbol{F}_{\mathrm{c}}^{\mathrm{cpl}} + \boldsymbol{v}_{\mathrm{int}}^{T}\boldsymbol{F}_{\mathrm{int}}^{\mathrm{cpl}} + \boldsymbol{v}_{\mathrm{RM}}^{T}\boldsymbol{\tau}_{\mathrm{RM}}^{\mathrm{cpl}} \\
&\quad - \boldsymbol{v}_{\mathrm{c}}^{T}\boldsymbol{F}_{\mathrm{c}}^{\mathrm{cpl}} + \boldsymbol{v}_{\mathrm{c,d}}^{T}\boldsymbol{\Omega}_{\mathrm{c}}^{T}\boldsymbol{F}_{\mathrm{c}}^{\mathrm{cpl}} - \Delta\boldsymbol{v}_{\mathrm{c}}^{T}\boldsymbol{D}_{\mathrm{c}}\Delta\boldsymbol{v}_{\mathrm{c}} \\
&\quad - \boldsymbol{v}_{\mathrm{int}}^{T}\boldsymbol{F}_{\mathrm{int}}^{\mathrm{cpl}} + \boldsymbol{v}_{\mathrm{int,d}}^{T}\boldsymbol{\Omega}_{\mathrm{int}}^{T}\boldsymbol{F}_{\mathrm{int}}^{\mathrm{cpl}} - \Delta\boldsymbol{v}_{\mathrm{int}}^{T}\boldsymbol{D}_{\mathrm{int}}\Delta\boldsymbol{v}_{\mathrm{int}} \\
&\quad - \boldsymbol{v}_{\mathrm{RM}}^{T}\boldsymbol{\tau}_{\mathrm{RM}}^{\mathrm{cpl}} - \boldsymbol{v}_{\mathrm{RM}}^{T}\boldsymbol{D}_{\mathrm{RM}}\boldsymbol{v}_{\mathrm{RM}} \\
&\quad + m\boldsymbol{g}_{0}^{T}\boldsymbol{v}_{\mathrm{c}} - \boldsymbol{v}_{\mathrm{sup}}^{T}\boldsymbol{F}_{\mathrm{sup}}^{\mathrm{opt}} + \boldsymbol{v}^{T}\boldsymbol{\tau}_{\mathrm{ext}} \\
&= \boldsymbol{v}_{\mathrm{c,d}}^{T}\boldsymbol{\Omega}_{\mathrm{c}}^{T}\boldsymbol{F}_{\mathrm{c}}^{\mathrm{cpl}} + \boldsymbol{v}_{\mathrm{int,d}}^{T}\boldsymbol{\Omega}_{\mathrm{int}}^{T}\boldsymbol{F}_{\mathrm{int}}^{\mathrm{cpl}} - \boldsymbol{v}_{\mathrm{RM}}^{T}\boldsymbol{D}_{\mathrm{RM}}^{\mathrm{var}}\boldsymbol{v}_{\mathrm{RM}} \\
&\quad + \sigma_{g}\, m\boldsymbol{g}_{0}^{T}\boldsymbol{v}_{\mathrm{c}} - \boldsymbol{v}_{\mathrm{sup}}^{T}\boldsymbol{F}_{\mathrm{sup}}^{\mathrm{opt}} + \boldsymbol{v}^{T}\boldsymbol{\tau}_{\mathrm{ext}} \\
&\quad \underbrace{-\Delta\boldsymbol{v}_{\mathrm{c}}^{T}\boldsymbol{D}_{\mathrm{c}}\Delta\boldsymbol{v}_{\mathrm{c}}}_{\leq 0}\,\underbrace{-\Delta\boldsymbol{v}_{\mathrm{int}}^{T}\boldsymbol{D}_{\mathrm{int}}\Delta\boldsymbol{v}_{\mathrm{int}}}_{\leq 0}\,\underbrace{-\boldsymbol{v}_{\mathrm{RM}}^{T}\boldsymbol{D}_{\mathrm{RM}}\boldsymbol{v}_{\mathrm{RM}}}_{\leq 0}\,.
\end{aligned}
\tag{6.72}
$$

As the last three terms are semi-negative definite, we can conclude that

$$
\begin{aligned}
\dot{V} &\leq \boldsymbol{v}_{\mathrm{c,d}}^{T}\boldsymbol{\Omega}_{\mathrm{c}}^{T}\boldsymbol{F}_{\mathrm{c}}^{\mathrm{cpl}} + \boldsymbol{v}_{\mathrm{int,d}}^{T}\boldsymbol{\Omega}_{\mathrm{int}}^{T}\boldsymbol{F}_{\mathrm{int}}^{\mathrm{cpl}} - \boldsymbol{v}_{\mathrm{RM}}^{T}\boldsymbol{D}_{\mathrm{RM}}^{\mathrm{var}}\boldsymbol{v}_{\mathrm{RM}} + \sigma_{g}\, m\boldsymbol{g}_{0}^{T}\boldsymbol{v}_{\mathrm{c}} \\
&\quad \underbrace{-\boldsymbol{v}_{\mathrm{sup}}^{T}\boldsymbol{F}_{\mathrm{sup}}^{\mathrm{opt}}}_{P_{\mathrm{act}}} + \underbrace{\boldsymbol{v}^{T}\boldsymbol{\tau}_{\mathrm{ext}}}_{\mathrm{port}}
\end{aligned}
\tag{6.73}
$$

must hold. The last term represents the port for exchanging energy between the environment and the subsystem containing the controller and the robot. The latter comprises the floor but also a human disturbing the robot. In order to make the closed-loop system passive, it is necessary to render (6.73) into $\dot{V} \leq \boldsymbol{v}^{T}\boldsymbol{\tau}_{\mathrm{ext}}$ by compensating with $\boldsymbol{v}_{\mathrm{c,d}}$, $\boldsymbol{v}_{\mathrm{int,d}}$, σ_{g}, and $\boldsymbol{D}_{\mathrm{RM}}^{\mathrm{var}}$ for the power $P_{\mathrm{act}} = -\boldsymbol{v}_{\mathrm{sup}}^{T}\boldsymbol{F}_{\mathrm{sup}}^{\mathrm{opt}}$, which is injected by the active element.

Note that the power P_{act} can be injected into the group motion as well as into the relative motion. The power injected into the group motion can be influenced by moving the setpoints of the CoM and interaction compliances via $\boldsymbol{v}_{\mathrm{c,d}}$ and $\boldsymbol{v}_{\mathrm{int,d}}$ and by disabling gravity compensation via the parameter σ_{g}. The power of the relative motion can be manipulated via the variable damper $\boldsymbol{D}_{\mathrm{RM}}^{\mathrm{var}}$, which is part of the relative compliance (6.67). Therefore, it is necessary to split the injected power P_{act} into a part corresponding to a group motion and into a relative motion of the support contacts. Based on the discussion in Sect. 3.3, this can be achieved by splitting the commanded support wrench $\boldsymbol{F}_{\mathrm{sup}}^{\mathrm{opt}} \in \mathbb{R}^{6\psi_{\mathrm{sup}}}$ into the overall support wrench $\boldsymbol{F}_{\mathrm{sup}}^{\mathrm{res}} \in \mathbb{R}^{6}$ and into internal stress via

$$
\boldsymbol{F}_{\mathrm{sup}}^{\mathrm{opt}} = \underbrace{(\boldsymbol{Ad}_{\mathrm{sup}}^{T})^{\boldsymbol{W}_{\mathrm{sup}}+}\boldsymbol{Ad}_{\mathrm{sup}}^{T}\boldsymbol{F}_{\mathrm{sup}}^{\mathrm{opt}}}_{\boldsymbol{F}_{\mathrm{sup}}^{\mathrm{res}}} + \underbrace{\left(\boldsymbol{I} - (\boldsymbol{Ad}_{\mathrm{sup}}^{T})^{\boldsymbol{W}_{\mathrm{sup}}+}\boldsymbol{Ad}_{\mathrm{sup}}^{T}\right)\boldsymbol{F}_{\mathrm{sup}}^{\mathrm{opt}}}_{\text{internal stress}}
\tag{6.74}
$$

with

$$(\boldsymbol{Ad}_{\text{sup}}^T)^{\boldsymbol{W}_{\text{sup}}+} = \boldsymbol{W}_{\text{sup}}^{-1}\,\boldsymbol{Ad}_{\text{sup}}\left(\boldsymbol{Ad}_{\text{sup}}^T\,\boldsymbol{W}_{\text{sup}}^{-1}\,\boldsymbol{Ad}_{\text{sup}}\right)^{-1} \tag{6.75}$$

denoting the weighted pseudoinverse of $\boldsymbol{Ad}_{\text{sup}}^T$. The positive definite weighting matrix $\boldsymbol{W}_{\text{sup}} \in \mathbb{R}^{6\psi_{\text{sup}}\times 6\psi_{\text{sup}}}$ specifies which portion of $\boldsymbol{F}_{\text{sup}}^{\text{opt}}$ is considered as part of the overall support wrench $\boldsymbol{F}_{\text{sup}}^{\text{res}}$ and which portion as internal stress. The logical choice is to choose $\boldsymbol{W}_{\text{sup}}$ equal to the weighting matrix $\boldsymbol{Q}_{\text{sup}}$, which is used in the constrained quadratic optimization (6.9) to (6.14) for solving the wrench distribution problem. This choice allows us to split the injected power

$$\begin{aligned} P_{\text{act}} = &\underbrace{(-\boldsymbol{v}_{\text{sup}})^T(\boldsymbol{Ad}_{\text{sup}}^T)^{\boldsymbol{Q}_{\text{sup}}+}\boldsymbol{F}_{\text{sup}}^{\text{res}}}_{P_{\text{GM}}} \\ &+ \underbrace{(-\boldsymbol{v}_{\text{sup}})^T\left(\boldsymbol{I} - (\boldsymbol{Ad}_{\text{sup}}^T)^{\boldsymbol{Q}_{\text{sup}}+}\,\boldsymbol{Ad}_{\text{sup}}^T\right)\boldsymbol{F}_{\text{sup}}^{\text{opt}}}_{P_{\text{RM}}} \end{aligned} \tag{6.76}$$

into P_{GM} and P_{RM} related to the group and the relative motion of the support contacts, respectively. Furthermore, P_{GM} can be split into

$$\begin{aligned} P_{\text{GM}} = &\underbrace{(-\boldsymbol{v}_{\text{sup}})^T(\boldsymbol{Ad}_{\text{sup}}^T)^{\boldsymbol{Q}_{\text{sup}}+}\sigma_g m\boldsymbol{g}_0}_{P_{\text{GM}}^g} \\ &+ \underbrace{(-\boldsymbol{v}_{\text{sup}})^T(\boldsymbol{Ad}_{\text{sup}}^T)^{\boldsymbol{Q}_{\text{sup}}+}(\boldsymbol{F}_{\text{c}}^{\text{cpl}} + \boldsymbol{Ad}_{\text{int}}^T\,\boldsymbol{F}_{\text{int}}^{\text{cpl}})}_{P_{\text{GM}}^{\text{cpl}}} \end{aligned} \tag{6.77}$$

by inserting the modified underactuation condition (6.65) with $\boldsymbol{F}_{\text{sup}}^{\text{res}} = \boldsymbol{Ad}_{\text{sup}}^T\,\boldsymbol{F}_{\text{sup}}$ (see Sect. 3.3). Here, P_{GM}^g and $P_{\text{GM}}^{\text{cpl}}$ denote the power of the group motion due to gravity compensation and the Cartesian compliances (CoM + Interaction), respectively. Inserting this segmentation into (6.73) yields

$$\begin{aligned} \dot{V} \le\ &(\boldsymbol{\Omega}_{\text{c}}\,\boldsymbol{v}_{\text{c,d}})^T(\boldsymbol{F}_{\text{c}}^{\text{cpl}}) + (\boldsymbol{\Omega}_{\text{int}}\,\boldsymbol{v}_{\text{int,d}})^T\boldsymbol{F}_{\text{int}}^{\text{cpl}} + \boldsymbol{v}_{\text{c}}^T\,\sigma_g\,m\boldsymbol{g}_0 + P_{\text{GM}}^{\text{cpl}} + P_{\text{GM}}^g \\ &- \boldsymbol{v}_{\text{RM}}^T\,\boldsymbol{D}_{\text{RM}}^{\text{var}}\,\boldsymbol{v}_{\text{RM}} + P_{\text{RM}} \\ &+ \boldsymbol{v}^T\boldsymbol{\tau}_{\text{ext}}, \end{aligned} \tag{6.78}$$

where the first line corresponds to the group motion, the second to the relative motion, and the third to the port with the environment.

In order to compensate for P_{GM}, we propose a combination of an "impedance" and an "admittance type" passivity control. The former is implemented by adjusting the gravity compensation of the controller via the parameter σ_g. The latter by moving the setpoints of the CoM and the interaction end-effectors appropriately. In general, there is an infinite amount of possibilities to combine a CoM motion with a motion of the hands. But for reasons of simplicity, in Henze et al. (2018) we proposed moving both in a coordinated manner, as if they were connected by a single rigid body. Other choices are also possible, although they can lead to a singular problem: For example, compensating for $P_{\mathrm{GM}}^{\mathrm{cpl}}$ is not possible if $\boldsymbol{F}_{\mathrm{c}}^{\mathrm{cpl}}$ is close to zero, because this would require an infinite velocity $\boldsymbol{v}_{\mathrm{c,d}}$. To avoid this problem, the setpoint velocities are chosen as if they were connected by a single rigid body. Incorporating this condition via $\boldsymbol{\Omega}_{\mathrm{int}}\, \boldsymbol{v}_{\mathrm{int,d}} = \boldsymbol{Ad}_{\mathrm{int}}\, \boldsymbol{\Omega}_{\mathrm{c}}\, \boldsymbol{v}_{\mathrm{c,d}}$ into (6.78) results in

$$\begin{aligned}\dot{V} \leq{} & (\boldsymbol{\Omega}_{\mathrm{c}}\, \boldsymbol{v}_{\mathrm{c,d}})^T (\boldsymbol{F}_{\mathrm{c}}^{\mathrm{cpl}} + \boldsymbol{Ad}_{\mathrm{int}}^T\, \boldsymbol{F}_{\mathrm{int}}^{\mathrm{cpl}}) + \boldsymbol{v}_{\mathrm{c}}^T\, \sigma_g\, m \boldsymbol{g}_0 + P_{\mathrm{GM}}^{g} + P_{\mathrm{GM}}^{\mathrm{cpl}} \\ & - \boldsymbol{v}_{\mathrm{RM}}^T\, \boldsymbol{D}_{\mathrm{RM}}^{\mathrm{var}}\, \boldsymbol{v}_{\mathrm{RM}} + P_{\mathrm{RM}} \\ & + \boldsymbol{v}^T \boldsymbol{\tau}_{\mathrm{ext}}.\end{aligned} \tag{6.79}$$

In order to compensate for the activity of the group motion P_{GM}, one has to choose $\boldsymbol{v}_{\mathrm{c,d}}$ and σ_g such that

$$\begin{aligned}(\boldsymbol{\Omega}_{\mathrm{c}}\, \boldsymbol{v}_{\mathrm{c,d}})^T (\boldsymbol{F}_{\mathrm{c}}^{\mathrm{cpl}} + \boldsymbol{Ad}_{\mathrm{int}}^T\, \boldsymbol{F}_{\mathrm{int}}^{\mathrm{cpl}}) + \boldsymbol{v}_{\mathrm{c}}^T\, \sigma_g\, m \boldsymbol{g}_0 + P_{\mathrm{GM}}^{g} + P_{\mathrm{GM}}^{\mathrm{cpl}} = & \\ \left(\boldsymbol{\Omega}_{\mathrm{c}}\, \boldsymbol{v}_{\mathrm{c,d}} - (\boldsymbol{Ad}_{\mathrm{sup}})^{\boldsymbol{Q}_{\mathrm{sup}}+} \boldsymbol{v}_{\mathrm{sup}}\right)^T \left(\boldsymbol{F}_{\mathrm{c}}^{\mathrm{cpl}} + \boldsymbol{Ad}_{\mathrm{int}}^T\, \boldsymbol{F}_{\mathrm{int}}^{\mathrm{cpl}}\right) + & \\ \left(\boldsymbol{v}_{\mathrm{c}} - (\boldsymbol{Ad}_{\mathrm{sup}})^{\boldsymbol{Q}_{\mathrm{sup}}+} \boldsymbol{v}_{\mathrm{sup}}\right) \left(\sigma_g\, m \boldsymbol{g}_0\right) \leq 0 &\end{aligned} \tag{6.80}$$

holds with P_{GM}^{g} and $P_{\mathrm{GM}}^{\mathrm{cpl}}$ according to (6.77). This condition can be ensured by selecting the setpoint velocity for the CoM and the interaction compliances according to

$$\boldsymbol{\Omega}_{\mathrm{c}}\, \boldsymbol{v}_{\mathrm{c,d}} = \begin{cases} \boldsymbol{0} & \text{if } P_{\mathrm{GM}}^{\mathrm{cpl}} \leq 0 \wedge \boldsymbol{d}^T \dot{\boldsymbol{d}} \geq 0 \\ (\boldsymbol{Ad}_{\mathrm{sup}})^{\boldsymbol{Q}_{\mathrm{sup}}+}\, \boldsymbol{v}_{\mathrm{sup}} & \text{if } P_{\mathrm{GM}}^{\mathrm{cpl}} \leq 0 \wedge \boldsymbol{d}^T \dot{\boldsymbol{d}} < 0 \\ (\boldsymbol{Ad}_{\mathrm{sup}})^{\boldsymbol{Q}_{\mathrm{sup}}+}\, \boldsymbol{v}_{\mathrm{sup}} & \text{if } P_{\mathrm{GM}}^{\mathrm{cpl}} > 0 \end{cases} \tag{6.81}$$

and by disabling the gravity compensation according to

$$\sigma_g = \begin{cases} 1 & \text{if } P_{\mathrm{GM}}^{g} \leq 0 \\ 0 & \text{if } P_{\mathrm{GM}}^{g} > 0. \end{cases} \tag{6.82}$$

Note that the expression $(\boldsymbol{Ad}_{\mathrm{sup}})^{\boldsymbol{Q}_{\mathrm{sup}}+}\, \boldsymbol{v}_{\mathrm{sup}}$ in (6.81) can be interpreted as a weighted mean value of the contact velocities. If $P_{\mathrm{GM}}^{\mathrm{cpl}} > 0$, then the active element injects power into the group motion. The controller compensates for the injected power by moving the setpoints at the computed mean velocity of the ground floor, as motivated in Sect. 6.4.1. If $P_{\mathrm{GM}}^{\mathrm{cpl}} \leq 0$, there is no need to move the setpoints, as the active element drains power from the group motion. However, if the setpoints have been moved before, the controller can use $P_{\mathrm{GM}}^{\mathrm{cpl}}$ to bring the setpoints back into their initial configuration via the second line of (6.81). For this, a measure $\boldsymbol{d} \in \mathbb{R}^6$ of the distance between the current and the initial setpoint of the CoM is introduced. But the setpoints are only moved if the motion decreases the distance $\boldsymbol{d}$. Otherwise, the power $P_{\mathrm{GM}}^{\mathrm{cpl}}$ is dissipated by the active element without commanding a motion of the setpoints (see first line of (6.81)). The strategy, which is given in (6.82) temporarily disables the gravity compensation of the controller if $P_{\mathrm{GM}}^{g} \leq 0$ in order to let the robot drop due to the existing gravitational field.

In order to compensate for the activity of the relative motion P_{RM} in (6.79), we proposed an "impedance type" passivity control in Henze et al. (2018) by adjusting the variable damping matrix $\boldsymbol{D}_{\mathrm{RM}}^{\mathrm{var}}$ according to

$$\boldsymbol{D}_{\mathrm{RM}}^{\mathrm{var}} \boldsymbol{v}_{\mathrm{RM}} = \begin{cases} \mathbf{0} & \text{if } P_{\mathrm{RM}} \leq 0 \\ -\left(\boldsymbol{I} - \boldsymbol{Ad}_{\mathrm{sup}} (\boldsymbol{Ad}_{\mathrm{sup}})^{\mathcal{Q}_{\mathrm{sup}}+}\right) \boldsymbol{F}_{\mathrm{sup}}^{\mathrm{opt}} & \text{if } P_{\mathrm{RM}} > 0. \end{cases} \tag{6.83}$$

If $P_{\mathrm{RM}} > 0$, then the active element injects power into the relative motion, which is compensated for by activating the damper $\boldsymbol{D}_{\mathrm{RM}}^{\mathrm{var}}$ accordingly. If $P_{\mathrm{RM}} \leq 0$ then the active element drains energy from the relative motion, which allows for a deactivation of the variable damper. Note that (6.83) never becomes singular because the relative power $P_{\mathrm{RM}} = 0$ if the relative velocity $\boldsymbol{v}_{\mathrm{RM}} = \mathbf{0}$ (see (6.76)). Thus, a suitable matrix $\boldsymbol{D}_{\mathrm{RM}}^{\mathrm{var}}$ always exists to compensate for P_{RM}.

Note that it is not possible to show the passivity of the subsystem containing the controller and the robot. Although (6.81), (6.82), and (6.83) ensure that

$$\dot{V} \leq \boldsymbol{\nu}^{T} \boldsymbol{\tau}_{\mathrm{ext}} \tag{6.84}$$

holds, the storage function V does not feature a lower bound due to $V_{g}(\boldsymbol{x}_{\mathrm{c,d}})$ in (6.69). For this reason, the passivity considerations must be extended to incorporate the properties of the ground floor. The latter is assumed to be passive itself, which means that there exists a positive definite storage function $V_{\mathrm{floor}}(\boldsymbol{H}_{\mathrm{sup},i}) \in \mathbb{R}$, which is a function of the homogeneous transformations of the support contacts $\boldsymbol{H}_{\mathrm{sup},i}$. Furthermore, $V_{\mathrm{floor}}(\boldsymbol{H}_{\mathrm{sup},i})$ satisfies

$$\dot{V}_{\mathrm{floor}} \leq -\boldsymbol{v}_{\mathrm{sup}}^{T} \boldsymbol{F}_{\mathrm{sup}}^{\mathrm{ext}} \tag{6.85}$$

with $\boldsymbol{v}_{\mathrm{sup}}^{T} \boldsymbol{F}_{\mathrm{sup}}^{\mathrm{ext}}$ representing the power port, which allows for an energy transfer between the floor and the robot. Combining $\dot{V}_{\mathrm{floor}}$ with (6.69) yields the total potential function

$$\begin{aligned} V_{\mathrm{tot}} &= V_{\mathrm{floor}} + V \\ &= V_{\mathrm{floor}} + V_{g}(\boldsymbol{x}_{\mathrm{c}}) + \frac{1}{2} \boldsymbol{\nu}^{T} \boldsymbol{M} \boldsymbol{\nu} + V_{\mathrm{c}}^{K} + V_{\mathrm{int}}^{K} + V_{\mathrm{RM}}^{K} \end{aligned} \tag{6.86}$$

comprising the floor, the robot, and the controller. An obvious requirement to stabilize the robot on a movable surface is the assumption that the environment can support the weight of the robot. This is formalized with the condition

$$V_{\mathrm{floor}}(\boldsymbol{H}_{\mathrm{sup},i}) + V_{g}(\boldsymbol{x}_{\mathrm{c}}) > -\kappa \in \mathcal{U} \tag{6.87}$$

with the scalar $\kappa > 0$. This condition is subject to the constraint $\mathcal{U}$, which limits the maximum distance between the support contacts and the CoM to be within the kinematic limits of the robot. Under this assumption, the total potential V_{tot} is bounded from below and qualifies as a candidate storage function for the passivity analysis.

In order to complete the passivity analysis, we have to show that $\dot{V}_{tot}$ is smaller or equal to the power transferred via a power port (Khalil 2014). Computing $\dot{V}_{tot}$ yields

$$\dot{V}_{tot} = \dot{V}_{floor} + \dot{V} \leq -\boldsymbol{v}_{sup}^T \boldsymbol{F}_{sup}^{ext} + \boldsymbol{\nu}^T \boldsymbol{\tau}_{ext} \tag{6.88}$$

considering (6.84) and (6.85). Let us split $\boldsymbol{\tau}_{ext} = \boldsymbol{J}_{sup}^T \boldsymbol{F}_{sup}^{ext} + \boldsymbol{\tau}^{dist}$ into the external support wrenches $\boldsymbol{F}_{sup}^{ext}$ and into the vector $\boldsymbol{\tau}_{dist} \in \mathbb{R}^{6+n}$. The latter comprises all remaining external disturbances, such as a human perturbing the robot. This partition allows for the isolation of the energy flow between the robot and the floor from the rest, which leads to

$$\dot{V}_{tot} \leq \boldsymbol{v}_{sup}^T \boldsymbol{F}_{sup}^{ext} + \boldsymbol{\nu}^T \boldsymbol{J}_{sup}^T \boldsymbol{F}_{sup}^{ext} + \boldsymbol{\nu}^T \boldsymbol{\tau}_{dist} = \boldsymbol{\nu}^T \boldsymbol{\tau}_{dist} \tag{6.89}$$

considering (3.21). Thus, one can conclude that the combined system containing the controller, the robot, and the floor is passive with respect to the power port $\boldsymbol{\nu}^T \boldsymbol{\tau}_{dist}$ based on the definition given by Khalil (2014).

6.4.3 *Simplification of the Control Strategy for the Group Motion*

The previous section uses two independent power ports to compensate for the activity of the group motion. The first power port is given by $(\boldsymbol{\Omega}_c\, \boldsymbol{v}_{c,d})^T (\boldsymbol{F}_c^{cpl} + \boldsymbol{Ad}_{int}^T \boldsymbol{F}_{int}^{cpl})$ and administered by (6.81) to compensate for P_{GM}^{cpl}. The second port is given by $(\boldsymbol{v}_c)^T (\sigma_g\, m \boldsymbol{g}_0)$ and administered by (6.82) to compensate for P_{GM}^{g}. The control strategy can be simplified under the assumption that the CoM and the interaction compliance show a sufficient tracking, such that only a single power port is required to passivate the group motion.

Besides this aspect, the strategy presented in the previous section results in a chattering of the contact wrenches $\boldsymbol{F}_{sup}$. Especially, the temporary deactivation of the gravity compensation according to (6.82) has a significant impact on the amplitude of the chattering. In the case of the humanoid robot *TORO*, the gravity compensation has to counteract a weight of 76.4 kg (see Chap. 5). In contrast to that, the simplified strategy presented in this section does not disable the gravity compensation at all (σ_g is always set to $\sigma_g = 1$), which results in a less “aggressive” behavior of the controller.

Let us revise (6.80) by first adding $\boldsymbol{\Omega}_c(\boldsymbol{v}_{c,d} - \boldsymbol{v}_{c,d}) = \boldsymbol{0}$ and then exploiting the modified underaction condition (6.65) with $\boldsymbol{Ad}_{sup}^T \boldsymbol{F}_{sup} = \boldsymbol{F}_{sup}^{res}$:

$$
\begin{aligned}
0 \geq\ & \left(\boldsymbol{\Omega}_{\mathrm{c}}\, \boldsymbol{v}_{\mathrm{c,d}} - (\boldsymbol{Ad}_{\mathrm{sup}})^{\mathcal{Q}_{\mathrm{sup}}+} \boldsymbol{v}_{\mathrm{sup}}\right)^T \left(\boldsymbol{F}_{\mathrm{c}}^{\mathrm{cpl}} + \boldsymbol{Ad}_{\mathrm{int}}^T\, \boldsymbol{F}_{\mathrm{int}}^{\mathrm{cpl}}\right) \\
& + \left(\boldsymbol{v}_{\mathrm{c}} + \boldsymbol{\Omega}_{\mathrm{c}}(\boldsymbol{v}_{\mathrm{c,d}} - \boldsymbol{v}_{\mathrm{c,d}}) - (\boldsymbol{Ad}_{\mathrm{sup}})^{\mathcal{Q}_{\mathrm{sup}}+} \boldsymbol{v}_{\mathrm{sup}}\right) \left(\sigma_g\, m \boldsymbol{g}_0\right) \\
=\ & \left(\boldsymbol{\Omega}_{\mathrm{c}}\, \boldsymbol{v}_{\mathrm{c,d}} - (\boldsymbol{Ad}_{\mathrm{sup}})^{\mathcal{Q}_{\mathrm{sup}}+} \boldsymbol{v}_{\mathrm{sup}}\right)^T \left(\boldsymbol{F}_{\mathrm{c}}^{\mathrm{cpl}} + \boldsymbol{Ad}_{\mathrm{int}}^T\, \boldsymbol{F}_{\mathrm{int}}^{\mathrm{cpl}} + \sigma_g\, m \boldsymbol{g}_0\right) \\
& + \left(\boldsymbol{v}_{\mathrm{c}} - \boldsymbol{\Omega}_{\mathrm{c}}\, \boldsymbol{v}_{\mathrm{c,d}}\right) \left(\sigma_g\, m \boldsymbol{g}_0\right) \\
=\ & \left(\boldsymbol{\Omega}_{\mathrm{c}}\, \boldsymbol{v}_{\mathrm{c,d}} - (\boldsymbol{Ad}_{\mathrm{sup}})^{\mathcal{Q}_{\mathrm{sup}}+} \boldsymbol{v}_{\mathrm{sup}}\right)^T \left(\boldsymbol{F}_{\mathrm{sup}}^{\mathrm{res}}\right) + \underbrace{\left(\boldsymbol{v}_{\mathrm{c}} - \boldsymbol{\Omega}_{\mathrm{c}}\, \boldsymbol{v}_{\mathrm{c,d}}\right)}_{\approx \boldsymbol{0}} \left(\sigma_g\, m \boldsymbol{g}_0\right).
\end{aligned}
\tag{6.90}
$$

If we assume a sufficient tracking of the CoM and interaction compliances, then the last term in (6.90) can be neglected. Therefore, it is only necessary to move the setpoints according to

$$
\boldsymbol{\Omega}_{\mathrm{c}}\, \boldsymbol{v}_{\mathrm{c,d}} = \begin{cases} \boldsymbol{0} & \text{if } P_{\mathrm{GM}} \leq 0 \wedge \boldsymbol{d}^T \dot{\boldsymbol{d}} \geq 0 \\ (\boldsymbol{Ad}_{\mathrm{sup}})^{\mathcal{Q}_{\mathrm{sup}}+}\, \boldsymbol{v}_{\mathrm{sup}} & \text{if } P_{\mathrm{GM}} \leq 0 \wedge \boldsymbol{d}^T \dot{\boldsymbol{d}} < 0 \\ (\boldsymbol{Ad}_{\mathrm{sup}})^{\mathcal{Q}_{\mathrm{sup}}+}\, \boldsymbol{v}_{\mathrm{sup}} & \text{if } P_{\mathrm{GM}} > 0 \end{cases}
\tag{6.91}
$$

to compensate for the power $P_{\mathrm{GM}} = \left((\boldsymbol{Ad}_{\mathrm{sup}})^{\mathcal{Q}_{\mathrm{sup}}+}\, \boldsymbol{v}_{\mathrm{sup}}\right)^T \left(\boldsymbol{F}_{\mathrm{sup}}^{\mathrm{res}}\right)$, which is injected into the group motion by the active element according to (6.76). Temporarily disabling the gravity compensation is no longer required, which means that the parameter σ_g can be set permanently to $\sigma_g = 1$.

6.4.4 Energy Tanks

The method proposed in Sect. 6.4.2 ensures the passivity of the closed-loop system (controller + robot + floor) at every time step the controller is executed. In order to be less conservative, in Henze et al. (2018) we proposed extending the control scheme with energy tanks (Franken et al. 2011), which allows the controller to become active for a short period of time, as long as the same amount of energy is dissipated afterwards. Note that the extension is designed for the simplified version of the control strategy given in Sect. 6.4.3.

Let us first address the power P_{GM} involved in the group motion of the support contacts. Based on the practical experience of Henze et al. (2018) it is recommended to use six separate energy tanks for every DoF of the group motion to ensure that no activity of one DoF is canceled or "masked" by the dissipation of another DoF. Of course implementing six separate tanks can make the controller more conservative than necessary. The energy level $T_{\mathrm{GM},k} \in \mathbb{R}$ of each tank features a lower and an upper bound such that $0 \leq T_{\mathrm{GM},k} < T_{\mathrm{GM},k}^{\max}$ $\forall k = 1 \ldots 6$ holds. If a tank is within these boundaries, the fill level changes according to

$$
\dot{T}_{\mathrm{GM},k} = (\beta_{\mathrm{GM}}^{\mathrm{act},k} + \beta_{\mathrm{GM}}^{\mathrm{drain},k}) \qquad \forall k = 1 \ldots 6
\tag{6.92}
$$

with $\beta_{\mathrm{GM}}^{\mathrm{act},k}, \beta_{\mathrm{GM}}^{\mathrm{drain},k} \in \mathbb{R}$ representing two power ports. The first one is directly connected to the active element via $\beta_{\mathrm{GM}}^{\mathrm{act},k} = \left((\boldsymbol{Ad}_{\mathrm{sup}})^{\boldsymbol{Q}_{\mathrm{sup}}+} \boldsymbol{v}_{\mathrm{sup}}\right)_k \left(\boldsymbol{F}_{\mathrm{sup}}^{\mathrm{res}}\right)_k$ to monitor the injected energy. If the group motion becomes passive according to the first line of (6.91), then the port is used to refill the tank. If the level reaches the upper limit $T_{\mathrm{GM},k}^{\max}$, the tank saturates, which means that the energy drained from the controller is dissipated. In case the tank reaches the lower limit of zero, the second port $\beta_{\mathrm{GM}}^{\mathrm{drain},k} = \left(-\boldsymbol{\Omega}_{\mathrm{c}}\, \boldsymbol{v}_{\mathrm{c,d}}\right)_k \left(\boldsymbol{F}_{\mathrm{sup}}^{\mathrm{res}}\right)_k$ opens, which is connected to the setpoints via

$$\left(\boldsymbol{\Omega}_{\mathrm{c}}\, \boldsymbol{v}_{\mathrm{c,d}}\right)_k = \begin{cases} 0 & \text{if } T_{\mathrm{GM},k} > 0 \\ 0 & \text{if } T_{\mathrm{GM},k} = 0 \wedge \beta_{\mathrm{GM}}^{\mathrm{act},k} \leq 0 \wedge d_k \dot{d}_k \geq 0 \\ \left((\boldsymbol{Ad}_{\mathrm{sup}})^{\boldsymbol{Q}_{\mathrm{sup}}+} \boldsymbol{v}_{\mathrm{sup}}\right)_k & \text{if } T_{\mathrm{GM},k} = 0 \wedge \beta_{\mathrm{GM}}^{\mathrm{act},k} \leq 0 \wedge d_k \dot{d}_k < 0 \\ \left((\boldsymbol{Ad}_{\mathrm{sup}})^{\boldsymbol{Q}_{\mathrm{sup}}+} \boldsymbol{v}_{\mathrm{sup}}\right)_k & \text{if } T_{\mathrm{GM},k} = 0 \wedge \beta_{\mathrm{GM}}^{\mathrm{act},k} > 0. \end{cases} \tag{6.93}$$

In other words, the second port drains as much power from the controller as the first port injects into it, such that the level of the tank does not sink below zero ensuring the overall passivity of the closed-loop system. Note that the setpoints of the CoM and the interaction end-effectors are only adjusted according to the second and third line of (6.91) if the energy level of the associated tank equals zero. Thus, the initial fill level of the tanks can be interpreted as a tuning parameter, which specifies how much energy the controller may inject via $\beta_{\mathrm{GM}}^{\mathrm{act},k}$ before passivity is enforced by opening the second port $\beta_{\mathrm{GM}}^{\mathrm{drain},k}$. In conclusion, the MCB-PC controller is identical to the MCB version presented in Sect. 6.1 as long as the energy tanks are filled. Thus, the energy tanks allow for full balancing capabilities for a short period of time until they are depleted and the robot switches to a "passive mode" by adjusting the setpoints.

Analogously, an energy tank $T_{\mathrm{RM},k} \in \mathbb{R}$ can be defined for each DoF involved in the relative motion of the support contacts. The saturation limits are given by $0 \leq T_{\mathrm{RM},k} < T_{\mathrm{RM},k}^{\max}$. If the energy level of the tank is within the boundaries, $T_{\mathrm{RM},k}$ changes according to

$$\dot{T}_{\mathrm{RM},k} = (\beta_{\mathrm{RM}}^{\mathrm{act},k} + \beta_{\mathrm{RM}}^{\mathrm{drain},k}) \qquad \forall k = 1 \ldots 6 \tag{6.94}$$

with the first port $\beta_{\mathrm{RM}}^{\mathrm{act},k} = \left(\boldsymbol{v}_{\mathrm{sup}}\right)_k \left((\boldsymbol{I} - (\boldsymbol{Ad}_{\mathrm{sup}}^T)^{\boldsymbol{Q}_{\mathrm{sup}}+} \boldsymbol{Ad}_{\mathrm{sup}}^T) \boldsymbol{F}_{\mathrm{sup}}^{\mathrm{opt}}\right)_k$ monitoring the active element. The second port $\beta_{\mathrm{RM}}^{\mathrm{drain},k} = (\boldsymbol{v}_{\mathrm{RM}})_k \left(\boldsymbol{D}_{\mathrm{RM}}^{\mathrm{var}}\, \boldsymbol{v}_{\mathrm{RM}}\right)_k$ is connected to the variable damper via

$$\left(\boldsymbol{D}_{\mathrm{RM}}^{\mathrm{var}} \boldsymbol{v}_{\mathrm{RM}}\right)_k = \begin{cases} 0 & \text{if } T_{\mathrm{RM},k} > 0 \\ 0 & \text{if } T_{\mathrm{RM},k} = 0 \wedge \beta_{\mathrm{RM}}^{\mathrm{drain},k} \leq 0 \\ \left(-\left(\boldsymbol{I} - \boldsymbol{Ad}_{\mathrm{sup}} (\boldsymbol{Ad}_{\mathrm{sup}})^{\boldsymbol{Q}_{\mathrm{sup}}+}\right) \boldsymbol{F}_{\mathrm{sup}}^{\mathrm{opt}}\right)_k & \text{if } T_{\mathrm{RM},k} = 0 \wedge \beta_{\mathrm{RM}}^{\mathrm{drain},k} > 0. \end{cases} \tag{6.95}$$

The first port is permanently connected to the active element, while the second port is used to drain energy from the controller via the variable damper $\boldsymbol{D}_{\mathrm{RM}}^{\mathrm{var}}$. Again, the energy is dissipated once the tank reaches the upper limit $T_{\mathrm{RM},k}^{\max}$. If the level of $T_{\mathrm{RM},k}$ reaches zero, the second port $\beta_{\mathrm{RM}}^{\mathrm{drain},k}$ opens in order to drain the same amount of power as injected via $\beta_{\mathrm{RM}}^{\mathrm{act},k}$.

6.4.5 *Experimental Evaluation*

The performance of the presented MCB-PC control framework for movable and deformable support surfaces was evaluated in Henze et al. (2018) using the torque-controlled humanoid robot *TORO*. A brief description of the robot hardware is given in Chap. 5. Although the control strategy presented in Sect. 6.4.2 provides a passivation of the group and the relative motion of the support contacts, previous experiments revealed that the group motion shows the biggest influence on passivity, while the relative motion can be neglected. Thus, the relative compliance (6.67) was permanently disabled for the presented experiments. Furthermore, the control strategy of the group motion was implemented on the basis of the simplifications in Sect. 6.4.3, as previous experiments revealed a sufficient tracking for the CoM and the interaction compliances.

6.4.5.1 Verification of the Control Strategy

The first experiment evaluates the effectiveness of the control strategy proposed in Sect. 6.4.3. For this, the robot *TORO* is placed with the left foot on a rocker board, which can be flipped manually by a human, as shown in Fig. 6.10a. The right foot of the robot rests on a pedestal. The algorithm for estimating the state of the base frame (see Sect. 5.2) was configured such that the information on the orientation is entirely provided by the on-board IMU ($\alpha_{\mathrm{IMU}} = 1$) and the one on the translation by the right foot ($\alpha_{\mathrm{FootR}} = 1$, $\alpha_{\mathrm{FootL}} = \alpha_{\mathrm{HandR}} = \alpha_{\mathrm{HandL}} = 0$). The rocker board can induce a vertical motion of the left foot, which corresponds to a superposition of a group and a relative motion.

Figure 6.10b compares the behavior of the robot using the MCB version of the control framework (see Sect. 6.1) with the extension for movable and deformable support surfaces (MCB-PC). For both, the rocker board is flipped twice inducing two consecutive up and down movements of the left foot. As can be seen in Fig. 6.10b, the MCB controller successfully stabilizes the height of the CoM and the hands above the ground, despite some minor deviations. However, if we take a look at the energy $\int \dot{\boldsymbol{q}}^T \boldsymbol{\tau}\, dt$, which is transfered between the controller and the robot, one can see that the MCB controller injects an overall amount of 11.4 J into the robot during the two motions. Note that the controller can only store energy in the potentials V_{c}^{K} and V_{int}^{K} of the Cartesian compliances, as the relative compliance was permanently disabled. As the control error of the Cartesian compliances is approximately zero before and after the two motions ($t = 0$ s and $t = 12$ s), one can conclude that the

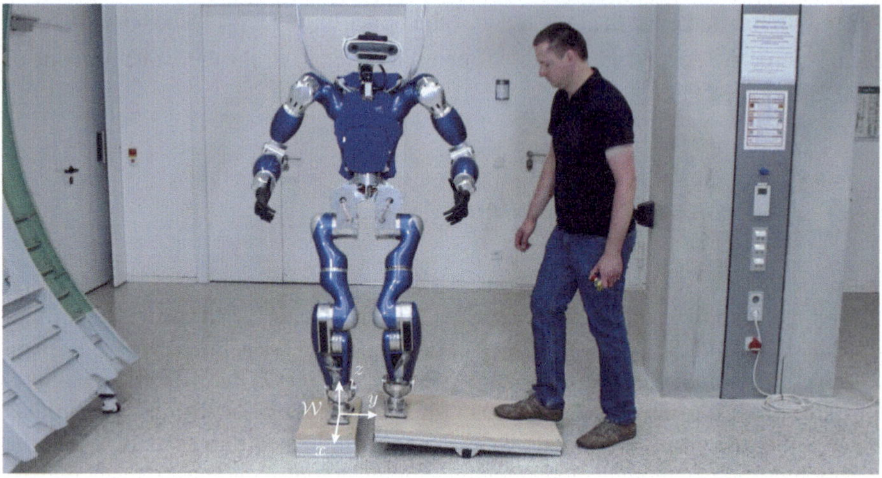

(a) Setup of the experiment.

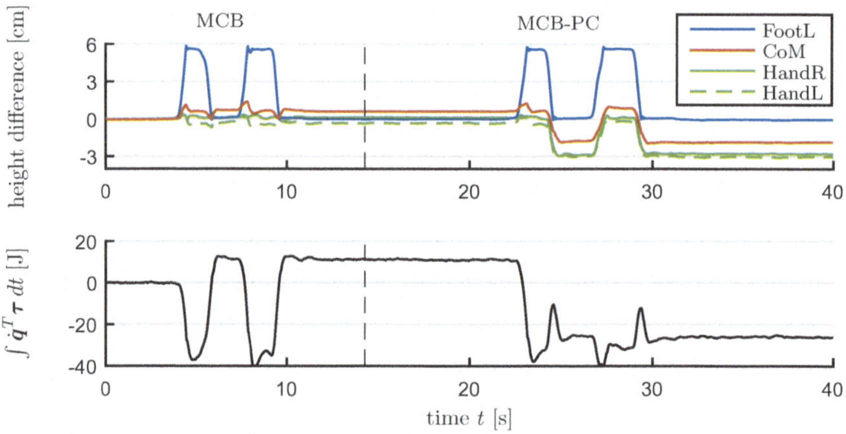

(b) Motion of the setpoints and the left foot.

Fig. 6.10 Comparison of the MCB and the MCB-PC controller for a vertical foot motion

injected energy must originate from the active element (6.76) inside the MCB controller.

If the experiment is repeated with the MCB-PC approach, Fig. 6.10b shows that the controller moves the setpoints of the CoM and the hands according to (6.91): During the first upward motion at $t = 22.9$ s, the active element P_{act} dissipates energy from the controller. Thus, the MCB-PC controller shows the same behavior as the MCB controller before by maintaining the height of the CoM and the hands (first line of (6.91)). But during the following downward motion, P_{act} is about to inject energy into the controller, which is prevented by lowering the CoM and the hands simultaneously (third line of (6.91)). During the second upward motion, P_{act} becomes dissipative again. Thus, the MCB-PC controller utilizes the power of P_{act} to lift the setpoints back again to their initial height (second line of (6.91)). It is noteworthy that the setpoints perform a vertical motion of 2.8 cm, which is exactly half of the amplitude of the left foot. This observation can be explained by interpreting $(\boldsymbol{Ad}_{\text{sup}})^{\boldsymbol{Q}_{\text{sup}}+}\boldsymbol{v}_{\text{sup}}$ in (6.91) as a weighted mean value of the foot velocities. As the right foot is not moving, only 50% of the motion of the left foot is mapped to the CoM and the hands. If we take a look at the energy transfered between controller and robot $\int \dot{\boldsymbol{q}}^T \boldsymbol{\tau}\, dt$, one can see that the MCB-PC controller drains an overall amount of 37 J from the robot. As the control error of the compliances within the controller are approximately zero before and after the experiment ($t = 20$ s and $t = 35$ s), one can conclude that the drained energy must be dissipated by the controller, which proves the effectiveness of the proposed MCB-PC control strategy. Note that strictly speaking, the assumption of a passive floor (see Sect. 6.4) does not hold in this particular scenario because of a human flipping the rocker board. However, this inadequacy does not diminish the main observation of the experiment that the MCB-PC controller drains energy from the robot and dissipates it to ensure passivity in contrast to the simple MCB approach.

6.4.5.2 Importance of Passivity Control

The next experiment demonstrates the importance of passivity control. Stramigioli (2015) recently showed that for every active controller one can find an environment that destabilizes the controller even if the environment is passive. In order to verify this statement, the robot *TORO* was placed with the left foot on three layers of gym mats, as shown in Fig. 6.11a. The right foot was used to estimate the base frame resting on a pedestal ($\alpha_{\text{FootR}} = 1$, $\alpha_{\text{FootL}} = \alpha_{\text{HandR}} = \alpha_{\text{HandL}} = 0$ in Sect. 5.2).

If the MCB version of the control framework (see Sect. 6.1) is used to operate the robot, the latter exhibits an increasing oscillation, as shown in Fig. 6.11b. The oscillation involves the translation of the CoM in x-direction and the orientation of the left foot about the y-axis of the world frame $\mathcal{W}$. As the amplitude keeps increasing up to 2.6 cm and 10° respectively, the experiment is aborted at $t = 20$ s. The observed instability can be explained by analyzing the energy transfer between controller and robot $\int \dot{\boldsymbol{q}}^T \boldsymbol{\tau}\, dt$. As can be seen in Fig. 6.11b, the MCB controller periodically injects energy into the robot and thereby destabilizes it.

Figure 6.11b also reveals that the energy transfer into the robot is negligible if the robot is operated using the MCB-PC controller, which features the extension for movable and deformable ground surfaces. Although there is also a slight oscillation, the amplitude stays within reasonable limits.

6.4.5.3 Application of Energy Tanks

This section evaluates the possibility to make the MCB-PC control approach less conservative by using energy tanks, as proposed in Sect. 6.4.4. For this purpose, the robot is placed on three layers of gym mats, as shown in Fig. 6.12a. At the beginning

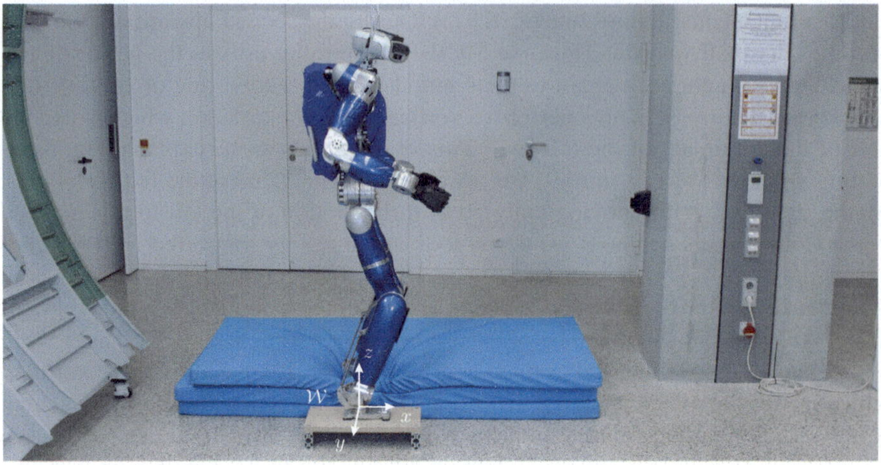

(a) Setup of the experiment.

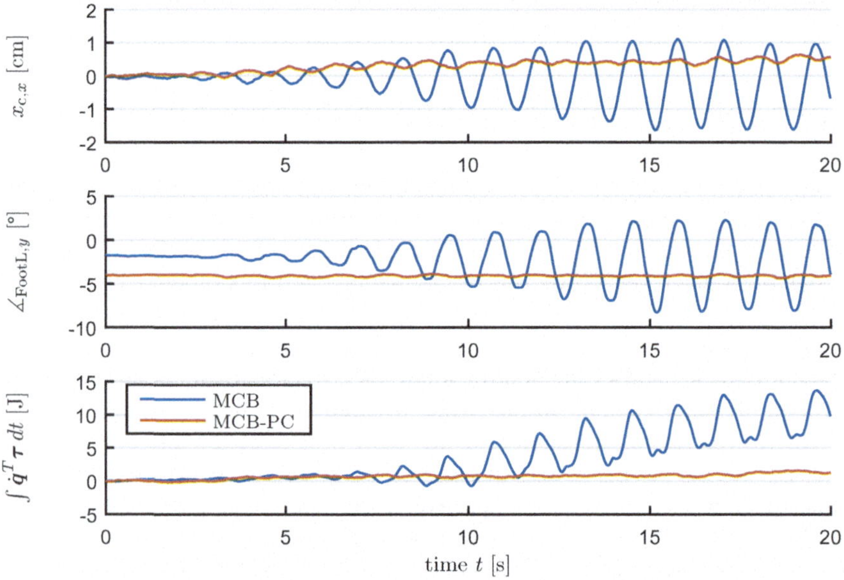

(b) Motion of the center of mass and the left foot in relation to the power transfer between controller and robot.

Fig. 6.11 Comparison of the MCB and the MCB-PC controller balancing on a mixed type of surfaces

of the experiment, the three energy tanks for the translation of the group motion are at their maximum level of 3 J, 0.2 J, and 5 J, respectively. Therefore, the MCB-PC controller exhibits the same balancing behavior as the basic MCB controller in

Sect. 6.1 until the tank in y-direction reaches zero at $t = 5$ s. After that, the power port $\beta_{\mathrm{GM}}^{\mathrm{drain},y}$ in (6.92) is opened, which causes the CoM and the CoP to move 1.1 cm to the side. However, the stance of the robot is wide enough, such that the movement of CoM and CoP does not compromise balancing. After $t = 25$ s, the energy tank in x-direction is depleted as well, which causes the CoM and the CoP to start drifting to the front. As soon as both reach the frontal boundary of the support polygon at $p_{\mathrm{Feet},x}^{\mathrm{max}} = 13$ cm, the robot shuts itself down and falls to the front.

In summary, the extended MCB-PC control concept ensures the passivity of the closed-loop system by moving the setpoints of the CoM and the hands in concert with the ground floor. However, this can also have negative effects, as the CoM can approach the border of the support polygon. In the presented experiment, the stance of the robot was wide enough, such that a motion of the CoM to the side does not pose a problem, in contrast to the x-direction where the size of the support polygon is rather limited.

6.5 Balancing While Performing High-Force Interaction Tasks (MCB-ACT)

As shown in Sect. 6.1.3, the basic version of the whole-body controller (MCB) can already carry a significant amount of load, such as a 12.2 kg box in this case. As can be seen in Fig. 6.4, the center of mass maintains its location during the lift, while the center of pressure moves to the front of the feet to counteract the weight of the box. As a consequence, the length of the feet limits the additional weight that the robot is able to carry without falling over. In more general terms, the magnitude of the interaction wrenches, which the robot is able to apply at the environment, is limited by the contact model (6.13) imposed on the support contacts. In order to overcome this restriction, we extended the basic MCB controller in Abi-Farraj et al. (2019) by allowing the controller to move the CoM automatically, such that the support wrenches $\boldsymbol{F}_{\mathrm{sup}}$ satisfy the contact model. For instance, this enables the robot to lean forward in order to push a heavy object, as demonstrated in Fig. 6.16a. The whole-body control approach presented in this section will be called Multi-Contact Balancer with Automatic CoM Task (MCB-ACT) for the remainder of the manuscript. A repetition of the experiment where *TORO* is carrying a box revealed that the payload can be increased from 12.2 to 25 kg, which corresponds to nearly a third of the robot's weight (see Fig. 6.15).

6.5.1 Controller Derivation

The MCB-ACT controller presented in Abi-Farraj et al. (2019) utilizes the polyhedron of feasible wrenches introduced in Sect. 3.6 to find an optimal CoM position as a function of the commanded interaction wrenches $\boldsymbol{F}_{\mathrm{int}}^{\mathrm{cpl}}$. Figure 6.13 offers a more detailed overview of the control architecture: The polyhedron $\mathcal{P}(\boldsymbol{A}, \boldsymbol{b})$ is used to

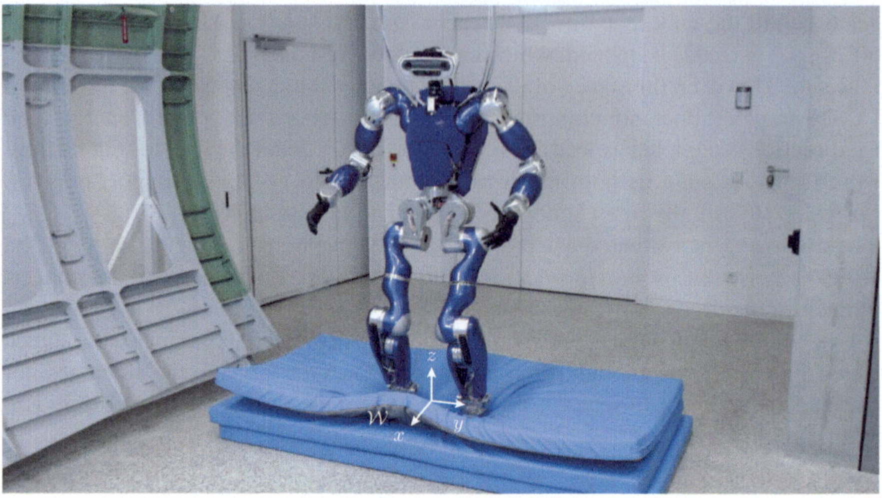

(a) Setup of the experiment.

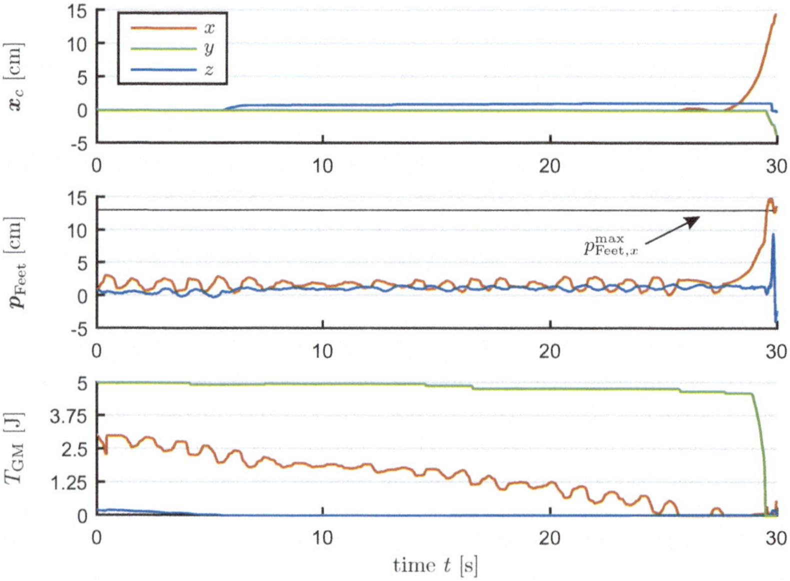

(b) Motion of the CoM and CoP in relation to the level of the energy tanks.

Fig. 6.12 Influence of the energy tanks while balancing on a compliant support

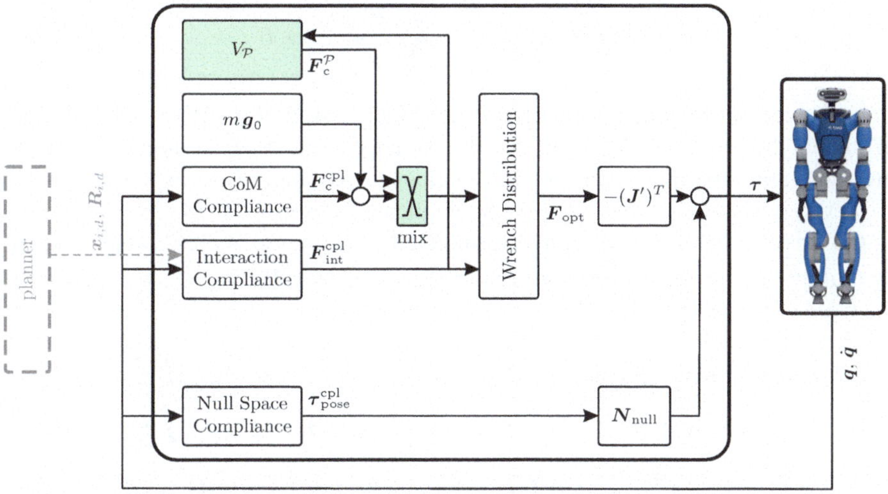

Fig. 6.13 Overview of the architecture of the MCB-ACT control framework extended for high-force interaction tasks. The "mix"-block combines both input signals by selecting the horizontal forces from $F_c^{\mathcal{P}}$ and the rest from $F_c^{cpl} + m g_0$ according to (6.101)

define a potential $V_{\mathcal{P}}(F_{int}, x_c)$, which drives the horizontal translation of the center of mass, such that the robot can utilize its own weight to counteract the interaction wrenches F_{int}^{cpl}. For instance, this allows the robot to lean against an object in order to push it, as demonstrated in Fig. 6.16. The remaining four CoM coordinates are governed by the Cartesian compliance (6.1). The wrenches generated by the potential $V_{\mathcal{P}}(F_{int}, x_c)$ and by the CoM compliance are combined and fed to the constrained quadratic optimization problem (6.9) to (6.14) in order to be distributed to the end-effectors. The mapping of F^{opt} to the joint torque τ is done via (6.17). The interaction end-effectors are again stabilized by the Cartesian compliance (6.4) generating F_{int}^{cpl}. In order to handle robots with a redundant kinematics or in singular configurations, the framework is also equipped with the null space controller presented in Sect. 6.2.

Under the assumption that all contact frames are stationary with respect to the world frame $\mathcal{W}$, the polyhedron of feasible wrenches $\mathcal{P}(A, b)$ is computed using the procedure depicted in Fig. 3.11. Note that the polyhedron is given in the $\mathcal{H}$-representation leading to (3.51). As the resulting support wrench $F_{sup}^{res} = Ad_{sup}^T F_{sup}$ acts on the CoM frame C, it is shifted to the world frame $\mathcal{W}$ via $Ad(I, x_c)^{-T}$ leading to $F_{sup,\mathcal{W}}^{res} = Ad(I, x_c)^{-T} F_{sup}^{res}$. Consequently, the polyhedron $\mathcal{P}(A, b) = \left\{ F_{sup,\mathcal{W}}^{res} : A\, F_{sup,\mathcal{W}}^{res} \leq b \right\}$ describes the set of feasible wrenches acting on the world frame. Note that $\mathcal{P}(A, b)$ is constant for a given contact configuration.

Let the kth face of $\mathcal{P}(A, b)$ be given by $\left\{ F_{sup,\mathcal{W}}^{res} : a_k^T\, F_{sup,\mathcal{W}}^{res} \leq b_k \right\}$. Based on the assumption that the robot starts from a feasible configuration (see Assumption 3.1), one can define a distance measure $d_k \in \mathbb{R}$

$$d_k(\boldsymbol{F}^{\text{res}}_{\text{sup},\mathcal{W}}) = \frac{b_k - \boldsymbol{a}_k^T \boldsymbol{F}^{\text{res}}_{\text{sup},\mathcal{W}}}{\|\boldsymbol{a}_k\|} \geq 0 \tag{6.96}$$

from the current resulting support wrench $\boldsymbol{F}^{\text{res}}_{\text{sup},\mathcal{W}}$ (shifted to $\mathcal{W}$) to the kth face of the polyhedron $\mathcal{P}(\boldsymbol{A}, \boldsymbol{b})$. Note that the distance measure d_k is unique and scale invariant, although there is an infinite number of possibilities to express $\mathcal{P}(\boldsymbol{A}, \boldsymbol{b})$. However, it is assumed that there are no duplicate faces in $\mathcal{P}(\boldsymbol{A}, \boldsymbol{b})$.

On the other hand, $\boldsymbol{F}^{\text{res}}_{\text{sup},\mathcal{W}}$ can be formulated as a function of the CoM position $\boldsymbol{x}_{\text{c}}$ by multiplying the underactuation condition (6.6) with $\boldsymbol{Ad}(\boldsymbol{I}, \boldsymbol{x}_{\text{c}})^{-T}$ to shift it from the CoM to the world frame $\mathcal{W}$:

$$\begin{aligned} \boldsymbol{Ad}(\boldsymbol{I}, \boldsymbol{x}_{\text{c}})^{-T} \boldsymbol{Ad}^T_{\text{sup}} \boldsymbol{F}_{\text{sup}} &= \boldsymbol{Ad}(\boldsymbol{I}, \boldsymbol{x}_{\text{c}})^{-T} \left(m\boldsymbol{g}_0 + \boldsymbol{F}^{\text{cpl}}_{\text{c}} + \boldsymbol{Ad}^T_{\text{int}} \boldsymbol{F}^{\text{cpl}}_{\text{int}} \right) \\ \boldsymbol{Ad}(\boldsymbol{I}, \boldsymbol{x}_{\text{c}})^{-T} \boldsymbol{F}^{\text{res}}_{\text{sup}} &= \boldsymbol{Ad}(\boldsymbol{I}, \boldsymbol{x}_{\text{c}})^{-T} \left(m\boldsymbol{g}_0 + \boldsymbol{F}^{\text{cpl}}_{\text{c}} + \boldsymbol{Ad}^T_{\text{int}} \boldsymbol{F}^{\text{cpl}}_{\text{int}} \right) \\ \boldsymbol{F}^{\text{res}}_{\text{sup},\mathcal{W}} &= \boldsymbol{Ad}(\boldsymbol{I}, \boldsymbol{x}_{\text{c}})^{-T} \left(m\boldsymbol{g}_0 + \boldsymbol{F}^{\text{cpl}}_{\text{c}} \right) + \boldsymbol{Ad}^T_{\text{int},\mathcal{W}} \boldsymbol{F}^{\text{cpl}}_{\text{int}} \end{aligned} \tag{6.97}$$

with $\boldsymbol{Ad}_{\text{int},\mathcal{W}} = \boldsymbol{Ad}_{\text{int}} \boldsymbol{Ad}(\boldsymbol{I}, \boldsymbol{x}_{\text{c}})$. Note that the potential $V_{\mathcal{P}}$ and the CoM compliance are designed independently of each other, which is why $\boldsymbol{F}^{\text{cpl}}_{\text{c}}$ can be neglected in (6.97). Although a combined design would account for a possible coupling between $V_{\mathcal{P}}$ and the compliance, it would also be considerably more cumbersome. Let us consider the case where the x- and y-axis of the world frame $\mathcal{W}$ span the horizontal plane, as shown in Fig. 3.12a. Then the above equation can be simplified by exploiting the structure of $\boldsymbol{g}_0 = (0\ 0\ \|\boldsymbol{g}_0\|\ 0\ 0\ 0)^T$ in combination with $\boldsymbol{x}_{\text{c}} = (x_{\text{c},x}\ x_{\text{c},y}\ x_{\text{c},z})^T$ and Definition 3.2:

$$\begin{aligned} \boldsymbol{F}^{\text{res}}_{\text{sup},\mathcal{W}} &= \boldsymbol{Ad}^T_{\text{int},\mathcal{W}} \boldsymbol{F}^{\text{cpl}}_{\text{int}} + \boldsymbol{Ad}(\boldsymbol{I}, \boldsymbol{x}_{\text{c}})^{-T} m\boldsymbol{g}_0 \\ &= \boldsymbol{Ad}^T_{\text{int},\mathcal{W}} \boldsymbol{F}^{\text{cpl}}_{\text{int}} + m \left\|\boldsymbol{g}_0\right\| \begin{pmatrix} 0 \\ 0 \\ 1 \\ x_{\text{c},y} \\ -x_{\text{c},x} \\ 0 \end{pmatrix} . \end{aligned} \tag{6.98}$$

Note that the wrench $\boldsymbol{F}^{\text{res}}_{\text{sup},\mathcal{W}}$ is a function of the commanded interaction wrenches $\boldsymbol{F}^{\text{cpl}}_{\text{int}}$ and the CoM position in the horizontal plane of the world frame $\mathcal{W}$. As the commanded interaction wrenches $\boldsymbol{F}^{\text{cpl}}_{\text{int}}$ are prescribed by the task at hand, the only remaining controllable variables are $x_{\text{c},x}$ and $x_{\text{c},y}$.

The support contacts can only generate the required wrench $\boldsymbol{F}_{\text{sup}}$ if and only if $d_k \geq 0\ \forall k$. In order to enforce this condition, a potential $V_{\mathcal{P}} \in \mathbb{R}$ can be defined encoding the proximity of the resulting support wrench $\boldsymbol{F}^{\text{res}}_{\text{sup}}$ to the faces of $\mathcal{P}(\boldsymbol{A}, \boldsymbol{b})$. The potential needs to be a monotone function that grows to infinity as d_k tends to zero. In Abi-Farraj et al. (2019), the potential was empirically chosen as

$$V_{\mathcal{P}}(x_{c,x}, x_{c,y}) = \sum_k \frac{1}{d_k(x_{c,x}, x_{c,y})}. \tag{6.99}$$

Due to the convexity of $\mathcal{P}(\boldsymbol{A}, \boldsymbol{b})$, a gradient control law

$$\boldsymbol{F}_c^{\mathcal{P}} = -k_{\mathcal{P}} \frac{\partial V_{\mathcal{P}}(x_{c,x}, x_{c,y})}{\partial \left(x_{c,x}\ x_{c,y}\right)^T} = -k_{\mathcal{P}}\, m\, \|\boldsymbol{g}_0\| \left(\sum_k \frac{\boldsymbol{a}_k^T}{d_k^2\, \|\boldsymbol{a}_k\|}\right) \begin{bmatrix} 0 & 0 \\ 0 & 0 \\ 0 & 0 \\ 0 & -1 \\ 1 & 0 \\ 0 & 0 \end{bmatrix} \tag{6.100}$$

can be formulated to regulate the center of mass location in the horizontal plane of $\mathcal{W}$ with $k_{\mathcal{P}} \in \mathbb{R}$ denoting a positive tuning parameter. Note that both $\boldsymbol{F}_c^{\mathcal{P}}$ and $\boldsymbol{F}_c^{\text{cpl}}$ in (6.4) are six-dimensional vectors. But $\boldsymbol{F}_c^{\mathcal{P}}$ is supposed to control the horizontal translation of the CoM, while $\boldsymbol{F}_c^{\text{cpl}}$ covers the remaining DoFs. Therefore, both need to be mixed into

$$\boldsymbol{F}_c = \text{diag}(1, 1, 0, 0, 0, 0)\, \boldsymbol{F}_c^{\mathcal{P}} - \text{diag}(0, 0, 1, 1, 1, 1)(\nabla\, V_c^K)^T - \boldsymbol{D}_c \boldsymbol{v}_c, \tag{6.101}$$

which is then fed to the constrained quadratic optimization problem (6.9) to (6.14), as shown in Fig. 6.13.

6.5.2 Reacting to Contact Transitions

In order to facilitate contact transitions, a humanoid robot must be capable of attaching and detaching end-effectors to/from the environment. A procedure for swapping end-effectors between the support and interaction task is discussed in Sect. 6.7.3. But as we know from Sect. 3.6, a change in the contact configuration has a direct influence on the size and shape of the support polygon $\mathcal{S}_{\text{SP}}$. In order to keep the CoM within the support polygon during a contact switch, the MCB-ACT controller in Sect. 6.5.1 can be adapted as shown in Abi-Farraj et al. (2019).

Let us consider a contact switch by changing the task assignment according to the procedure described in Sect. 6.7.3 without actually moving the respective end-effector. Furthermore, let us denote the polyhedron of feasible wrenches before and after the contact switch with $\mathcal{P}_{\text{before}} = \mathcal{P}(\boldsymbol{A}_{\text{before}}, \boldsymbol{b}_{\text{before}})$ and $\mathcal{P}_{\text{after}} = \mathcal{P}(\boldsymbol{A}_{\text{after}}, \boldsymbol{b}_{\text{after}})$, respectively. The potentials for both polyhedra are analogously given by $V_{\mathcal{P},\text{before}}(x_{c,x}, x_{c,y})$ and $V_{\mathcal{P},\text{after}}(x_{c,x}, x_{c,y})$ based on (6.99). One can obtain an optimal CoM position for both polyhedra by solving

$$\begin{pmatrix} x_{c,x}^{\text{before}} \\ x_{c,y}^{\text{before}} \end{pmatrix} = \underset{(x_{c,x}\ x_{c,y})^T}{\text{argmin}}\ V_{\mathcal{P},\text{before}}(x_{c,x}, x_{c,y}) \tag{6.102}$$

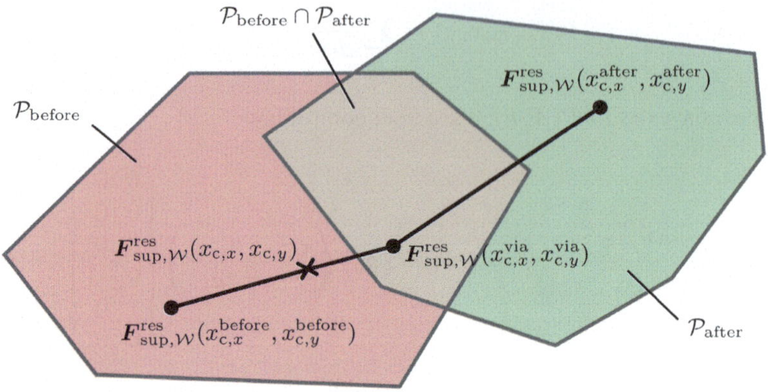

Fig. 6.14 The overall support wrench $\boldsymbol{F}^{\text{res}}_{\text{sup}}$ moves from $\mathcal{P}_{\text{before}}$ to $\mathcal{P}_{\text{after}}$ during a contact transition

and

$$\begin{pmatrix} x^{\text{after}}_{\text{c},x} \\ x^{\text{after}}_{\text{c},y} \end{pmatrix} = \underset{(x_{\text{c},x} \ x_{\text{c},y})^T}{\text{argmin}} \ V_{\mathcal{P},\text{after}}(x_{\text{c},x}, x_{\text{c},y}). \tag{6.103}$$

Note that both optimal positions depend on the interaction wrenches $\boldsymbol{F}^{\text{cpl}}_{\text{int}}$, as can be seen from (6.99) with (6.96) and (6.98). Thus, the optimal CoM positions must be recomputed online if there is a change of $\boldsymbol{F}^{\text{cpl}}_{\text{int}}$.

If the CoM moves quasi-statically from one location (6.102) to another (6.103), the contacts must generate the overall support wrench $\boldsymbol{F}^{\text{res}}_{\text{sup},\mathcal{W}}(x_{\text{c},x}, x_{\text{c},y})$ according to (6.98) in order to support the weight of the robot. To ensure the feasibility of $\boldsymbol{F}^{\text{res}}_{\text{sup},\mathcal{W}}(x_{\text{c},x}, x_{\text{c},y})$ during the transition, both polyhedra $\mathcal{P}_{\text{before}}$ and $\mathcal{P}_{\text{after}}$ must overlap, as shown in Fig. 6.14. However, it is possible that the start and/or the end point of the transition are outside of the intersection $\mathcal{P}_{\text{before}} \cap \mathcal{P}_{\text{after}}$. Therefore, the potentials $V_{\mathcal{P},\text{before}}$ and $V_{\mathcal{P},\text{after}}$ are not properly defined during the entire transition, which prohibits a direct interpolation between $V_{\mathcal{P},\text{before}}$ and $V_{\mathcal{P},\text{after}}$ to move the CoM from the optimal position (6.102) to (6.103). Instead, in Abi-Farraj et al. (2019) we proposed interpolating between (6.102) and (6.103) and stabilizing the resulting CoM position with another potential.[4] Note that the polyhedra $\mathcal{P}_{\text{before}}$ and $\mathcal{P}_{\text{after}}$ are convex, but not necessarily the union $\mathcal{P}_{\text{before}} \cup \mathcal{P}_{\text{after}}$. Therefore, a linear interpolation between (6.102) and (6.103) could cause the required overall support wrench $\boldsymbol{F}^{\text{res}}_{\text{sup},\mathcal{W}}(x_{\text{c},x}, x_{\text{c},y})$ to leave the union $\mathcal{P}_{\text{before}} \cup \mathcal{P}_{\text{after}}$, which in turn would lead to the robot falling. In order to prevent that, a via-point $(x^{\text{via}}_{\text{c},x} \ x^{\text{via}}_{\text{c},y})$ is introduced with $\boldsymbol{F}^{\text{res}}_{\text{sup},\mathcal{W}}(x^{\text{via}}_{\text{c},x}, x^{\text{via}}_{\text{c},y}) \in \mathcal{P}_{\text{before}} \cap \mathcal{P}_{\text{after}}$ (see Fig. 6.14).

In order to drive the center of mass from the optimal position (6.102) to (6.103), one can define a quadratic potential

[4] The interpolation proposed in Abi-Farraj et al. (2019) is formulated for the overall support wrench, but is equivalent to the algorithm presented in this section thanks to (6.98).

$$V_{\text{sw}} = \frac{1}{2}\left\{\begin{pmatrix} x_{\text{c},x}^{\text{ipol}} \\ x_{\text{c},y}^{\text{ipol}} \end{pmatrix} - \begin{pmatrix} x_{\text{c},x} \\ x_{\text{c},y} \end{pmatrix}\right\}^T \left\{\begin{pmatrix} x_{\text{c},x}^{\text{ipol}} \\ x_{\text{c},y}^{\text{ipol}} \end{pmatrix} - \begin{pmatrix} x_{\text{c},x} \\ x_{\text{c},y} \end{pmatrix}\right\}, \tag{6.104}$$

which stabilizes the CoM at $(x_{\text{c},x}^{\text{ipol}} \; x_{\text{c},y}^{\text{ipol}})^T$. The latter is a position that is obtained from a piecewise linear interpolation starting from the optimal position (6.102) via $(x_{\text{c},x}^{\text{via}} \; x_{\text{c},y}^{\text{via}})$ to (6.103). Analogously to (6.101), the wrench to be applied at the center of mass is given by

$$\boldsymbol{F}_{\text{c}}^{\text{sw}} = -k_{\text{sw}} \frac{\partial V_{\text{sw}}(x_{\text{c},x}, x_{\text{c},y})}{\partial \left(x_{\text{c},x} \; x_{\text{c},y}\right)^T}, \tag{6.105}$$

which is then mixed with the CoM compliance (6.1) into

$$\boldsymbol{F}_{\text{c}}^{\text{sw}} = -\operatorname{diag}(1, 1, 0, 0, 0, 0)\, \boldsymbol{F}_{\text{c}}^{\text{sw}} - \operatorname{diag}(0, 0, 1, 1, 1, 1)(\nabla V_{\text{c}}^{K})^T - \boldsymbol{D}_{\text{c}}\boldsymbol{v}_{\text{c}}. \tag{6.106}$$

If we furthermore assume that (6.101) successfully stabilizes the CoM at the optimal position before the contact transition $(x_{\text{c},x}^{\text{opt,before}} \; x_{\text{c},y}^{\text{opt,before}})^T$, then switching from the controller (6.101) to (6.106) will not cause a discontinuity in the wrench applied at the center of mass. The same holds for switching back to (6.101) after the interpolation process.

6.5.3 *Experimental Evaluation*

We evaluated the performance of the presented MCB-ACT control framework, which is designed for interaction tasks with high forces, in Abi-Farraj et al. (2019) using the torque-controlled humanoid robot *TORO* developed at *DLR*. A brief description of the robot hardware is given in Chap. 5.

6.5.3.1 Balancing While Lifting a Box

This experiment is a repetition of the one presented in Fig. 6.4, where *TORO* lifts a heavy box with both hands. Its purpose is to discuss the difference between the basic MCB controller (see Sect. 6.1) and the MCB-ACT extension to large interaction forces (see Sect. 6.5). As in the previous experiment, the robot *TORO* balances with both feet on the ground floor, as shown in Fig. 6.15a. But instead of *TORO* picking up the box from a table, it is handed over by a human operator. Furthermore, the weight of the box was increased from 12.2 to 25 kg, which corresponds to nearly a third of the robot's weight of 76.4 kg.

At the beginning and at the end of the experiment, the center of mass coincides with the center of pressure, as predicted by theory. However, as soon as the box is handed to the robot (3.3 s $< t <$ 6.5 s), it induces a static displacement of the hand compliances, which causes the generated interaction wrenches to

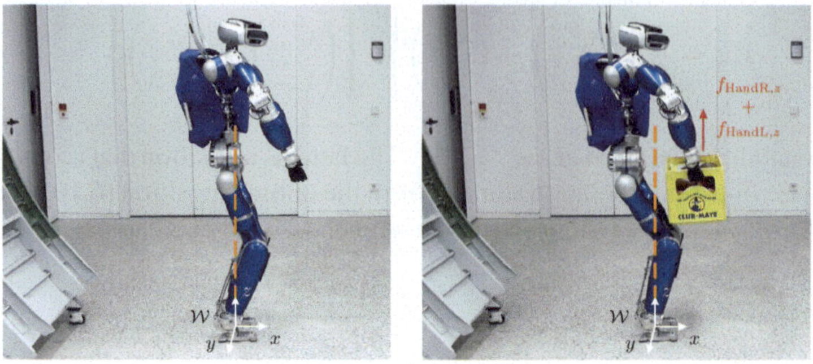

(a) Setup of the experiment.

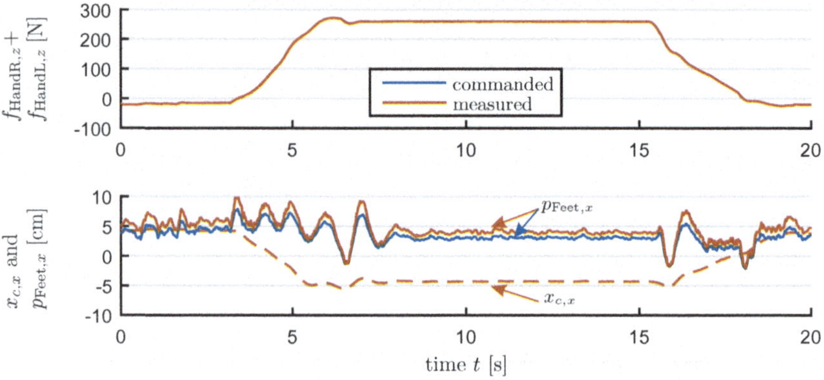

(b) Motion of the CoM and the CoP in relation to the hand wrenches.

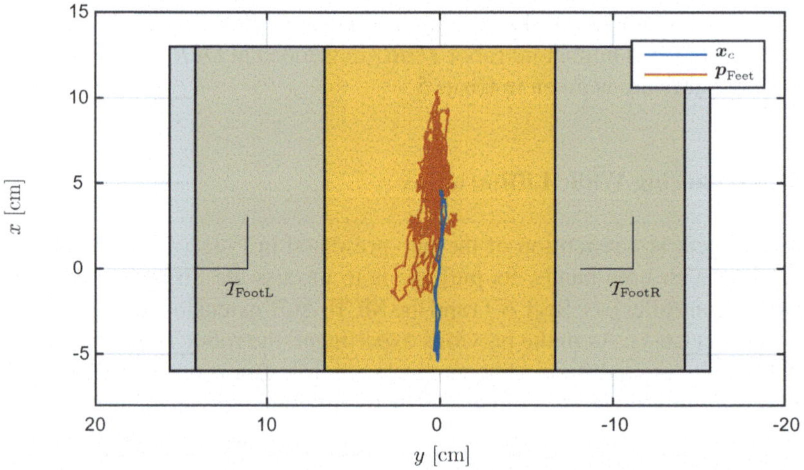

(c) Trajectory of the CoM and the CoP relative to the support polygon $\mathcal{S}_{SP}$ (yellow). The contact areas of the feet $\mathcal{S}_{FootR}$ and $\mathcal{S}_{FootL}$ are given in gray.

Fig. 6.15 Lifting a 25 kg box handed over by a human: the robot shifts the CoM to the back to counteract the additional load

rise to $f_{\mathrm{HandR},z} + f_{\mathrm{HandL},z} = 262\,\mathrm{N}$ in total. In order to counteract the interaction wrench, the MCB-ACT controller moves the CoM by 10 cm to the back, as shown in Fig. 6.15a–c, while the CoP maintains its location in the center of the feet. Note that the measured and commanded CoP are remarkably close to each other, which emphasizes the quality of the control concept.

In contrast to the basic MCB controller (see Sect. 6.1), the MCB-ACT extension moves the CoM instead of the CoP, which increases the producible interaction wrenches significantly. Here, the limitations are no longer given by the size of the support polygon $\mathcal{S}_{\mathrm{SP}}$ but by the strength of the arms. Note that in this scenario, the arms of the robot must be rather stretched to carry this amount of weight to avoid overloading the elbow joints.

6.5.3.2 Balancing While Pushing a Table

This experiment shows another application of whole-body control for pushing a heavy table, which combines balancing, support, and interaction. For this purpose, the robot *TORO* stands in front of a table, as shown in Fig. 6.16a, with an additional weight on top of it. The push is initiated by commanding the setpoints for the hand compliances by 25 cm to the front. Figure 6.16b shows the hand forces during the push, which reach a magnitude of 79.5 N and 60.8 N. The resulting interaction wrenches are counteracted by the robot by shifting its CoM by 16.8 cm to the front. In other words, the robot leans against the table in order to push it. Other than the CoM, the combined CoP of the feet roughly remains at its initial location in x-direction. Note that the posture of the robot is asymmetric, which causes the CoM and CoP to move slightly to the side as well.

Figure 6.16c shows the motion of the CoM $\boldsymbol{x}_{\mathrm{c}}$ and the CoP $\boldsymbol{p}_{\mathrm{Feet}}$ in relation to the support polygon $\mathcal{S}_{\mathrm{SP}}$. As can be seen, the CoP roughly stays in the middle of the support polygon. The CoM moves to the front and even leaves the support polygon to allow the robot to lean against the table during the push.

6.5.3.3 Contact Switching

This experiment illustrates the reaction of the robot to contact switches using the algorithm presented in Sect. 6.5.2. As shown in Fig. 6.17, the humanoid robot *TORO* is placed next to a support structure for the right hand. At the beginning of the experiment ($t = 0$ s), both feet are assigned to the support task, while both hands are operated in interaction mode. The right arm is slowly stretched until the hand touches the handle of the support structure. At $t = 2.35$ s, the task assignment of the right hand is changed from interaction to support, which causes the support polygon $\mathcal{S}_{\mathrm{SP}}$ to change as depicted in Fig. 6.17. As a consequence, the algorithm presented in Sect. 6.5.2 shifts the center of mass $\boldsymbol{x}_{\mathrm{c}}$ by 6.6 cm to the right in order to reach the new optimal position $\boldsymbol{x}_{\mathrm{c}}^{\mathrm{after}}$ (see Sect. 6.5.2). From $t = 8.4$ s to $t = 13$ s, the algorithm prepares the robot to lift the right foot by shifting the CoM back to the

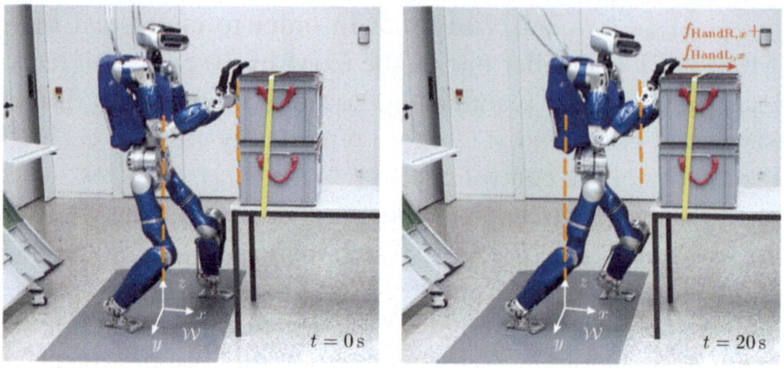

(a) Setup of the experiment.

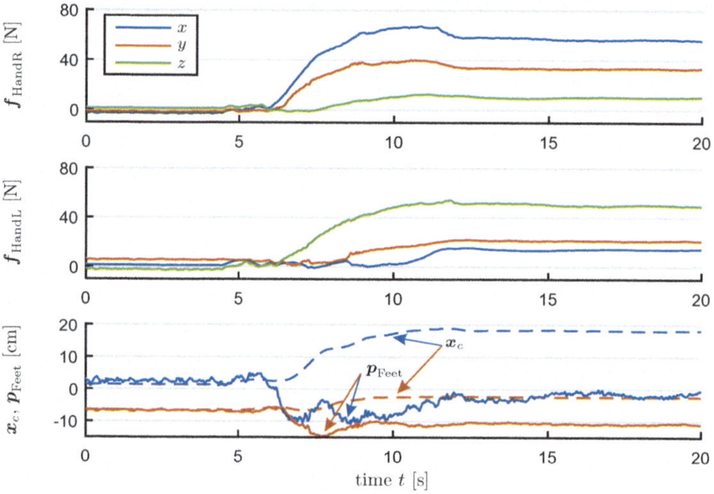

(b) Motion of the CoM and the CoP in relation to the hand wrenches.

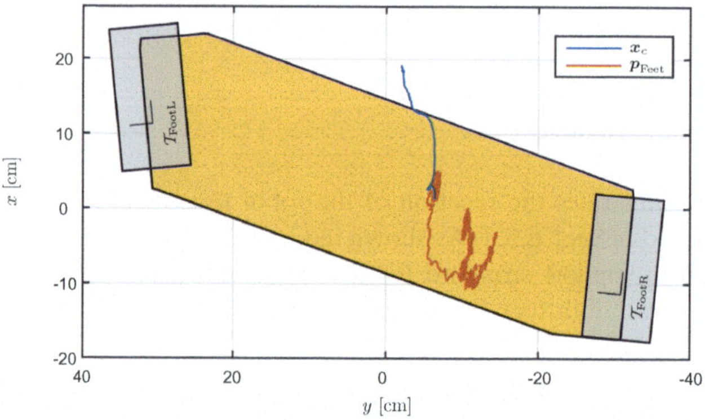

(c) Trajectory of the CoM and the CoP relative to the support polygon $\mathcal{S}_{\text{SP}}$ (yellow). The contact areas of the feet $\mathcal{S}_{\text{FootR}}$ and $\mathcal{S}_{\text{FootL}}$ are given in gray.

Fig. 6.16 Pushing a heavy table: the robot shifts the CoM to the front in order to lean against the table

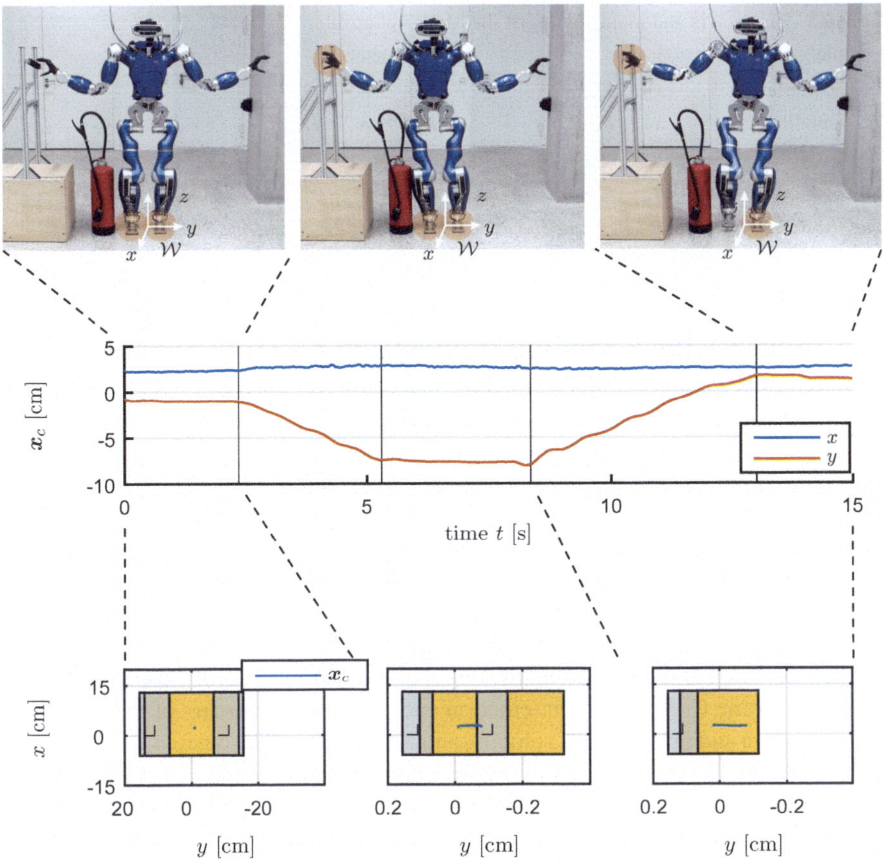

Fig. 6.17 Motion of the CoM in consequence of a contact switch: the top row shows the contact configuration of the robot (red markers). The row at the bottom visualizes the contact configuration assumed by the algorithm to determine the optimal CoM position (yellow: support polygon $\mathcal{S}_{\text{SP}}$; gray: contact areas $\mathcal{S}_{\text{FootR}}$ and $\mathcal{S}_{\text{FootL}}$)

left ($x_{c,y} = 1.5$ cm). This motion is triggered by providing the algorithm at $t = 8.4$ s with the upcoming task assignment specifying the left foot and the right hand as support contacts. The actual contact switch takes place at $t = 13$ s. Although the contact configuration at the beginning (FootR + FootL) and at the end of the experiment (FootL + HandR) are fundamentally different, they lead to a similar support polygon $\mathcal{S}_{\text{SP}}$ as shown at the bottom of Fig. 6.17. Therefore, one could say that in this particular scenario, the right hand can be used to replace the support of the right foot.

Table 6.2 Parameters and input signals of the MCB+ controller

Module	Input	Parameter	Comment
CoM compliance	$\boldsymbol{x}_{\mathrm{c,d}}, \boldsymbol{R}_{\mathrm{c,d}}$	$\boldsymbol{K}_{\mathrm{c}}^{\mathrm{trans}}, \boldsymbol{K}_{\mathrm{c}}^{\mathrm{rot}}, \boldsymbol{D}_{\mathrm{c}}'$	
Int. compliances	$\boldsymbol{x}_{i,d}, \boldsymbol{R}_{i,d}$	$\boldsymbol{K}_{i}^{\mathrm{trans}}, \boldsymbol{K}_{i}^{\mathrm{rot}}, \boldsymbol{D}_{i}'$	$\forall\, \psi_{\mathrm{sup}} < i \leq \psi$
Feedforward terms	$\boldsymbol{v}_{\mathrm{c,d}}, \dot{\boldsymbol{v}}_{\mathrm{c,d}}\ \boldsymbol{v}_i, \dot{\boldsymbol{v}}_i$		$\forall\, \psi_{\mathrm{sup}} < i \leq \psi$
Null space controller	$\boldsymbol{q}_d$		
cQP	$\boldsymbol{F}_{\mathrm{sup}}^{\mathrm{def}}$	$\boldsymbol{Q}_{\mathrm{sup}}, \boldsymbol{A}_i, \boldsymbol{b}_i$	$\forall\, 1 \leq i \leq \psi_{\mathrm{sup}}$

6.6 Interface for Providing Inputs from External Sources

We have demonstrated several times that the framework for whole-body balancing on multiple contacts can be interfaced with various external components such as algorithms for path planning (Werner et al. 2016; Sundaram et al. 2018) or input devices for teleoperation (Abi-Farraj et al. 2019; Porges et al. 2019). Of course, the variants of the multi-contact balancer presented in Sects. 6.1–6.5 require a different interface depending on the provided features and internal architecture. However, the interface for the multi-contact balancer extended to the tracking case (MCB+, see Sect. 6.3) can be considered as the most general one.

The interface for the MCB+ controller is given in Table 6.2: It expects a desired trajectory for the CoM and the interaction end-effectors consisting of a desired set-point (position and orientation), velocity, and acceleration. The desired configuration of the robot can be provided to the null space controller detailed in Sect. 6.2 via the vector of desired joint angles $\boldsymbol{q}_d$. If there is no information available on the desired configuration in joint space, $\boldsymbol{q}_d$ can also be set to the initial configuration, although this can reduce the performance of the robot, as $\boldsymbol{q}_d$ might not match the setpoints for the Cartesian compliances. In order to adjust the controller to the task at hand, the compliances can be parameterized via their respective stiffness and damping matrices (see Table 6.2 and Chap. 4). For example, if the robot is supposed to pick up a sphere from a table, they should be chosen rather high to provide a sufficient accuracy of the hand position. In contrast, the rotational stiffness can be lowered, as the grasp orientation is irrelevant for this particular scenario.

The constrained quadratic optimization problem expects the vector $\boldsymbol{F}_{\mathrm{sup}}^{\mathrm{def}}$, which specifies a default wrench distribution for the support contacts. It can be used in combination with the weighting matrix $\boldsymbol{Q}_{\mathrm{sup}}$ to shift the load between the support end-effectors. Ideally, $\boldsymbol{F}_{\mathrm{sup}}^{\mathrm{def}}$ is also provided by an external planning algorithm, but it can also be set manually if there is no input source available. The latter might slightly decrease the performance of the robot in some situations but will not compromise the overall functionality. Besides $\boldsymbol{Q}_{\mathrm{sup}}$, the cQP also requires the parameters $\boldsymbol{A}_i$ and $\boldsymbol{b}_i$ describing the model of each support contact. If the contact properties change over time, for example, due to locomotion, the contact model must be repetitively updated by the external planning tool.

6.7 Implementation

This section presents and discusses several aspects of the constrained quadratic optimization (6.9) to (6.14) and (6.44) to (6.49), which facilitate a simple but efficient implementation. For instance, Sect. 6.7.1 proposes a method for increasing the robustness of the support contacts. Section 6.7.2 discusses the advantages of formulating the optimization problem using soft constraints regarding the computational effort and the possibility to adjust the task assignment online. A procedure for changing the task assignment is presented in Sect. 6.7.3.

6.7.1 *Increasing the Robustness of the Supporting Contacts*

Assumption 3.2 states that the support end-effectors do not move relative to their respective contact area $\mathcal{S}_i$ to simplify the discussion on the controller properties. However, the possibility that the support end-effectors unintentionally lose contact can never be entirely ruled out in practice. One of the reasons can be uncertainties in the contact model or the fact that the contact model is invoked on the commanded instead of the actual contact wrenches. Consequently, there are situations in which $\boldsymbol{F}^{\text{ext}}_{\text{sup}}$ can deviate from $\boldsymbol{F}_{\text{sup}}$, as discussed in Sects. 6.1.2 and 6.3.2.

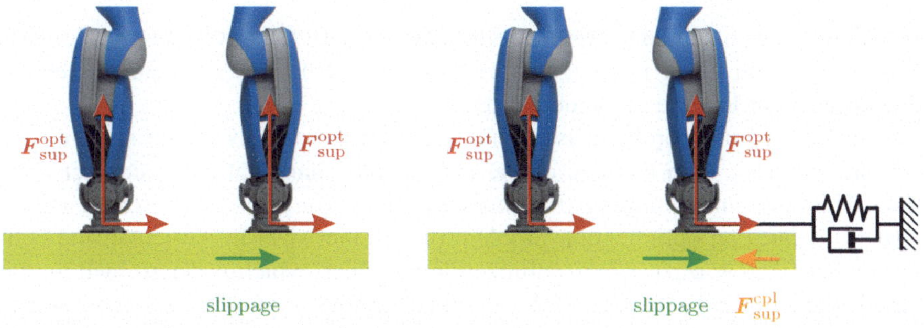

Fig. 6.18 Influence of a Cartesian compliance at the support contact on the wrench distribution. Left: wrench distribution without compliance. Right: wrench distribution with compliance

In order to increase the robustness of the contacts, in Henze et al. (2016b) we proposed subjecting not only the interaction end-effectors but also the support end-effectors to a Cartesian compliance, which can be achieved by modifying (6.11) or (6.46) according to

$$\boldsymbol{\delta}_{\text{sup}} = \boldsymbol{F}_{\text{sup}} - (\boldsymbol{F}^{\text{def}}_{\text{sup}} + \boldsymbol{F}^{\text{cpl}}_{\text{sup}}). \tag{6.107}$$

The wrench $\boldsymbol{F}^{\text{cpl}}_{\text{sup}} \in \mathbb{R}^{6\psi_{\text{sup}}}$ is computed analogously to (6.4) and superimposed with the default wrench distribution $\boldsymbol{F}^{\text{def}}_{\text{sup}} \in \mathbb{R}^{6\psi_{\text{sup}}}$. As discussed in Remark 6.1, the vector

$(\boldsymbol{F}^{\text{def}}_{\text{sup}} + \boldsymbol{F}^{\text{cpl}}_{\text{sup}})$ serves as a regularization term, which influences the wrench distribution in the null space of the contact map $\boldsymbol{Ad}^T_{\text{sup}}$. An example of the effect is shown in Fig. 6.18: Let us consider the case in which the robot uses both legs as support end-effectors. Furthermore, let the left foot be slipping over the ground floor due to an uncertainty in the contact model. In other words, the commanded wrench $\boldsymbol{F}_{\text{sup}}$ overloads the left contact causing the latter to move. The compliance issues a wrench $\boldsymbol{F}^{\text{cpl}}_{\text{sup}}$ to counteract the resulting control error. As $\boldsymbol{F}^{\text{cpl}}_{\text{sup}}$ functions as a regularization term, it causes the optimization to shift some of the load from the left to the right foot. This means that the compliance reduces the load on the left foot to prevent it from slipping any further. Of course, this method is only applicable if the remaining contacts (here the right foot) can take over the additional load.

Note that the described method is not able to stabilize the contact if the connection with the floor is completely lost and the end-effector is in midair. However, from an empirical point of view, it has proven itself to increase the robustness against lift-off, slipping, and tilting, especially in challenging situations such as stair climbing (see e.g. Werner et al. 2016). The stabilizing effect of the compliance also increases the robustness during contact transitions.

6.7.2 *Formulation of the Constrained Quadratic Optimization*

Formulating the contained quadratic optimization (cQP) for solving the wrench distribution problem given in (6.9) to (6.14) or in (6.44) to (6.49) features several advantages regarding the implementation:

If the optimization problem is formulated such that it only comprises the support end-effectors, then the size of the cQP would change at every contact transition. However, incorporating the interaction task via the soft constraint (6.12) or (6.47) creates an optimization problem of constant size. In particular, the cQP comprises $6(\psi_{\text{sup}} + \psi_{\text{int}}) = 6\psi$ optimization variables, which is determined by the overall number of end-effectors ψ. Besides this, the notion in this work assumes a certain numbering of the end-effectors for reasons of simplicity, such that the first ψ_{sup} end-effectors are assigned to the support task and the remaining ψ_{int} to the interaction task (see Sect. 3.2.1). Therefore, a change in the task assignment would require a reordering of the end-effectors, which is rather cumbersome to implement. However, introducing the interaction task via the soft constraint (6.12) or (6.47) allows for a formulation of the optimization problem using a fixed (task independent) numbering of the end-effectors. Thus, the task assignment of the end-effectors can be swapped easily between the support and the interaction task without causing any structural changes to the cQP. The latter is important to enable an attaching and detaching of the end-effectors from the environment to facilitate contact transitions. This effect becomes visible if the optimization (6.9) to (6.14) or (6.44) to (6.49) is expressed in a more general form

$$\boldsymbol{F}^{\text{opt}} = \underset{\boldsymbol{F}}{\operatorname{argmin}} \left(\frac{1}{2} \boldsymbol{\delta}_{\text{c}}^T \boldsymbol{Q}_{\text{c}} \boldsymbol{\delta}_{\text{c}} + \frac{1}{2} \boldsymbol{\delta}^T \boldsymbol{Q} \boldsymbol{\delta} \right) \tag{6.108}$$

minimizing the residua

$$\boldsymbol{\delta}_{\text{c}} = \boldsymbol{Ad}^T \boldsymbol{F} - m\boldsymbol{g}_0 - \boldsymbol{F}_{\text{c}}^{\text{cpl}} - \boldsymbol{F}_{\text{ff}} \tag{6.109}$$

$$\boldsymbol{\delta} = \boldsymbol{F} - (\boldsymbol{F}^{\text{def}} + \boldsymbol{F}^{\text{cpl}}) \tag{6.110}$$

with respect to the inequality constraints

$$\boldsymbol{A}_i \boldsymbol{F}_i \leq \boldsymbol{b}_i \qquad \forall i = 1 \ldots \psi \tag{6.111}$$

$$\boldsymbol{B}_\tau \boldsymbol{F} \leq \boldsymbol{\tau}^{\max}. \tag{6.112}$$

The residuum $\boldsymbol{\delta} \in \mathbb{R}^{6\psi}$ comprises all end-effectors of the robot based on a fixed numbering. In order to specify the assignment of a particular end-effector, the corresponding values within the weighting matrix $\boldsymbol{Q} \in \mathbb{R}^{6\psi \times 6\psi}$ must be set either to the values of the support or the interaction task. The vector of default wrenches $\boldsymbol{F}^{\text{def}} \in \mathbb{R}^{6\psi}$ can be used to specify the default wrench distribution for the support contacts or a feedforward wrench for the interaction end-effectors, respectively. The compliance wrench $\boldsymbol{F}^{\text{cpl}} \in \mathbb{R}^{6\psi}$ comprises the Cartesian compliances for the interaction task (6.4) as well as the ones in Sect. 6.7.1 for increasing the robustness of the support contacts. The contact model (6.111) is computed for all end-effectors ($i = 1 \ldots \psi$) although it is rendered ineffective for the interaction task by setting the respective vector to $\boldsymbol{b}_i \rightarrow \infty$ (or an unrealistic high value). In summary, the above formulation features the same functionality as (6.9) to (6.14) or (6.44) to (6.49). However, a change in the task assignment does not require any structural changes of the cQP. Instead, the formulation allows for an online adaptation of the task assignment by modifying the parameters of the cQP. An algorithm for updating the task assignment to switch contacts is presented in Sect. 6.7.3.

Another advantage of formulating (6.109) and (6.110) as soft constraints instead of hard constraints is a reduction of computation time. In Henze et al. (2016b), we were able to reduce the computation time by 30% using *qpOASES* (version 2.0) by Ferreau et al. (2008, 2014) to solve the cQP on the real-time computer of the humanoid robot *TORO* (see Chap. 5).

6.7.3 Procedure for Contact Transition

As already mentioned in Sect. 6.7.2, the task assignment can be changed by modifying the parameters of the constrained quadratic optimization problem (6.108) to (6.112). But, as the support and the interaction task require a different wrench at the end-effector, an instantaneous update of the parameters would result in a jump of the joint torques $\boldsymbol{\tau}$, which is undesirable from a hardware perspective. In order to obtain

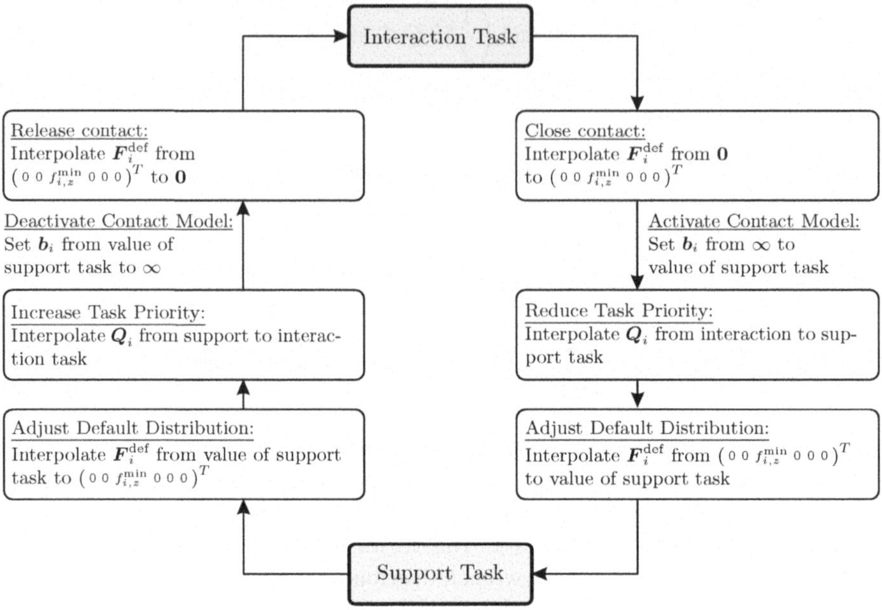

Fig. 6.19 Procedure for switching contacts

a smooth transition, Fig. 6.19 sketches an algorithm for the continuous fading of the parameters of the optimization problem.

For instance, let us consider the case that the ith end-effector is transfered from the interaction to the support task. As a first step, the corresponding default wrench $\boldsymbol{F}_i^{\text{def}} \in \mathbb{R}^6$ is interpolated from $\boldsymbol{F}_i^{\text{def}} = \mathbf{0}$ to $\boldsymbol{F}_i^{\text{def}} = (0\ 0\ f_{i,z}^{\text{min}}\ 0\ 0\ 0)^T$ in order to respect the unilaterality of the contact (see Sect. 3.4). As a result, the end-effector is pressed against the environment thanks to the minimum perpendicular force $f_{i,z}^{\text{min}} \geq 0$. As $\boldsymbol{F}_i^{\text{def}}$ is superimposed with the compliance wrench $\boldsymbol{F}_i^{\text{cpl}} \in \mathbb{R}^6$ according to (6.110), the end-effector is stabilized until the default wrench has reached $f_{i,z}^{\text{min}}$. After that, the contact model (6.111) can be activated instantaneously by setting $\boldsymbol{b}_i$ from $\boldsymbol{b}_i = \infty$ to the value specified by Sect. 3.4. If we assume that the compliance $\boldsymbol{F}_i^{\text{cpl}}$ only shows a minimal control error, activating the contact model on the end-effector wrench $\boldsymbol{F}_i^{\text{opt}}$ will only cause an insignificant discontinuity in $\boldsymbol{F}_i^{\text{opt}}$. The third step is used to continuously lower the task weight $\boldsymbol{Q}_i \in \mathbb{R}^{6\times 6}$ from the values for the interaction task to the ones for the support task (see Table 6.1), which allows for an increasing deviation of the end-effector wrench $\boldsymbol{F}_i^{\text{opt}}$ from $\boldsymbol{F}_i^{\text{def}} = (0\ 0\ f_{i,z}^{\text{min}}\ 0\ 0\ 0)^T$. As a last step, the $\boldsymbol{F}_i^{\text{def}}$ is continuously ramped from $\boldsymbol{F}_i^{\text{def}} = (0\ 0\ f_{i,z}^{\text{min}}\ 0\ 0\ 0)^T$ to the desired wrench distribution for the support task. As the default wrench $\boldsymbol{F}_i^{\text{def}}$ is superimposed with the Cartesian compliance $\boldsymbol{F}_i^{\text{cpl}}$, the end-effector is stabilized during all phases of the transition leading to an increased robustness.

References

Abi-Farraj, F., Henze, B., Ott, C., Giordano, P.R., Roa, M.A.: Torque-based balancing for a humanoid robot performing high-force interaction tasks. IEEE Robot. Autom. Lett. **4**(2), 2023–2030 (2019)

Dietrich, A.: Whole-Body Impedance Control of Wheeled Humanoid Robots. Springer Tracts in Advanced Robotics, vol. 116, 1st edn. Springer International Publishing, Switzerland (2016)

Escande, A., Mansard, N., Wieber, P.B.: Hierarchical quadratic programming: fast online humanoid-robot motion generation. Int. J. Robot. Res. **33**(7), 1522–1543 (2014)

Ferreau, H.J., Bock, H.G., Diehl, M.: An online active set strategy to overcome the limitations of explicit MPC. Int. J. Robust Nonlinear Control **18**(8), 816–830 (2008)

Ferreau, H.J., Kirches, C., Potschka, A., Bock, H.G., Diehl, M.: qpOASES: a parametric active-set algorithm for quadratic programming. Math. Program. Comput. **6**(4), 327–363 (2014)

Franken, M., Stramigioli, S., Misra, S., Secchi, C., Macchelli, A.: Bilateral telemanipulation with time delays: a two-layer approach combining passivity and transparency. IEEE Trans. Robot. **27**(4), 741–756 (2011)

Hannaford, B., Ryu, J.H.: Time-domain passivity control of haptic interfaces. IEEE Trans. Robot. Autom. **18**(1), 1–10 (2002)

Henze, B., Dietrich, A., Ott, C.: An approach to combine balancing with hierarchical whole-body control for legged humanoid robots. IEEE Robot. Autom. Lett. **1**(2), 700–707 (2016a)

Henze, B., Roa, M.A., Ott, C.: Passivity-based whole-body balancing for torque-controlled humanoid robots in multi-contact scenarios. Int. J. Robot. Res. **35**(12), 1522–1543 (2016b)

Henze, B., Balachandran, R., Roa, M.A., Ott, C., Albu-Schäffer, A.: Passivity analysis and control of humanoid robots on movable ground. IEEE Robot. Autom. Lett. **3**(4), 3457–3464 (2018)

Henze, B., Roa, M.A., Werner, A., Dietrich, A., Ott, C., Albu-Schäffer, A.: Experiments with human-inspired behaviors in a humanoid robot: quasi-static balancing using toe-off motion and stretched knees. In: IEEE International Conference on Robotics and Automation, pp. 2510–2516 (2019)

Herzog, A., Rotella, N., Mason, S., Grimminger, F., Schaal, S., Righetti, L.: Momentum control with hierarchical inverse dynamics on a torque-controlled humanoid. Auton. Robot. **40**(3), 473–491 (2016)

Hyon, S.H., Hale, J.G., Cheng, G.: Full-body compliant human-humanoid interaction: balancing in the presence of unknown external forces. IEEE Trans. Robot. **23**(5), 884–898 (2007)

Khalil, H.K.: Nonlinear Systems, 3rd edn. Pearson Eduction, Essex (2014)

Mistry, M., Buchli, J., Schaal, S.: Inverse dynamics control of floating base systems using orthogonal decomposition. In: IEEE International Conference on Robotics and Automation, pp. 3406–3412 (2010)

Ott, C.: Cartesian Impedance Control of Redundant and Flexible-Joint Robots. Springer Tracts in Advanced Robotics, vol. 49, 1st edn. Springer, Berlin (2008)

Ott, C., Roa, M.A., Hirzinger, G.: Posture and balance control for biped robots based on contact force optimization. In: IEEE-RAS International Conference on Humanoid Robots, pp. 26–33 (2011)

Paden, B., Panja, R.: Globally asymptotically stable 'PD+' controller for robot manipulators. Int. J. Control **47**(6), 1697–1712 (1988)

Porges, O., Connan, M., Henze, B., Gigli, A., Castellini, C., Roa, M.A.: A wearable, ultralight interface for bimanual teleoperation of a compliant, whole-body-controlled humanoid robot. In: IEEE International Conference on Robotics and Automation (2019). www.youtube.com/watch?v=YLEUBFu5qgI. Accessed 13 Sept 2019

Righetti, L., Buchli, J., Mistry, M., Kalakrishnan, M., Schaal, S.: Optimal distribution of contact forces with inverse dynamics control. Int. J. Robot. Res. **32**(3), 280–298 (2013)

RMC, YouTube channel of the Institute of Robotics and Mechatronics at the German Aerospace Center (DLR) (2019). www.youtube.com/user/DLRRMC. Accessed 13 Sept 2019

Sentis, L.: Compliant control of whole-body multi-contact behaviors in humanoid robots. Motion Planning for Humanoid Robots, pp. 29–66. Springer, London (2010)

Slotine, J.J.E., Lee, W.: Applied Nonlinear Control. Prentice Hall International, Upper Saddle River (1991)
Stramigioli, S.: Energy-aware robotics. Mathematical Control Theory I. Lecture Notes in Control and Information Sciences, vol. 461, pp. 37–50. Springer International Publishing, Cham (2015)
Sundaram, A.M., Henze, B., Porges, O., Marton, Z.C., Roa, M.A.: Autonomous bipedal humanoid grasping with base repositioning and whole-body control. In: IEEE-RAS International Conference on Humanoid Robots, pp. 395–402 (2018)
Wensing, P.M., Hammam, G.B., Dariush, B., Orin, D.E.: Optimizing foot centers of pressure through force distribution in a humanoid robot. Int. J. Humanoid Robot. **10**(03), 1–21 (2013)
Werner, A., Henze, B., Rodriguez, D.A., Gabaret, J., Porges, O., Roa, M.A.: Multi-contact planning and control for a torque-controlled humanoid robot. In: IEEE/RSJ International Conference on Intelligent Robots and Systems, pp. 5708–5715 (2016)
Whitcomb, L.L., Rizzi, A.A., Koditschek, D.E.: Comparative experiments with a new adaptive controller for robot arms. IEEE Trans. Robot. Autom. **9**(1), 59–70 (1993)
Williams, D., Khatib, O.: The virtual linkage: a model for internal forces in multi-grasp manipulation. In: IEEE International Conference on Robotics and Automation, pp. 1025–1030 (1993)

Chapter 7
Combining Multi-contact Balancing with Hierarchical Whole-Body Control

Dietrich (2016) developed a framework for hierarchical multi-objective control of wheeled humanoid robots. A representative of this class of robots is the humanoid robot *Rollin' Justin* developed by *DLR* (Borst et al. 2009), which consists of a moving platform with an upper body on top, as shown in Fig. 7.1a. The upper body comprises 17 DoFs not counting the hands and the neck. The platform is governed by an underlying velocity controller, which provides a torque interface for controlling the Cartesian motion of the base (two translational and one rotational DoFs). This abstraction results in a formulation of the system dynamics, which shows the same structure as a fixed base robot, although strictly speaking *Rollin' Justin* features a moving base. Considering the three DoFs of the platform and the 17 DoFs of the upper body creates a highly redundant serial kinematic chain. In order to deal with the kinematic redundancy, the framework by Dietrich (2016) allows for the definition of a stack of multiple control objectives for the upper body, such as the Cartesian end-effector pose, the pose in joint space, or singularity avoidance. Often the control objectives or control tasks comprise more DoFs than the robot, which is resolved by assigning a priority level to each control objective resulting in a hierarchical task execution.

The humanoid robot *TORO* (see Chap. 5) was developed to conduct research on the field of balancing and locomotion, which requires a floating base description of the robot. If the robot is balancing on multiple contacts, the supporting limbs form a closed kinematic chain, as shown in Fig. 7.1b. The resulting redundancy in the space of the contact wrenches is also known as the wrench distribution problem, as introduced in Sect. 3.3. The control framework presented in Chap. 6 allows for the solving of the WDP via a constrained quadratic optimization. The latter features a hierarchy as well; but the objectives are defined in wrench space, which affects the task execution in the case of problems with the contact model or the actuator constraints. A comparison of both approaches is given in Table 7.1.

B. Henze, *Whole-Body Control for Multi-Contact Balancing of Humanoid Robots*,
Springer Tracts in Advanced Robotics 143,
https://doi.org/10.1007/978-3-030-87212-0_7

(a) Humanoid robot *Rollin' Justin* featuring a serial kinematic chain originating from a fixed base.

(b) Humanoid robot *TORO* featuring a closed kinematic chain due to multiple contacts.

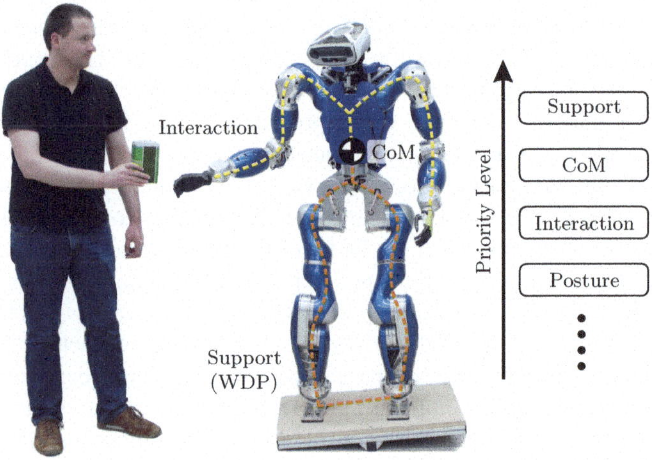

(c) Combining multi-objective control with multi-contact balancing.

Fig. 7.1 Basic concept of the presented framework

Table 7.1 Overview of both control frameworks

	Multi-objective control (Dietrich 2016)	Multi-contact control (Chap. 6)
Purpose	Multiple control objectives	Balancing on multiple contacts
Target of hierarchy	Serial kinematics	Closed kinematic chain
Type of robot	Fixed base	Floating base
Implementation	Null space projectors	Numerical optimization

Both control approaches are combined in Henze et al. (2016a) into a unified framework, which handles serial as well as closed kinematic chains (see Fig. 7.1c). The framework features two hierarchies of different types: The first one addresses the serial kinematics by defining multiple control objectives. The objectives are prioritized via null space projectors, which can be computed analytically. To incorporate the inequality constraints describing the contact model and the actuator limits, the cQP is solved numerically. Thus, the combined framework minimizes the computational effort by reducing the usage of numerical methods. The combined framework will be called Hierarchical Multi-Contact Balancing controller (HMCB) for the remainder of the manuscript.

Combining the multi-contact balancing controller presented in Chap. 6 with hierarchical multi-objective control adds two more features to the abilities of the robot:

1. The combined framework implements a dynamic decoupling of the tasks (Henze et al. 2016a), such that low priority tasks do not disturb tasks with a high priority level. For instance, this feature can be beneficial if the robot carries a glass of water while performing a motion with the CoM while walking.
2. Furthermore, the combined framework allows for the employment of humanoid robots in confined spaces, where the available space limits the motion capabilities of the rather bulky limbs. For this reason, the robot should be able to support itself with contacts scattered all over the body instead of relying only on the feet and the hands. Adding possible contact points to the knees, elbows, the pelvis, or the backpack provides the robot with an enhanced support polygon and therefore with a more robust stance for performing the interaction task. The challenge of balancing in confined spaces arises from the additional contacts, which constrain the motion of the robot to a submanifold of the contact constraint. The information on the contact constraints is incorporated into the controller by projecting all other tasks onto the null space of the Jacobian matrix describing the support task, which in turn enables the support task to provide a suitable set of contact wrenches (Henze et al. 2017). For instance, another projection technique for two priority levels is presented in Aghili and Su (2016), which leads to similar results.

This chapter is structured as follows: Sect. 7.1 revisits the approach presented in Dietrich (2016), Dietrich et al. (2018) for multi-objective control of robots with a fixed base. The latter is combined in Sect. 7.2 with the framework for multi-contact balancing presented in Chap. 6. Section 7.3 presents the application to confined spaces.

7.1 Multi-objective Control for Fixed Base Robots

This section gives a brief introduction to the framework presented in Dietrich (2016), Dietrich et al. (2018) for hierarchical multi-objective control of fixed base robots. Based on $n \in \mathbb{N}$ joint angles, the dynamics of a robot with a fixed base is given by

$$\tilde{\boldsymbol{M}}\ddot{\boldsymbol{q}} + \tilde{\boldsymbol{C}}\dot{\boldsymbol{q}} + \tilde{\boldsymbol{g}} = \underbrace{\boldsymbol{\tau}}_{\tilde{\boldsymbol{u}}} + \tilde{\boldsymbol{\tau}}_{\text{ext}} \tag{7.1}$$

with $\tilde{\boldsymbol{M}} \in \mathbb{R}^{n \times n}$ denoting the inertia and $\tilde{\boldsymbol{C}} \in \mathbb{R}^{n \times n}$ the Coriolis/centrifugal matrix. Note that $\dot{\tilde{\boldsymbol{M}}} - 2\tilde{\boldsymbol{C}}$ is a skew-symmetric matrix (Ott 2008), which is strongly related to the passivity properties of the robot. The influence of gravity is considered by $\tilde{\boldsymbol{g}} \in \mathbb{R}^n$. The joint torques generated by the actuators are denoted by $\boldsymbol{\tau} \in \mathbb{R}^n$, while the generalized external forces are given by $\tilde{\boldsymbol{\tau}}_{\text{ext}} \in \mathbb{R}^n$. In preparation of the next section, $\tilde{\boldsymbol{u}} = \boldsymbol{\tau}$ is defined as a virtual control input.

The framework can handle an arbitrary number of $r \in \mathbb{N}$ tasks or control objectives. Each one of them is characterized by a set of task coordinates $\boldsymbol{x}_{\text{task}_k} = \boldsymbol{f}_{\text{task}_k}(\boldsymbol{q}) \in \mathbb{R}^{m_k}\ \forall 1 \leq k \leq r$, which are a function of $\boldsymbol{q}$ and show a task dimension of $m_k \in \mathbb{N}$. In total, the tasks can feature more DoFs than the robot ($\sum_{k=1}^{r} m_k \geq n$), which requires a prioritization of the tasks or control objectives. The notation in Dietrich (2016) is chosen such that $k = 1$ represents the task with the highest and $k = r$ the task with the lowest priority level. Each task outputs a desired control action in the form of a generalized task wrench $\boldsymbol{F}_{\text{task}_k}^{\text{ctrl}}(\boldsymbol{q}, \dot{\boldsymbol{q}}) \in \mathbb{R}^{m_k}$, which can be used to implement various functionalities, such as a joint compliance, a Cartesian compliance, or a repulsive potential field for singularity avoidance. The associated task velocities are given by

$$\boldsymbol{v}_{\text{task}_k} = \dot{\boldsymbol{x}}_{\text{task}_k} = \boldsymbol{J}_{\text{task}_k} \dot{\boldsymbol{q}} \qquad \forall 1 \leq k \leq r \tag{7.2}$$

with the task Jacobian matrix $\boldsymbol{J}_{\text{task}_k} = \partial \boldsymbol{f}_{\text{task}_k}(\boldsymbol{q}) / \partial \boldsymbol{q}$.

The basic idea of the hierarchy is that a task at priority level k may disturb all lower priority tasks $(k + 1 \ldots r)$ but not the tasks with a higher level $(1 \ldots k - 1)$. Thus, task k must be executed in the null space of the Jacobian matrices of the tasks $1 \ldots (k - 1)$, which can be formalized via

$$\boldsymbol{J}_{\text{task}_{k-1}}^{\text{aug}} \boldsymbol{Z}_{\text{task}_k}^T = \boldsymbol{0} \qquad \forall 2 \leq k \leq r. \tag{7.3}$$

Here, $\boldsymbol{J}_{\text{task}_{k-1}}^{\text{aug}}$ represents the so-called "augmented Jacobian matrix"

$$\boldsymbol{J}_{\text{task}_{k-1}}^{\text{aug}} = \begin{bmatrix} \boldsymbol{J}_{\text{task}_1} \\ \vdots \\ \boldsymbol{J}_{\text{task}_{k-1}} \end{bmatrix}, \tag{7.4}$$

which gathers the Jacobian matrices of the tasks 1 to $k - 1$. Note that the augmented Jacobian of the last task level $\boldsymbol{J}_{\text{task}_r}^{\text{aug}}$ is assumed to be of full column-rank, such that the sum of all tasks covers the whole workspace of the robot (Dietrich et al. 2018). The matrix $\boldsymbol{Z}_{\text{task}_k} \in \mathbb{R}^{m_k \times n}$ in (7.3) represents a full row-rank "null space base matrix", which spans the null space of $\boldsymbol{J}_{\text{task}_{k-1}}^{\text{aug}}$ (see Dietrich (2016) for details on the computation of $\boldsymbol{Z}_{\text{task}_k}$).

For each task, the set of task velocities $\boldsymbol{v}_{\text{task}_k}$ corresponds to a subset of the joint velocities $\dot{\boldsymbol{q}}$. The subsets of two or more tasks can overlap with each other if the sum of the task dimensionalities exceeds the number of joints ($\sum_{k=1}^{r} m_k > n$). In other words, there can be kinematic conflicts between two or more tasks. In order to remedy this overlap, Dietrich (2016) introduced the concept of local null space velocities

$$\breve{\boldsymbol{v}}_{\text{task}_k} = \breve{\boldsymbol{J}}_{\text{task}_k} \dot{\boldsymbol{q}} \qquad \forall 1 \leq k \leq r, \tag{7.5}$$

which span a subspace of the original task velocities $\boldsymbol{v}_{\text{task}_k}$. The mapping

$$\breve{\boldsymbol{J}}_{\text{task}_k} = \begin{cases} \boldsymbol{J}_1^{\text{task}} & \text{if } k = 1 \\ \left(\boldsymbol{Z}_{\text{task}_k} \tilde{\boldsymbol{M}} \boldsymbol{Z}_{\text{task}_k}^T \right)^{-1} \boldsymbol{Z}_{\text{task}_k} \tilde{\boldsymbol{M}} & \text{if } 2 \leq k \leq r \end{cases} \tag{7.6}$$

provides a dynamically consistent null space projection according to Dietrich (2016). As task $k = 1$ is not restricted by any other task, $\breve{\boldsymbol{J}}_1^{\text{task}}$ equals $\boldsymbol{J}_1^{\text{task}}$. Note that the "hierarchy-consistent Jacobian matrix" $\breve{\boldsymbol{J}} \in \mathbb{R}^{n \times n}$ in

$$\breve{\boldsymbol{v}} = \begin{pmatrix} \breve{\boldsymbol{v}}_{\text{task}_1} \\ \vdots \\ \breve{\boldsymbol{v}}_{\text{task}_r} \end{pmatrix} = \breve{\boldsymbol{J}} \dot{\boldsymbol{q}} = \begin{bmatrix} \breve{\boldsymbol{J}}_{\text{task}_1} \\ \vdots \\ \breve{\boldsymbol{J}}_{\text{task}_r} \end{bmatrix} \dot{\boldsymbol{q}} \tag{7.7}$$

is of full rank, which means that the vector $\breve{\boldsymbol{v}} \in \mathbb{R}^n$ describes the full velocity state of the robot. According to Dietrich (2016), the inverse is given by

$$\breve{\boldsymbol{J}}^{-1} = \left[\boldsymbol{J}_{\text{task}_1}^{\tilde{\boldsymbol{M}}+} \ \boldsymbol{Z}_{\text{task}_2}^T \ \ldots \ \boldsymbol{Z}_{\text{task}_r}^T \right] \tag{7.8}$$

with

$$\boldsymbol{J}_{\text{task}_1}^{\tilde{\boldsymbol{M}}+} = \tilde{\boldsymbol{M}}^{-1} \boldsymbol{J}_{\text{task}_1}^T \left(\boldsymbol{J}_{\text{task}_1} \tilde{\boldsymbol{M}}^{-1} \boldsymbol{J}_{\text{task}_1}^T \right)^{-1} \tag{7.9}$$

denoting the dynamically consistent pseudo-inverse of $\boldsymbol{J}_{\text{task}_1}$.

In order to implement the task hierarchy, (Dietrich 2016) proposes the following control law:

$$\begin{aligned} \tilde{\boldsymbol{u}} &= \tilde{\boldsymbol{g}} + \tilde{\boldsymbol{\tau}}_\mu + \sum_{k=1}^{r} \breve{\boldsymbol{J}}_{\text{task}_k}^T \boldsymbol{Z}_{\text{task}_k} \boldsymbol{J}_{\text{task}_k}^T \boldsymbol{F}_{\text{task}_k}^{\text{ctrl}} \\ &= \tilde{\boldsymbol{g}} + \tilde{\boldsymbol{\tau}}_\mu + \sum_{k=1}^{r} \boldsymbol{N}_{\text{task}_k} \boldsymbol{J}_{\text{task}_k}^T \boldsymbol{F}_{\text{task}_k}^{\text{ctrl}}. \end{aligned} \tag{7.10}$$

The task torques are first mapped to joint space via the transpose of $\boldsymbol{J}_{\text{task}_k}$ and then via the null space projector

$$\boldsymbol{N}_{\text{task}_k} = \boldsymbol{I} - \boldsymbol{J}_{\text{aug}_{k-1}}^T \boldsymbol{J}_{\text{aug}_{k-1}}^{\tilde{\boldsymbol{M}}+,T} \qquad \forall 2 \leq k \leq r \tag{7.11}$$

to the control input $\tilde{\boldsymbol{u}}$. The vector $\tilde{\boldsymbol{\tau}}_\mu \in \mathbb{R}^n$ is used to cancel out several coupling terms within the Coriolis/centrifugal matrix $\tilde{\boldsymbol{C}}$ (see Dietrich (2016) for details). If the motion of the robot is sufficiently slow, $\tilde{\boldsymbol{\tau}}_\mu$ may also be omitted.

Inserting the control law (7.10) with (7.7) into the model of the fixed base robot (7.1) yields to the closed-loop dynamics:

$$\underbrace{\begin{bmatrix} \breve{\boldsymbol{\Lambda}}_1 & \mathbf{0} & \dots & \mathbf{0} \\ \mathbf{0} & \breve{\boldsymbol{\Lambda}}_2 & \dots & \mathbf{0} \\ \vdots & \vdots & \ddots & \vdots \\ \mathbf{0} & \mathbf{0} & \dots & \breve{\boldsymbol{\Lambda}}_r \end{bmatrix}}_{\breve{\boldsymbol{\Lambda}}} \dot{\tilde{\boldsymbol{v}}} + \underbrace{\begin{bmatrix} \breve{\boldsymbol{\mu}}_1 & \mathbf{0} & \dots & \mathbf{0} \\ \mathbf{0} & \breve{\boldsymbol{\mu}}_2 & \dots & \mathbf{0} \\ \vdots & \vdots & \ddots & \vdots \\ \mathbf{0} & \mathbf{0} & \dots & \breve{\boldsymbol{\mu}}_r \end{bmatrix}}_{\breve{\boldsymbol{\mu}}} \tilde{\boldsymbol{v}} + \begin{pmatrix} \boldsymbol{F}^{\text{ctrl}}_{\text{task}_1} \\ \boldsymbol{Z}_{\text{task}_2} \boldsymbol{J}^T_{\text{task}_2} \boldsymbol{F}^{\text{ctrl}}_{\text{task}_2} \\ \vdots \\ \boldsymbol{Z}_{\text{task}_r} \boldsymbol{J}^T_{\text{task}_r} \boldsymbol{F}^{\text{ctrl}}_{\text{task}_r} \end{pmatrix} = \breve{\boldsymbol{J}}^{-T} \tilde{\boldsymbol{\tau}}_{\text{ext}}. \tag{7.12}$$

Note that (7.12) indicates a decoupled behavior of the tasks due to the block-diagonal structure of the transformed inertia $\breve{\boldsymbol{\Lambda}}$ and Coriolis matrix $\breve{\boldsymbol{\mu}}$. As the local velocities $\breve{\boldsymbol{v}}_{\text{task}_k}$ depend on $\boldsymbol{J}^{\text{aug}}_{\text{task}_{k-1}}$, the velocities of a task can be disturbed by task velocities with a higher priority level, but not vice versa. The equilibrium of the closed-loop dynamics (7.12) is asymptotically stable, as shown in Dietrich (2016) by means of conditional stability. The experiments presented in Dietrich (2016) suggest that the closed-loop system exhibits a cascaded convergence starting with the convergence of the task on priority level $k = 1$, followed by level $k = 2$, through to priority level $k = r$.

Another interesting aspect of (7.12) is the influence of external forces. If all external forces $\tilde{\boldsymbol{\tau}}_{\text{ext}}$ act solely on the task coordinates, then $\breve{\boldsymbol{J}}^{-T} \tilde{\boldsymbol{\tau}}_{\text{ext}}$ becomes

$$\breve{\boldsymbol{J}}^{-T} \tilde{\boldsymbol{\tau}}_{\text{ext}} = \breve{\boldsymbol{J}}^{-T} \left(\boldsymbol{J}^{\text{aug}}_{\text{task}_r}\right)^T \tilde{\boldsymbol{F}}_{\text{ext}} = \begin{bmatrix} \boldsymbol{I} & \boldsymbol{J}^{\tilde{\boldsymbol{M}}+,T}_{\text{task}_1} \boldsymbol{J}^T_{\text{task}_2} & \dots & \boldsymbol{J}^{\tilde{\boldsymbol{M}}+,T}_{\text{task}_1} \boldsymbol{J}^T_{\text{task}_r} \\ \mathbf{0} & \boldsymbol{Z}_{\text{task}_2} \boldsymbol{J}^T_{\text{task}_2} & \dots & \boldsymbol{Z}_{\text{task}_2} \boldsymbol{J}^T_{\text{task}_r} \\ \vdots & \vdots & \ddots & \vdots \\ \mathbf{0} & \mathbf{0} & \dots & \boldsymbol{Z}_{\text{task}_r} \boldsymbol{J}^T_{\text{task}_r} \end{bmatrix} \tilde{\boldsymbol{F}}^{\text{ext}}_{\text{task}} \tag{7.13}$$

with $\tilde{\boldsymbol{F}}^{\text{ext}}_{\text{task}} \in \mathbb{R}^{\sum_{k=1}^{r} m_k}$. It can be verified that $\breve{\boldsymbol{J}}^{-T} \left(\boldsymbol{J}^{\text{aug}}_{\text{task}_r}\right)^T$ represents an upper triangular matrix, which implies that the external forces of task k can disturb tasks with a higher priority level. This coupling is contrary to the actual task hierarchy, but a direct consequence of the dynamic decoupling. Hypothetically speaking, this effect can be remedied by explicitly measuring the external forces $\tilde{\boldsymbol{F}}^{\text{ext}}_{\text{task}}$ and compensating for the coupling. But this would cause the well-known problems regarding causality and additional measurement noise. The effect of this coupling on the closed-loop system is studied in Sect. 7.2.3 in the form of an experiment with the humanoid robot *TORO*.

7.2 Transfer to Robots with a Floating Base (HMCB)

The hierarchical control framework by Dietrich (2016) is combined in Henze et al. (2016a) with the approach for multi-contact balancing (MCB) presented in Sect. 6.1. A schematic overview of the resulting HMCB controller is given in Fig. 7.2. The

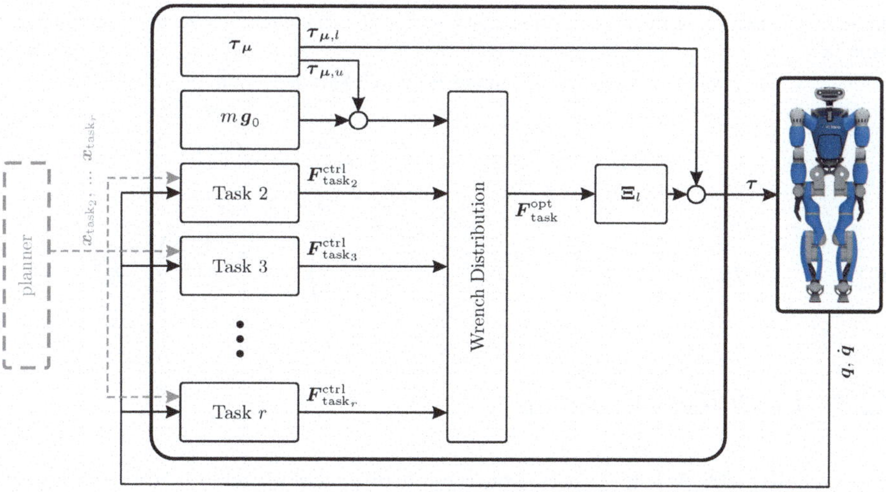

Fig. 7.2 Overview of the architecture of the HMCB control framework

combined framework (HMCB) can comprise an arbitrary number of user-defined tasks, although there are two restrictions that are mandatory for balancing: The hierarchy must contain a task

- for stabilizing the center of mass and
- for generating suitable contact wrenches $\boldsymbol{F}_{\text{sup}}$ to support the robot.

It is important that the support task is located at the highest priority level, such that no other task can interfere with the generation of the necessary overall support wrench $\boldsymbol{F}^{\text{res}}_{\text{sup}} = \boldsymbol{Ad}^{T}_{\text{sup}} \boldsymbol{F}_{\text{sup}}$. Table 7.2 presents three examples for potential task hierarchies. The respective task wrenches and task Jacobian matrices are defined in Table 7.3. The variants "CoM over Int" and "Int over CoM" were studied in Henze et al. (2016a) regarding the dynamic decoupling of the tasks. Another variant was proposed in Henze et al. (2017) to operate the robot in confined spaces (see Sect. 7.3). As detailed in Sect. 7.2.2, the last column of Table 7.3 provides a hierarchy that is equivalent to the control approach presented in Chap. 6 for multi-contact balancing.

Table 7.2 Exemplary task hierarchies for whole-body balancing (Henze et al. 2016a, b, 2017)

Priority level	"CoM over Int" (HMCB)	"Int over CoM" (HMCB)	Confined spaces (HMCB)	Multi-contact balancing (MCB)
1	Support	Support	Support	Sup. + CoM + Int.
2	CoM	Interaction	CoM + Int.	"Posture"
3	Interaction	CoM	Posture	–
4	Posture	Posture	–	–

Table 7.3 Definition of the task Jacobian matrices and the task wrenches for the tasks listed in Table 7.2

	Support	CoM	Interaction	CoM + Int.	Sup. + CoM + Int.	Posture
$\boldsymbol{J}_{\text{task}_k}$	$\boldsymbol{J}_{\text{sup}}$	$\boldsymbol{J}_{\text{c}}$	$\boldsymbol{J}_{\text{int}}$	$\begin{bmatrix}\boldsymbol{J}_{\text{c}}\\ \boldsymbol{J}_{\text{int}}\end{bmatrix}$	$\begin{bmatrix}\boldsymbol{J}_{\text{sup}}\\ \boldsymbol{J}_{\text{c}}\\ \boldsymbol{J}_{\text{int}}\end{bmatrix}$	$\boldsymbol{J}_{\text{pose}}$
$\boldsymbol{F}^{\text{ctrl}}_{\text{task}_k}$	$-\boldsymbol{F}_{\text{sup}}$	$\boldsymbol{F}^{\text{cpl}}_{\text{c}}$	$\boldsymbol{F}^{\text{cpl}}_{\text{int}}$	$\begin{pmatrix}\boldsymbol{F}^{\text{cpl}}_{\text{c}}\\ \boldsymbol{F}^{\text{cpl}}_{\text{int}}\end{pmatrix}$	$\begin{pmatrix}-\boldsymbol{F}_{\text{sup}}\\ \boldsymbol{F}^{\text{cpl}}_{\text{c}}\\ \boldsymbol{F}^{\text{cpl}}_{\text{int}}\end{pmatrix}$	$\boldsymbol{\tau}^{\text{cpl}}_{\text{pose}}$

7.2.1 Controller Derivation

Let us recall (3.20), which describes the dynamic model of a humanoid robot featuring a floating base. Applying (7.10) to the dynamics (3.20) results in the control law

$$
\begin{aligned}
\boldsymbol{u} = \begin{pmatrix}\boldsymbol{0}\\ \boldsymbol{\tau}\end{pmatrix} &= \boldsymbol{\tau}_{\mu} + \boldsymbol{g} + \boldsymbol{J}^T_{\text{sup}}(-\boldsymbol{F}_{\text{sup}}) + \sum_{k=2}^{r} \boldsymbol{N}_{\text{task}_k}\boldsymbol{J}^T_{\text{task}_k}\boldsymbol{F}^{\text{ctrl}}_{\text{task}_k} \\
&= \boldsymbol{\tau}_{\mu} + \boldsymbol{g} + \underbrace{\left[\boldsymbol{J}^T_{\text{sup}}\;\; \boldsymbol{N}_{\text{task}_2}\boldsymbol{J}^T_{\text{task}_2}\;\cdots\; \boldsymbol{N}_{\text{task}_r}\boldsymbol{J}^T_{\text{task}_r}\right]}_{\boldsymbol{\Xi}} \underbrace{\begin{pmatrix}-\boldsymbol{F}_{\text{sup}}\\ \boldsymbol{F}^{\text{ctrl}}_{\text{task}_2}\\ \vdots \\ \boldsymbol{F}^{\text{ctrl}}_{\text{task}_r}\end{pmatrix}}_{\boldsymbol{F}^{\text{ctrl}}_{\text{task}}}
\end{aligned}
\tag{7.14}
$$

with $\boldsymbol{J}_{\text{task}_1} = \boldsymbol{J}_{\text{sup}}$ representing the task Jacobian matrix of the support task at priority level $k = 1$. The negative sign in front of $\boldsymbol{F}_{\text{sup}}$ accounts for the definition of contact wrenches, as explained in Sect. 6.1.1. The Jacobian matrices of the other tasks can be chosen according to Table 7.2, for instance. The CoM task listed in Table 7.3 features a compliance $\boldsymbol{F}^{\text{cpl}}_{\text{c}}$ according to (6.1). The associated task Jacobian is given by $\boldsymbol{J}_{\text{c}} = [\, \boldsymbol{I}_{6\times 6}\; \boldsymbol{0}_{6\times n}\,]$ mapping $\boldsymbol{F}^{\text{cpl}}_{\text{c}}$ to the CoM coordinates. The posture task comprises a joint compliance $\boldsymbol{\tau}^{\text{cpl}}_{\text{pose}}$ according to (6.33) to stabilize the posture of the robot in joint space. The corresponding task Jacobian matrix is given by $\boldsymbol{J}_{\text{pose}} = [\, \boldsymbol{0}_{n\times 6}\; \boldsymbol{I}_{n\times n}\,]$. Note that the presented framework features a generic method to define the tasks, which allows for an arbitrary location of the frames of the support and interaction task on the body of the robot (knees, elbows, shoulders etc.). The mapping between the stacked task wrenches $\boldsymbol{F}^{\text{ctrl}}_{\text{task}} \in \mathbb{R}^{\sum_{k=1}^{r} m_k}$ and the virtual control input $\boldsymbol{u}$ is provided by the matrix $\boldsymbol{\Xi} \in \mathbb{R}^{(6+n)\times \sum_{k=1}^{r} m_k}$.

Analogously to (6.8), the control law (7.14) can be partitioned into the CoM and the joint coordinates according to

$$
\begin{pmatrix}\boldsymbol{0}\\ \boldsymbol{\tau}\end{pmatrix} = \begin{pmatrix}\boldsymbol{\tau}_{\mu,u}\\ \boldsymbol{\tau}_{\mu,l}\end{pmatrix} + \begin{pmatrix}m\boldsymbol{g}_0\\ \boldsymbol{0}\end{pmatrix} + \begin{bmatrix}\boldsymbol{\Xi}_u\\ \boldsymbol{\Xi}_l\end{bmatrix}\boldsymbol{F}^{\text{ctrl}}_{\text{task}} \tag{7.15}
$$

with $\boldsymbol{\tau}_{\mu,u} \in \mathbb{R}^6, \boldsymbol{\tau}_{\mu,l} \in \mathbb{R}^n$ and $\boldsymbol{\Xi}_u \in \mathbb{R}^{6\times\sum_{k=1}^{r} m_k}$, $\boldsymbol{\Xi}_l \in \mathbb{R}^{n\times\sum_{k=1}^{r} m_k}$. In order to account for the underactuation of the base, the support wrenches $\boldsymbol{F}_{\text{sup}}$ in $\boldsymbol{F}_{\text{task}}^{\text{ctrl}}$ must be chosen such that

$$\mathbf{0} = \boldsymbol{\tau}_{\mu,u} + m\boldsymbol{g}_0 + \boldsymbol{\Xi}_u\, \boldsymbol{F}_{\text{task}}^{\text{ctrl}} \tag{7.16}$$

holds at all times. Thus, (7.16) can be regarded as the underactuation condition for the hierarchical case analogously to (6.6).

As introduced in Sect. 3.3, the support task comprises the wrench distribution problem since the contact map $\boldsymbol{Ad}_{\text{sup}}^T \in \mathbb{R}^{6\times 6\psi_{\text{sup}}}$ within $\boldsymbol{\Xi}_u$ is subjective but not injective for more than one contact ($m_{\text{sup}} > 6$). Therefore, the controller can select $\boldsymbol{F}_{\text{sup}}$ analogously to Chap. 6 from a set of feasible solutions to generate the required overall wrench $\boldsymbol{F}_{\text{sup}}^{\text{res}}$ at the center of mass. This redundancy in the space of the contact wrenches $\boldsymbol{F}_{\text{sup}}$ can be resolved by computing the following constrained quadratic optimization problem (cQP):

$$\boldsymbol{F}_{\text{task}}^{\text{opt}} = \underset{\boldsymbol{F}_{\text{task}}}{\text{argmin}} \left(\frac{1}{2}\boldsymbol{\delta}_{\text{sup}}^T \boldsymbol{Q}_{\text{sup}} \boldsymbol{\delta}_{\text{sup}} + \frac{1}{2}\sum_{k=2}^{r} \boldsymbol{\delta}_{\text{task}_k}^T \boldsymbol{Q}_{\text{task}_k} \boldsymbol{\delta}_{\text{task}_k} \right) \tag{7.17}$$

minimizing the residua

$$\boldsymbol{\delta}_{\text{sup}} = \boldsymbol{F}_{\text{sup}} - \boldsymbol{F}_{\text{sup}}^{\text{def}} \tag{7.18}$$

$$\boldsymbol{\delta}_{\text{task}_k} = \boldsymbol{F}_{\text{task}_k} - \boldsymbol{F}_{\text{task}_k}^{\text{ctrl}} \qquad \forall k = 2\ldots r \tag{7.19}$$

with respect to the underactuation condition (7.16) and the inequality constraints

$$\boldsymbol{A}_i \boldsymbol{F}_i \leq \boldsymbol{b}_i \qquad \forall i = 1\ldots\psi_{\text{sup}} \tag{7.20}$$

$$\boldsymbol{B}_\tau \boldsymbol{F} \leq \boldsymbol{\tau}^{\max}. \tag{7.21}$$

The residua within the cost function (7.17) are weighted with the symmetric and positive semi-definite matrices $\boldsymbol{Q}_{\text{sup}} \in \mathbb{R}^{6\psi_{\text{sup}}\times 6\psi_{\text{sup}}}$ and $\boldsymbol{Q}_{\text{task},k} \in \mathbb{R}^{m_k\times m_k}$. The first residuum (7.18) regularizes the contact wrenches $\boldsymbol{F}_{\text{sup}}$ to a default wrench distribution $\boldsymbol{F}_{\text{sup}}^{\text{def}} \in \mathbb{R}^{6\psi_{\text{sup}}}$, which is ideally provided by an external planning algorithm as part of the reference trajectory. If there is no trajectory available, then $\boldsymbol{F}_{\text{sup}}^{\text{def}}$ can also be set manually to shift the load between the support contacts. The remaining residua (7.19) substitute $\boldsymbol{F}_{\text{task}_k}^{\text{ctrl}}$ in the control law (7.14) with $\boldsymbol{F}_{\text{ctrl}_k}^{\text{ctrl}} = \boldsymbol{F}_{\text{task}_k}$. The inequality constraint (7.20) ensures that all contact wrenches $\boldsymbol{F}_{\text{sup}}$ satisfy the contact model detailed in Sect. 3.4. A model of the actuators is given by (7.21) limiting the maximum joint torque to $\boldsymbol{\tau}^{\max} \in \mathbb{R}^n$ using $\boldsymbol{B}_\tau \in \mathbb{R}^{n\times\psi}$.

The presented HMCB approach implements two different hierarchies: The first one addresses the serial kinematic chain of the robot by prioritizing multiple control objectives using null space projectors. As shown in Sects. 7.2.3 and 7.3, this feature can be used to achieve a dynamic decoupling of the tasks as well as to resolve kinematic conflicts, as they occur, for example, in overconstrained contact configurations. The second hierarchy originates from the wrench distribution problem of the support task dealing with the closed kinematic chain. The employed optimization tries to keep the support wrenches $\boldsymbol{F}_{\text{sup}}$ as close as possible to $\boldsymbol{F}_{\text{sup}}^{\text{def}}$ by minimizing (7.18).

But the optimization also comprises the residua of the remaining tasks (7.19). The values within $\boldsymbol{Q}_{\text{task}_2}$ to $\boldsymbol{Q}_{\text{task}_r}$ must be chosen significantly higher than the ones in $\boldsymbol{Q}_{\text{sup}}$ such that the residua (7.19) are always sufficiently fulfilled unless they are in conflict with the contact model (7.20) or the actuator limitations (7.20). In that case, the weighting matrices $\boldsymbol{Q}_{\text{task}_2}$ to $\boldsymbol{Q}_{\text{task}_r}$ specify which tasks will be sacrificed first in order to allow the controller to generate the required overall support wrench $\boldsymbol{F}_{\text{sup}}^{\text{res}}$ nevertheless. In summary, the presented framework combines the two hierarchies from Chap. 6 and Dietrich (2016). The one responsible for the serial kinematics is resolved via null space projectors, which is a rather efficient method in terms of computational effort. The hierarchy addressing the closed kinematic chain involves a numeric optimization in order to deal with the inequality constraints of the contact and the actuator model. Thus, the computational effort is minimized by reducing the usage of numerical methods. For the remainder of this work, we make the following assumption to simplify the discussion of the controller properties:

Assumption 7.1 *(Fulfillment of $\boldsymbol{\delta}_{\text{task}_k} \approx \boldsymbol{0}\ \forall 2 \leq k \leq r$)* The residua of the cQP are always sufficiently fulfilled, such that $\boldsymbol{\delta}_{\text{task}_k} \approx \boldsymbol{0}\ \forall 2 \leq k \leq r$ holds.

After the task wrenches $\boldsymbol{F}_{\text{task}}^{\text{opt}}$ are computed via the constrained quadratic optimization using the upper part of the control law (7.14), the lower part is exploited to compute the control torques via

$$\boldsymbol{\tau} = \boldsymbol{\tau}_{\mu,l} + \boldsymbol{\Xi}_l\, \boldsymbol{F}_{\text{task}}^{\text{opt}}. \tag{7.22}$$

Inserting the controller (7.15) into the model (3.20) yields the closed-loop dynamics

$$\breve{\boldsymbol{\Lambda}} \begin{pmatrix} \dot{\boldsymbol{v}}_{\text{sup}} \\ \dot{\breve{\boldsymbol{v}}}_{\text{task}_2} \\ \vdots \\ \dot{\breve{\boldsymbol{v}}}_{\text{task}_r} \end{pmatrix} + \breve{\boldsymbol{\mu}} \begin{pmatrix} \boldsymbol{v}_{\text{sup}} \\ \breve{\boldsymbol{v}}_{\text{task}_2} \\ \vdots \\ \breve{\boldsymbol{v}}_{\text{task}_r} \end{pmatrix} + \begin{pmatrix} -\boldsymbol{F}_{\text{sup}} \\ \boldsymbol{Z}_{\text{task}_2} \boldsymbol{J}_{\text{task}_2}^T \boldsymbol{F}_{\text{task}_2}^{\text{ctrl}} \\ \vdots \\ \boldsymbol{Z}_{\text{task}_r} \boldsymbol{J}_{\text{task}_r}^T \boldsymbol{F}_{\text{task}_r}^{\text{ctrl}} \end{pmatrix} = \breve{\boldsymbol{J}}^{-T} \boldsymbol{\tau}_{\text{ext}} \tag{7.23}$$

according to (7.12). If we consider the static case under the assumption that all external loads $\boldsymbol{\tau}_{\text{ext}}$ exclusively act on the task wrenches, then (7.23) becomes

$$\begin{pmatrix} -\boldsymbol{F}_{\text{sup}} \\ \boldsymbol{Z}_{\text{task}_2} \boldsymbol{J}_{\text{task}_2}^T \boldsymbol{F}_{\text{task}_2}^{\text{ctrl}} \\ \vdots \\ \boldsymbol{Z}_{\text{task}_r} \boldsymbol{J}_{\text{task}_r}^T \boldsymbol{F}_{\text{task}_r}^{\text{ctrl}} \end{pmatrix} = \begin{bmatrix} \boldsymbol{I} & \boldsymbol{J}_{\text{sup}}^{M+,T} \boldsymbol{J}_{\text{task}_2}^T & \cdots & \boldsymbol{J}_{\text{sup}}^{M+,T} \boldsymbol{J}_{\text{task}_r}^T \\ \boldsymbol{0} & \boldsymbol{Z}_{\text{task}_2} \boldsymbol{J}_{\text{task}_2}^T & \cdots & \boldsymbol{Z}_{\text{task}_2} \boldsymbol{J}_{\text{task}_r}^T \\ \vdots & \vdots & \ddots & \vdots \\ \boldsymbol{0} & \boldsymbol{0} & \cdots & \boldsymbol{Z}_{\text{task}_r} \boldsymbol{J}_{\text{task}_r}^T \end{bmatrix} \begin{pmatrix} \boldsymbol{F}_{\text{sup}}^{\text{ext}} \\ \boldsymbol{F}_{\text{task}_2}^{\text{ext}} \\ \vdots \\ \boldsymbol{F}_{\text{task}_r}^{\text{ext}} \end{pmatrix} \tag{7.24}$$

based on (7.13). Note that (7.23) and (7.24) only hold if the soft constraints (7.19) are exactly fulfilled as demanded by Assumption 7.1. An analysis of the case where the soft constraints are not exactly fulfilled would be considerably more cumbersome. In this case, the constraints will affect the quality of the task fulfillment according to their corresponding weights $\boldsymbol{Q}_{\text{task}_k}$ leading to a different closed-loop dynamics.

7.2.2 Link to the MCB Control Approach for Multi-contact Balancing

The MCB control scheme, which is presented in Sect. 6.1 for whole-body balancing on multiple contacts, can be considered a special case of the hierarchical balancing controller (HMCB) introduced in Sect. 7.2. As shown in Table 7.2, the MCB controller can be replicated by combining the CoM, the support, and the interaction task together at priority level $k = 1$. This is possible because the MCB controller assumes that the frames of the support and interaction tasks are solely located at the end-effectors. Thus, there are no kinematic conflicts between the CoM, the support, and the interaction task if the robot is in a non-singular configuration, which facilitates a kinematic coexistence of all Cartesian tasks.

The task at level $k = 2$ is given by the joint compliance $\boldsymbol{\tau}_{\text{pose}}^{\text{cpl}}$ according to (6.33) in order to stabilize the posture of the robot in joint space. But instead of using

$$\boldsymbol{u} = \boldsymbol{\tau}_{\mu} + \boldsymbol{g} + \begin{bmatrix} \boldsymbol{J}_{\text{sup}} \\ \boldsymbol{J}_{\text{c}} \\ \boldsymbol{J}_{\text{int}} \end{bmatrix}^T \begin{pmatrix} -\boldsymbol{F}_{\text{sup}} \\ \boldsymbol{F}_{\text{c}}^{\text{cpl}} \\ \boldsymbol{F}_{\text{int}}^{\text{cpl}} \end{pmatrix} + \boldsymbol{N}_{\text{pose}} \boldsymbol{J}_{\text{pose}}^T \boldsymbol{\tau}_{\text{pose}}^{\text{cpl}} \qquad \text{(Sect. 7.25)} \qquad (7.25)$$

according to (7.14), the MCB-controller from Sect. 6.1 employs

$$\boldsymbol{u} = \boldsymbol{g} + \begin{bmatrix} \boldsymbol{J}_{\text{sup}} \\ \boldsymbol{J}_{\text{c}} \\ \boldsymbol{J}_{\text{int}} \end{bmatrix}^T \begin{pmatrix} -\boldsymbol{F}_{\text{sup}} \\ \boldsymbol{F}_{\text{c}}^{\text{cpl}} \\ \boldsymbol{F}_{\text{int}}^{\text{cpl}} \end{pmatrix} + \boldsymbol{J}_{\text{pose}}^T \boldsymbol{N}_{\text{null}} \boldsymbol{\tau}_{\text{pose}}^{\text{cpl}} \qquad \text{(Sect. 7.26)} \qquad (7.26)$$

based on (6.8) and (6.31). As can be seen, the MCB-approach does not feature the generalized torque $\boldsymbol{\tau}_{\mu}$ for canceling several coupling terms within the Coriolis/centrifugal matrix $\boldsymbol{C}$. Thus, the MCB controller does not offer a dynamic decoupling between task level $k = 1$ and $k = 2$. Furthermore, the mapping of the posture task $\boldsymbol{\tau}_{\text{pose}}^{\text{cpl}}$ to the virtual control input $\boldsymbol{u}$ is different: Despite the order of the task Jacobian matrix and the null space projector, the MCB controller uses

$$\boldsymbol{N}_{\text{null}} = \boldsymbol{I} - \begin{bmatrix} \boldsymbol{J}_{\text{sup}} \\ \boldsymbol{J}_{\text{int}} \end{bmatrix}^T \begin{bmatrix} \boldsymbol{J}_{\text{sup}} \\ \boldsymbol{J}_{\text{int}} \end{bmatrix}^{\boldsymbol{M}_{\text{qq}}+,T} \qquad (7.27)$$

according to (6.31), instead of

$$\boldsymbol{N}_{\text{pose}} = \boldsymbol{I} - \begin{bmatrix} \boldsymbol{J}_{\text{sup}} \\ \boldsymbol{J}_{\text{c}} \\ \boldsymbol{J}_{\text{int}} \end{bmatrix}^T \begin{bmatrix} \boldsymbol{J}_{\text{sup}} \\ \boldsymbol{J}_{\text{c}} \\ \boldsymbol{J}_{\text{int}} \end{bmatrix}^{\boldsymbol{M}+,T} \qquad (7.28)$$

according to (7.14).

In summary, the MCB control approach presented in Sect. 6.1 can be considered a special case of the hierarchical framework (HMCB) presented in Sect. 7.2, where the tasks of priority level $k = 1$ are defined such that they can coexist. The main difference lies in the posture task at priority level $k = 2$. Although the hierarchical framework (HMCB) provides more features, several experiments revealed that the MCB controller is slightly more robust, especially in extreme and difficult situations (Henze et al. 2019; Werner et al. 2016).

7.2.3 *Experimental Evaluation*

The performance of the combined control framework (HMCB) was tested in Henze et al. (2016a) using the torque-controlled humanoid robot *TORO* developed at *DLR*. A brief description of the robot hardware is given in Chap. 5. The evaluation is based on two exemplary hierarchies, which are listed as "CoM over Int" and "Int over CoM" in Table 7.2. Both hierarchies consist of four priority levels comprising a task for support, CoM, interaction, and the pose in joint space. The difference between both is that the priority of the CoM and the interaction task is swapped. The hierarchy "Int over CoM" can be motivated with a scenario in which a humanoid robot is supposed to walk while carrying an object, such as a glass of water. In order to prevent the motion of the CoM from dynamically affecting the interaction task, the latter is assigned to a higher priority level ($k = 2$) than the CoM task ($k = 3$). The other hierarchy "CoM over Int" is motivated by scenarios that require a high accuracy of the CoM task, for instance, when the robot is supposed to traverse a narrow bridge without falling. Therefore, the CoM task is given a higher priority ($k = 2$) than the interaction task at priority level $k = 3$. The MCB control framework for multi-contact balancing (see Sect. 6.1) serves as a reference for the evaluation.

7.2.3.1 Dynamic Decoupling

Two pairs of simulation and experiment were conducted in Henze et al. (2016a) to verify the dynamic decoupling offered by the HMCB approach. The setup for the simulations and the experiments is shown in Fig. 7.3a, b. The first pair implements a motion of the CoM to study the effect on the interaction task represented by the hands. According to theory, the hierarchical approach using the task order "Int over CoM" should prevent the interaction task from being disturbed by a motion of the CoM. In contrast to that, the hierarchy "CoM over Int" as well as the MCB approach (see Sect. 6.1) should exhibit a coupling from the CoM onto the interaction task. For evaluation, a simulation was conducted in which a vertical step of 5 cm is commanded for the desired CoM position. As can be seen in Fig. 7.3c, the hierarchical controller "Int over CoM" shows, as expected, a significantly smaller control error for the position and orientation of the right hand. The performance of the hierarchical controller "CoM over Int" is worse than the multi-contact balancer (MCB), but still comparable. Furthermore, we conducted an experiment where a continuous trajectory was commanded to the robot to trigger a dynamic motion rotating the CoM frame C. Note that a rotation of the CoM frame corresponds to a rotation of the hip due to the definition of C in Sect. 3.2.2. The commanded CoM rotation consists of a sinusoidal trajectory with a frequency of 0.5 Hz. As shown in Fig. 7.3e, the amplitude was linearly increased from 0° to 15°, held constant, and decreased again within 5 s each. The difference between the hierarchical approach "Int over CoM" and the multi-contact balancer (MCB) is not as evident as in the simulation. But both perform better than the controller "CoM over Int".

The second pair of simulation and experiment addresses the inverse coupling by commanding a motion of the interaction end-effectors and studying the effect on the CoM (see Fig. 7.3a, b). Here, it is expected that the controller "CoM over Int" performs best due to the high priority of the CoM task. In simulation, a step was commanded to the robot to increase the desired height of both hands by 10 cm to cause a vertical disturbance for the CoM task. As can be seen in Fig. 7.3d, the hierarchical approach "CoM over Int" exhibits a significantly smaller control error than the other controllers. In the conducted experiment, a continuous trajectory was commanded to the interaction compliances causing a complementary vertical motion of the hands to inflict a rotational coupling on the hip or the CoM frame C, respectively. The sinusoidal signal features a frequency of 0.6 Hz and an amplitude of 25 cm, as shown in Fig. 7.3f. Again, the amplitude was linearly increased, held constant, and decreased within 5 s each. As a result, the controller "CoM over Int" outperforms the other ones regarding the translational error of the CoM. For the orientation error of the CoM frame C, all three approaches show a similar performance. The reason is that the inertial effect that the hands have on the robot torso is relatively small compared to the effect of joint friction.

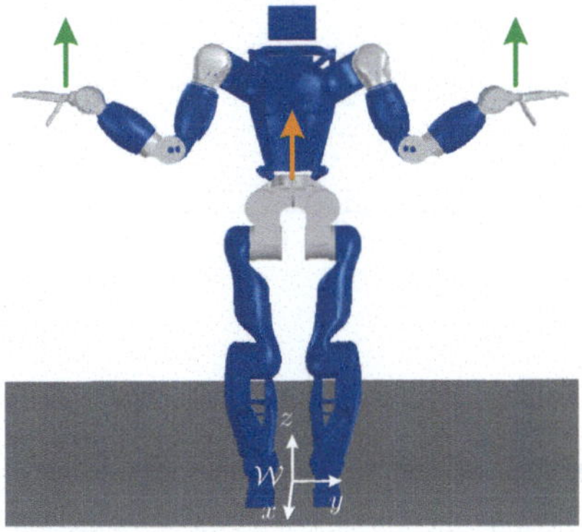

(a) Setup of the simulation: Jump in the CoM position (orange) and jump in the hand positions (green).

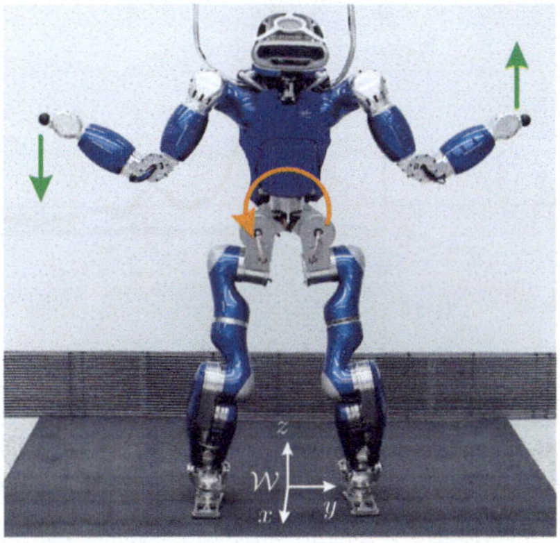

(b) Setup of the experiment: Trajectory of the CoM orientation (orange) and trajectory of the hand positions (green).

Fig. 7.3 Simulations and experiments comparing the dynamic decoupling of the MCB and the HMCB framework

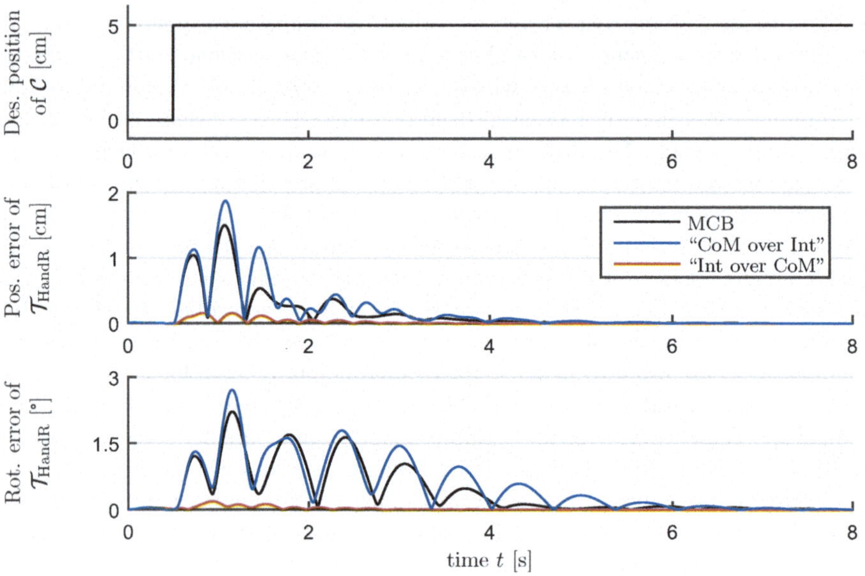

(c) Simulation of the response to a step in the desired CoM position, evaluated at the right hand.

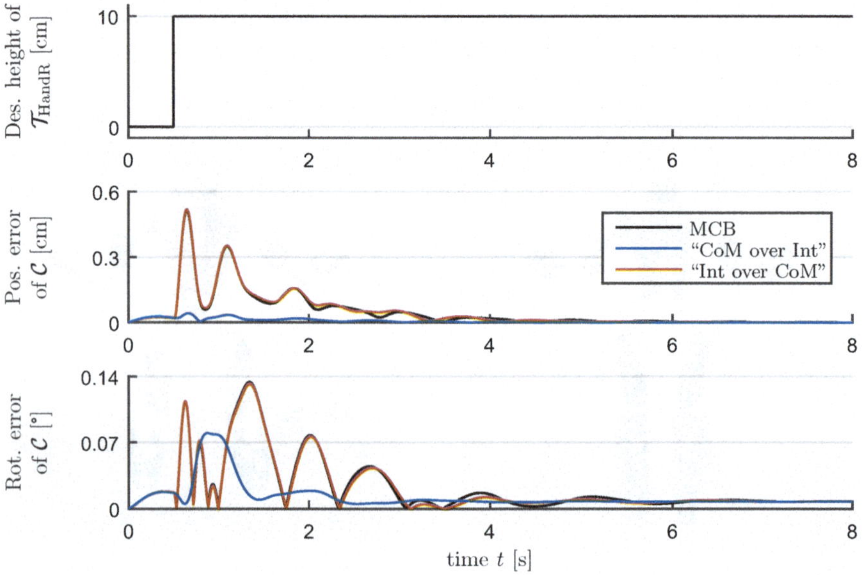

(d) Simulation of the response to a step in the desired hand positions, evaluated at the CoM frame $\mathcal{C}$.

Fig. 7.3 (continued)

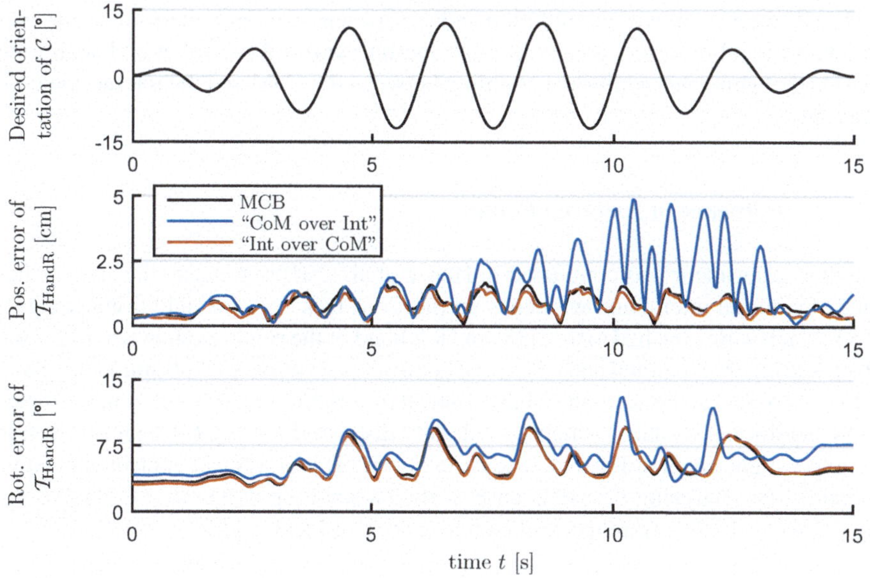

(e) Experiment on the response to a trajectory of the desired orientation of the COM frame $\mathcal{C}$, evaluated at the right hand.

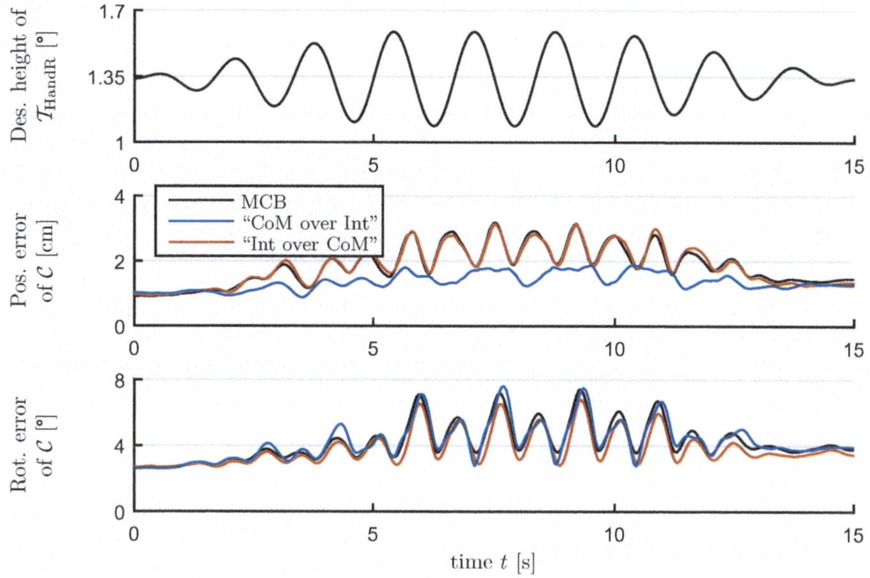

(f) Experiment on the response to trajectory of the desired hand positions, evaluated at the CoM frame $\mathcal{C}$.

Fig. 7.3 (continued)

The difference between simulation and experiment can be explained, among others, with modeling errors concerning the inertia matrix of the robot and with joint friction that generates additional coupling between the CoM and the interaction end-effectors.

7.2.3.2 Influence of External Forces

Another experiment was conducted in Henze et al. (2016a) to study the static influence of external interaction wrenches on the CoM task. For this, additional weights of 5 kg each were attached to the right and left hand of the robot, as shown in Fig. 7.4a. The resulting translational and rotational control errors of the CoM frame C are given in Fig. 7.4b. As can be seen, the MCB controller presented in Sect. 6.1 is not affected by the additional weights, which was already discussed in Sect. 6.1.2 and proven in the experiment given in Fig. 6.4. The behavior of the hierarchical controllers can be explained by analyzing the static case of the closed-loop dynamics (7.24). Applying (7.24) for both hierarchies and isolating the CoM task yields

$$\boldsymbol{Z}_{\mathrm{c}}\,\boldsymbol{J}_{\mathrm{c}}^T\,\boldsymbol{F}_{\mathrm{c}}^{\mathrm{cpl}} = \boldsymbol{Z}_{\mathrm{c}}\,\boldsymbol{J}_{\mathrm{c}}^T\,\boldsymbol{F}_{\mathrm{c}}^{\mathrm{ext}} + \boldsymbol{Z}_{\mathrm{c}}\,\boldsymbol{J}_{\mathrm{int}}^T\,\boldsymbol{F}_{\mathrm{int}}^{\mathrm{ext}} \qquad \text{(“CoM over Int”)} \tag{7.29}$$

and

$$\boldsymbol{Z}_{\mathrm{c}}\,\boldsymbol{J}_{\mathrm{c}}^T\,\boldsymbol{F}_{\mathrm{c}}^{\mathrm{cpl}} = \boldsymbol{Z}_{\mathrm{c}}\,\boldsymbol{J}_{\mathrm{c}}^T\,\boldsymbol{F}_{\mathrm{c}}^{\mathrm{ext}} \qquad \text{(“Int over CoM”).} \tag{7.30}$$

If we consider the hierarchy “Int over CoM”, the wrench of the CoM compliance $\boldsymbol{F}_{\mathrm{c}}^{\mathrm{cpl}}$ is not affected by the external interaction wrench $\boldsymbol{F}_{\mathrm{int}}^{\mathrm{ext}}$. As a consequence, the hierarchy “Int over CoM” shows a negligible control error, similar to the MCB controller. If we consider the hierarchy “CoM over Int”, then $\boldsymbol{F}_{\mathrm{int}}^{\mathrm{ext}}$ is mapped to $\boldsymbol{F}_{\mathrm{c}}^{\mathrm{cpl}}$ according to (7.29). Thus, the CoM compliance shows a steady state error of 2.3 cm and 1.5°, respectively. This observation is caused by $\breve{\boldsymbol{J}}^{-T}\,(\boldsymbol{J}_{\mathrm{task}_r}^{\mathrm{aug}})^T$ in (7.13) being an upper triangular matrix, which is a direct consequence of the dynamic decoupling, as discussed in Sect. 7.1. Hypothetically speaking, this effect can be remedied by explicitly measuring the external interaction forces $\boldsymbol{F}_{\mathrm{int}}^{\mathrm{ext}}$ and compensating for the coupling; but this would cause the well-known problems regarding causality and additional measurement noise.

7.3 Transfer to Confined Spaces

In Henze et al. (2017), we addressed the challenge of operating humanoid robots in confined spaces by extending the HMCB framework presented in Sect. 7.2. So far, we have only considered cases where the frames of the support and the interaction

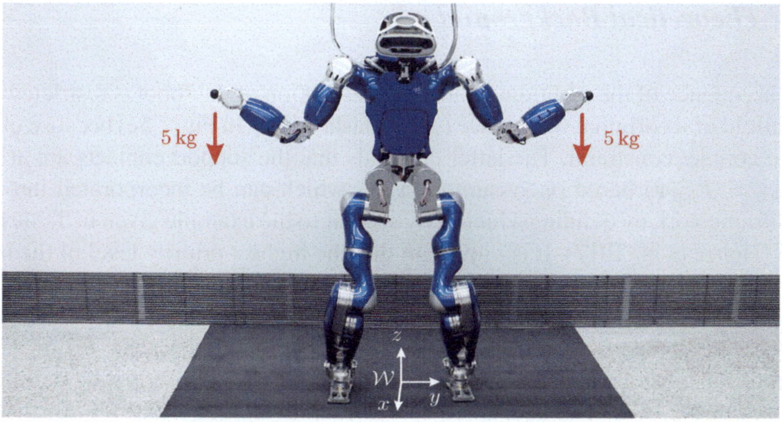

(a) Setup of the experiment: The red arrows represent the applied external forces.

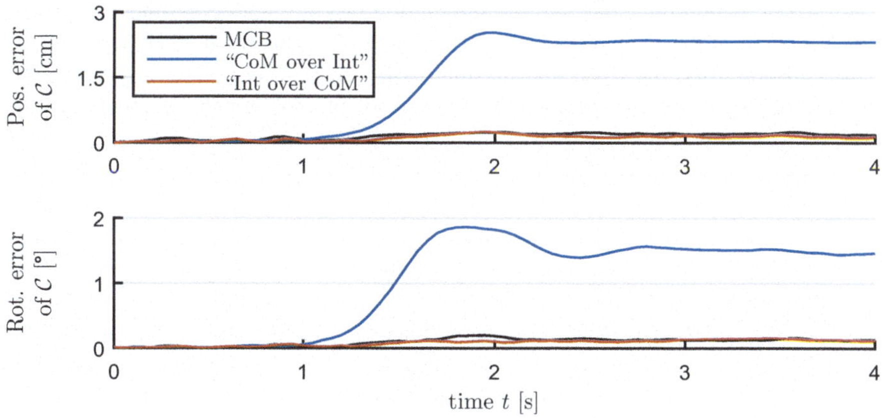

(b) Control error of the CoM task.

Fig. 7.4 Experiment comparing the influence of external disturbances of the MCB and the HMCB framework

task are located at the end-effectors of the robot (see Chap. 6). But due to the limited maneuverability in confined spaces, the robot needs to be able to exploit contacts (or task frames in general) that might be located arbitrarily on the robot structure and not only at the end-effectors. Thus, this section extends our considerations to scenarios where the contact and interaction frames can be scattered all over the surface of the robot, as schematically depicted in Fig. 7.5.

7.3.1 Theoretical Background

As a consequence of the additional contacts, the motion of the robot is restricted to a submanifold of its original workspace (green dashed lines in Fig. 7.5c) because of the physical contact constraint. The latter demands that the support contacts are at rest ($\mathbf{0} = \boldsymbol{v}_{\text{sup}} = \boldsymbol{J}_{\text{sup}}\, \boldsymbol{\nu}$) based on Assumption 3.2, which can be incorporated into the HMCB framework by defining a hierarchy similar to the example given in Tables 7.2 and 7.3 (Henze et al. 2017). It is important that the highest priority level of the hierarchy consists of the support task. Previously we have justified this decision with the accessibility of the support wrenches $\boldsymbol{F}_{\text{sup}}$ (see Sect. 7.2). But the support task can also be interpreted as the physical contact constraint given by $\mathbf{0} = \boldsymbol{v}_{\text{sup}} = \boldsymbol{J}_{\text{sup}}\, \boldsymbol{\nu}$. Thus, the support task can be considered a task, which is always converged by definition due to $\dot{\boldsymbol{v}}_{\text{sup}} = \boldsymbol{v}_{\text{sup}} = \mathbf{0}$. The decoupled closed-loop dynamics (7.23) reveals that all other tasks are mapped onto the null space of the physical contact constraint via their respective null space base matrix $\boldsymbol{Z}_{\text{task}_k}$. The remaining tasks can be chosen arbitrarily according to the target application of the robot. We proposed an example in Henze et al. (2017), where the CoM and the interaction task are both combined at priority level $k = 2$ (see Table 7.2). As the interaction frames in Henze et al. (2017) are defined at the end-effectors, the CoM and the interaction task do not cause any kinematic conflict, which allows for pooling at the same priority level. The lowest priority level ($k = 3$) is given by a joint compliance $\boldsymbol{\tau}_{\text{pose}}^{\text{cpl}}$ to stabilize the configuration of the robot in joint space in the case of a redundant kinematics and/or a singular configuration.

Another important aspect of balancing in confined spaces is that the additional contacts might result in an overdetermined contact configuration. In other words, the robot might not feature a sufficient number of joints to generate all contact wrenches independently of one another. Thus, it can happen that the external contact wrenches $\boldsymbol{F}_{\text{sup}}^{\text{ext}}$ deviate from the commanded ones $\boldsymbol{F}_{\text{sup}}$ and, in the worst case, violate the contact model given in Sect. 3.4, which ultimately can cause the robot to fall. In order to avoid that, let us discuss the contact configurations in which the robot is able to command a suitable set of contact wrenches by studying the static case of the desired closed-loop dynamics (7.24) for the task hierarchy given in Table 7.2. Isolating the first line of (7.24) with $\boldsymbol{F}_{\text{int}}^{\text{ext}} = \mathbf{0}$ and $\boldsymbol{\tau}_{\text{pose}}^{\text{ext}} = \mathbf{0}$ yields

$$\boldsymbol{F}_{\text{sup}} = \boldsymbol{J}_{\text{sup}}^{\boldsymbol{M}+,T}\, \boldsymbol{J}_{\text{sup}}^{T}\, \boldsymbol{F}_{\text{sup}}^{\text{ext}}. \tag{7.31}$$

As can be seen, $\boldsymbol{J}_{\text{sup}}$ must be of full rank to ensure $\boldsymbol{F}_{\text{sup}} = \boldsymbol{F}_{\text{sup}}^{\text{ext}}$. Fig. 7.5a, b show two examples of an overdetermined contact configuration leading to a rank deficiency of $\boldsymbol{J}_{\text{sup}}$. In Fig. 7.5a, the contacts II and III are redundant since both act on the shank of the robot. In Fig. 7.5b, the redundancy is given between the contacts I and II in the form of forces parallel to the shank. Such an overdetermined contact configuration can be remedied via two methods:

1. The contacts can be redefined as shown in Fig. 7.5c, such that $\boldsymbol{J}_{\text{sup}}$ is of full rank representing a statically indeterminate contact configuration.

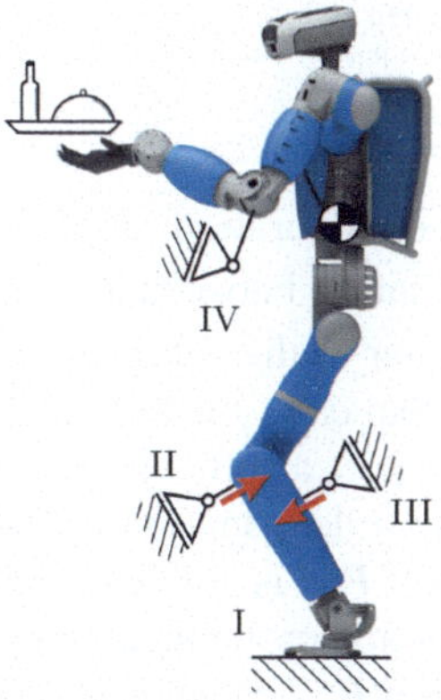

(a) Redundant support wrenches (red) due to contacts II and III.

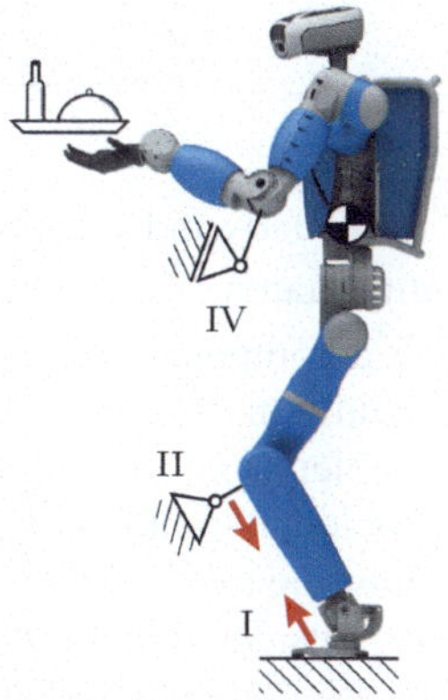

(b) Redundant support wrenches (red) due to contacts I and II.

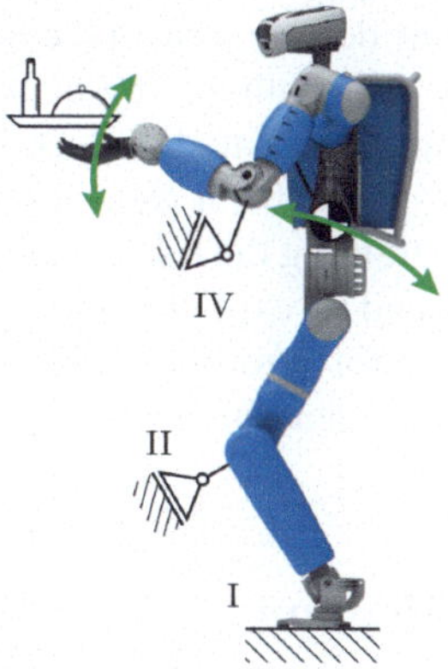

(c) Statically indeterminate contact configuration.

Fig. 7.5 Examples of contact configurations with external wrenches acting all over the body of the robot. The green lines represent the manifold to which the CoM and the interaction points are constrained by the support contacts

2. The knowledge of the controller can be augmented with a model for the elasticity of the environment and/or the robot structure, such that it is mathematically possible to distinguish between the individual contact wrenches. However, the knowledge on the environment can be limited depending on the application scenario.

For the remainder of this work, we always assume a statically indeterminate contact configuration leading to $\boldsymbol{J}_{\text{sup}}$ being of full rank.

A similar problem can arise from the external interaction wrenches $\boldsymbol{F}_{\text{int}}^{\text{ext}}$. Analyzing again the first line of the static closed-loop system (7.24) for $\boldsymbol{\tau}_{\text{pose}}^{\text{ext}} = \mathbf{0}$ yields

$$\boldsymbol{F}_{\text{sup}} = \underbrace{\boldsymbol{J}_{\text{sup}}^{M+,T} \boldsymbol{J}_{\text{sup}}^{T}}_{\boldsymbol{I}} \boldsymbol{F}_{\text{sup}}^{\text{ext}} + \boldsymbol{J}_{\text{sup}}^{M+,T} \boldsymbol{J}_{\text{int}}^{T} \boldsymbol{F}_{\text{int}}^{\text{ext}}. \tag{7.32}$$

In the case of a statically indeterminate contact configuration, $\boldsymbol{J}_{\text{sup}}^{M+,T} \boldsymbol{J}_{\text{sup}}^{T}$ simplifies to the identity matrix. As can be seen, a deviation between $\boldsymbol{F}_{\text{sup}}$ and $\boldsymbol{F}_{\text{sup}}^{\text{ext}}$ can occur if $\boldsymbol{F}_{\text{int}}^{\text{ext}}$ has a component that is perpendicular to the submanifold introduced by the contact condition ($\boldsymbol{0} = \boldsymbol{v}_{\text{sup}} = \boldsymbol{J}_{\text{sup}} \boldsymbol{\nu}$). In that case, the term $\boldsymbol{J}_{\text{sup}}^{M+,T} \boldsymbol{J}_{\text{int}}^{T} \boldsymbol{F}_{\text{int}}^{\text{ext}}$ is nonzero. In simpler terms, the robot does not feature a sufficient number of DoFs, such that the problematic part of $\boldsymbol{F}_{\text{int}}^{\text{ext}}$ can cause a deformation of the interaction compliances. Thus, the external interaction wrench $\boldsymbol{F}_{\text{int}}^{\text{ext}}$ is not detected by the robot and therefore not considered in the wrench distribution problem, although $\boldsymbol{F}_{\text{int}}^{\text{ext}}$ influences $\boldsymbol{F}_{\text{sup}}^{\text{ext}}$. Thus, the computed wrench distribution $\boldsymbol{F}_{\text{sup}}$ can deviate from the actual one $\boldsymbol{F}_{\text{sup}}^{\text{ext}}$. Such a situation is shown in Fig. 7.5c involving the interaction wrench $\boldsymbol{F}_{\text{int}}^{\text{ext}}$ and contact IV: The component of $\boldsymbol{F}_{\text{int}}^{\text{ext}}$ that is parallel to the forearm is directly routed to contact IV without causing any deformation of the robot. Thus, the contact wrench in IV is directly influenced by $\boldsymbol{F}_{\text{int}}^{\text{ext}}$ without any notion of the robot. A similar situation is studied in an experiment presented in Sect. 7.3.2. Of course, this problem does not only exist in confined spaces, but it gains more and more importance with a higher number of contacts constraining the robot, which is characteristic of confined spaces. Remedy could be provided by explicitly measuring the interaction wrenches via additional sensors, such as a force/torque sensor or a tactile skin, although this would cause secondary problems regarding causality or additional sensor noise.

7.3.2 Experimental Evaluation

The capabilities of the HMCB control framework in the context of confined spaces was demonstrated in Henze et al. (2017) using the torque-controlled humanoid robot *TORO* developed at *DLR*. A brief description of the robot hardware is given in Chap. 5. The evaluation is based on the hierarchy listed in Table 7.2 consisting of three priority levels. The first level ($k = 1$) comprises the support contacts, while the second level ($k = 2$) combines the CoM and the interaction compliances. The third level consists of a compliance in joint space in order to stabilize the posture of the robot. The first three experiments utilize the knees as point contacts based on the contact model given in Sect. 3.4.3 (point contact). The latter enables the knees to generate a unilateral contact force perpendicular to the shin of the robot. The last experiment demonstrates balancing exclusively on knees, which requires the exploitation of all six DoFs of the knee contacts according to the contact model given in Sect. 3.4.2.

7.3.2.1 Extending the Support Polygon

The first experiment presented in Henze et al. (2017) demonstrates an application from the field of industrial manufacturing, where the robot is supposed to perform an assembly task within the fuselage of an aircraft. As shown in Fig. 7.6a, the robot stands in front of a section of the fuselage and has to reach for the hull with the hands. However, the robot is only able to shift the CoM to the frontal edge of the support polygon $\mathcal{S}_{\mathrm{SP}}$ given by the convex hull of the feet, which limits the reach of the arms. In order to overcome this restriction, the robot lowers the center of mass until the knees establish contact with a support structure mounted to the hull. As previously mentioned, the knees are modeled as point contacts generating only unilateral forces ${}^{\mathrm{KneeR}}f_{\mathrm{KneeR},z}$ and ${}^{\mathrm{KneeL}}f_{\mathrm{KneeL},z}$ perpendicular to the shin. (Note that the leading superscript indicates that the forces are specified in the knee frames $\mathcal{T}_{\mathrm{KneeR}}$ and $\mathcal{T}_{\mathrm{KneeL}}$ according to the notation introduced in Sect. 3.1.) Because of the additional contacts, the robot can now shift the CoM even further to the front, leaving the convex hull spanned by the feet, which significantly increases the reach of the arms.

The recorded telemetry data is given in Fig. 7.6b. Note that the fuselage is equipped with additional uniaxial force sensors to measure the contact forces at the knees, which are only used for validation purposes and by the feedback controller. The recorded data reveals that the contact forces computed by the optimization ($\boldsymbol{f}_{\mathrm{KneeR}}$ and $\boldsymbol{f}_{\mathrm{KneeL}}$) match quite well with the forces obtained by the sensors ($\boldsymbol{f}^{\mathrm{ext}}_{\mathrm{KneeR}}$ and $\boldsymbol{f}^{\mathrm{ext}}_{\mathrm{KneeL}}$), which is an indicator for the validity of the presented control approach. The knees of the robot start to touch the fuselage at $t = 4.6$ s, which causes a spike in the measured force of the left knee due to the impact. At $t = 6$ s, the knees are added to the set of the support contacts, which causes the knee forces to rise from 0 N (deactivated state) to their minimum contact force of ${}^{\mathrm{KneeR}}f^{\min}_{\mathrm{KneeR},z} = {}^{\mathrm{KneeL}}f^{\min}_{\mathrm{KneeL},z} =$ 50 N given by the unilaterality constraint according to Sect. 3.4.3. The procedure for contact switching is sketched in Sect. 6.7.3. Afterwards, the robot starts to move the CoM to the front, which causes the knee forces to rise up to $\|\boldsymbol{f}_{\mathrm{KneeR}}\| = 138$ N and $\|\boldsymbol{f}_{\mathrm{KneeL}}\| = 136$ N, respectively. Note that also the position of the combined center of pressure $\boldsymbol{p}_{\mathrm{Feet}}$, which is obtained from the measured foot wrenches, changes throughout the experiment. First, the CoP of the feet moves 4.5 cm to the front following the motion of the CoM. After the knee contacts are added to the cQP, the CoP moves to the back to counteract the minimum contact forces for the knees. As soon as the forces in the knees rise, the CoP of the feet moves back to the origin of the world frame $\mathcal{W}$ located underneath the ankles (see Fig. 7.6b). This effect is caused by the cQP (7.17)–(7.21), which is configured such that mainly contact forces are generated instead of contact torques. Despite supporting the weight of the robot, the additional knee contacts generate a torque about the y-axis of the world frame $\mathcal{W}$, which the robot no longer needs to counteract with the feet. The additional knee contacts allow the robot to shift the CoM up to $x_{\mathrm{c},x} = 20.5$ cm, which is 7.5 cm in front of the convex hull of the feet characterized by $p^{\max}_{\mathrm{Feet},x} = 13$ cm. The slight difference between the commanded and measured CoM position can be explained by the fact that the CoM is commanded on a straight line, whereas the actual CoM can only move on the submanifold determined by the contact constraint (see Fig. 7.6a).

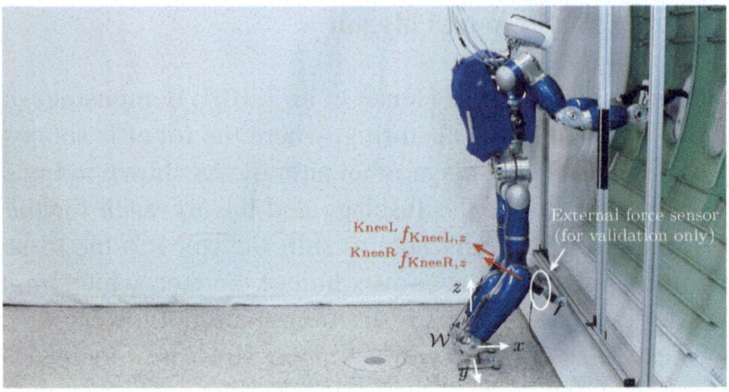

(a) Setup of the experiment. Additional force sensors are only mounted on the aircraft fuselage for validation purposes.

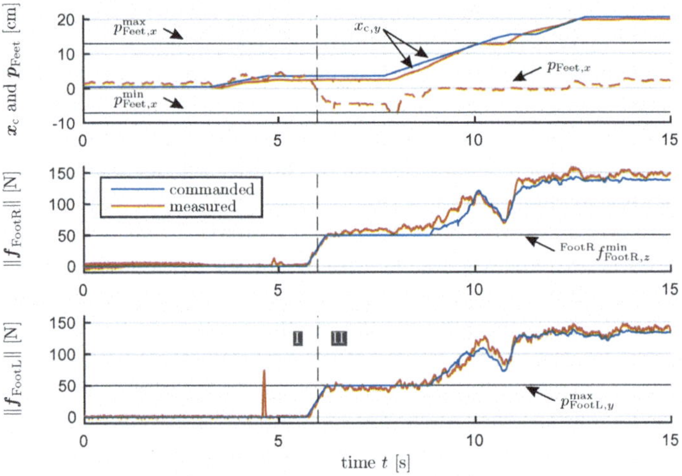

(b) Position of the CoM, the CoP of the feet, and the knee forces.

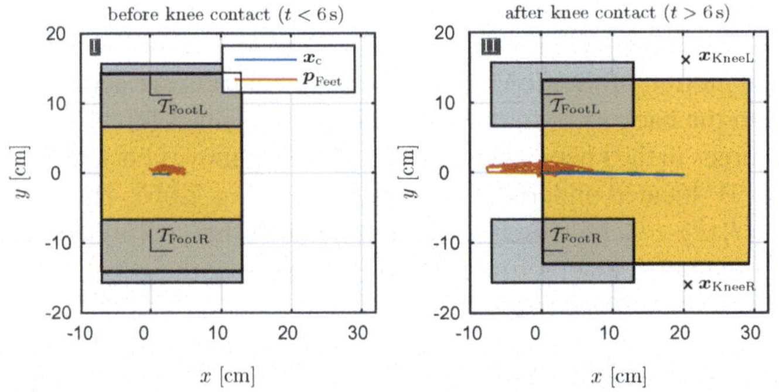

(c) Support polygon $\mathcal{S}_{\text{SP}}$ (yellow) before and after establishing contact with the knees. The contact areas $\mathcal{S}_{\text{FootR}}$ and S_{FootL} of the feet are given in gray.

Fig. 7.6 The support polygon is extended by the additional contacts at the knees in order to increase the reach of the arms

The effect of the additional knee contacts on the support polygon $\mathcal{S}_{\mathrm{SP}}$ is shown in Fig. 7.6c. Before the knee contacts are added to the cQP at $t = 6$ s, the support polygon comprises the full length of the feet. The width is smaller than the outer edge of the feet because of the minimum contact force $f_{\mathrm{FootR},z} = f_{\mathrm{FootL},z} = 50$ N. As soon as the knee contacts are added to the wrench distribution problem, $\mathcal{S}_{\mathrm{SP}}$ extends significantly to the front, such that the leading edge is almost 30 cm in front of the world frame $\mathcal{W}$. Note that the leading edge even extends beyond the knees $\boldsymbol{x}_{\mathrm{KneeR}}$ and $\boldsymbol{x}_{\mathrm{KneeL}}$ because of the friction in the foot surfaces of $\tilde{\mu} = 0.4$. The trailing edge also moves to the front due to the minimum contact force of the knees (${}^{\mathrm{KneeR}}f_{\mathrm{KneeR},z} = {}^{\mathrm{KneeR}}f_{\mathrm{KneeL},z} = 50$ N). But this effect does not compromise balancing, as the controller precautionarily moves the CoM to a position that lies inside the support polygon both before and after the contact transition. Note that the width of the support polygon is further reduced by the minimum knee forces ${}^{\mathrm{KneeR}}f_{\mathrm{KneeR},z} = {}^{\mathrm{KneeR}}f_{\mathrm{KneeL},z} = 50$ N. In summary, the additional knee contacts extend the support polygon to the front and therefore increase the operating range of the CoM and the hands, which ultimately enables the robot to fulfill the designated assembly task. The advantage of this feature was demonstrated in the context of aircraft manufacturing, as presented in Sect. 9.3.

7.3.2.2 Influence of External Forces

The second experiment presented in Henze et al. (2017) illustrates the influence of external interaction forces $\boldsymbol{F}_{\mathrm{int}}^{\mathrm{ext}}$ on the contact wrenches $\boldsymbol{F}_{\mathrm{sup}}^{\mathrm{ext}}$, as discussed in Sect. 7.3.1. The robot is placed in front of the fuselage with both knees in contact, as in the experiment before. Instead of performing an autonomous motion, several external disturbances are manually applied to the robot (see Fig. 7.7a). In the first half of the experiment ($t < 11$ s), a vertical force is applied twice to the hands of the robot (1st : 71N, 2nd : 78N). As can be seen from Fig. 7.7b, the wrenches generated by the interaction compliances $\boldsymbol{F}_{\mathrm{HandR}}^{\mathrm{cpl}}$ and $\boldsymbol{F}_{\mathrm{HandL}}^{\mathrm{cpl}}$ are immediately counteracted by the support wrenches comprising the feet and the knees. Note that the commanded contact wrenches match the measured ones quite well despite minor discrepancies due to sensor calibration. As explained in Sect. 7.3.1, the additional knee contacts constrain the motion of the hip but not of the hands. Thus, the applied external force results in a deflection of the hand compliances. The produced compliance wrench is in turn taken into account by the optimization computing the wrench distribution for the support contacts. As a consequence, the difference between the computed contact wrenches $\boldsymbol{F}_{\mathrm{sup}}$ and the actual ones $\boldsymbol{F}_{\mathrm{sup}}^{\mathrm{ext}}$ is sufficiently small.

In the second half of the experiment ($t > 11$ s), the external force is applied twice at the hip in the direction parallel to the thigh. As the robot cannot move in this particular direction, the HMCB controller does not become aware of the disturbance and thus does not adjust the commanded wrench distribution. But the actual contact forces change significantly since the thigh transfers the external disturbance directly to the knee contacts. As a consequence, there is a significant difference of up to 236 N between the commanded contact wrenches $\boldsymbol{F}_{\mathrm{sup}}$ and the actual ones $\boldsymbol{F}_{\mathrm{sup}}^{\mathrm{ext}}$. This illustrates that the problem of non-detectable external forces increases in the case of confined spaces, as discussed in Sect. 7.3.1.

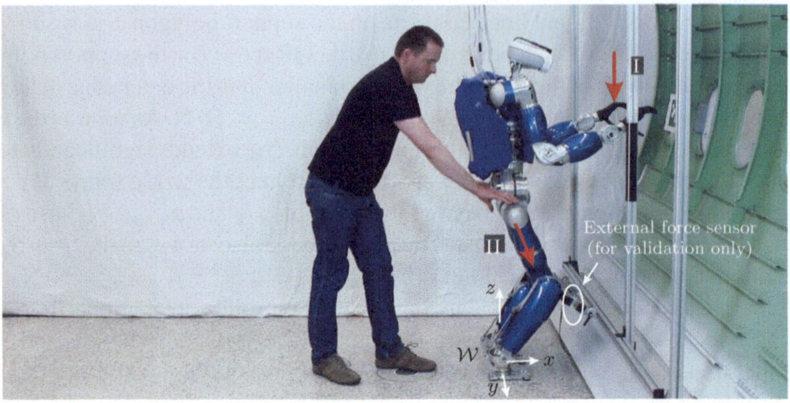

(a) Setup of the experiment. The red arrows represent the applied external forces.

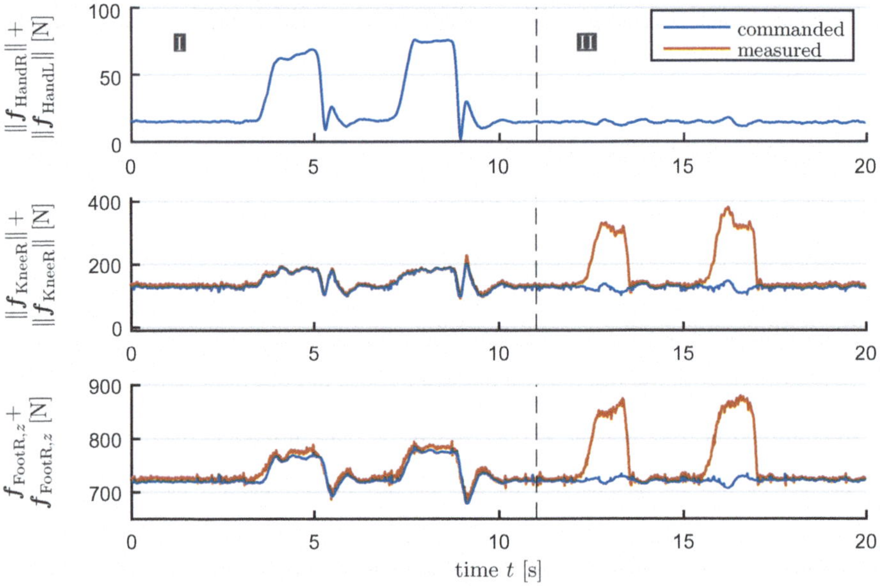

(b) Contact forces at the hands, the knees, and the feet.

Fig. 7.7 Influence of external wrenches on the support contacts

7.3.2.3 Replacing one Foot with a Knee

The third experiment presented in Henze et al. (2017) brings the robot into a highly asymmetric pose, as shown in Fig. 7.8a. The robot starts with both feet on the ground (phase I) and brings the right hand in contact with a support structure during phase II, which is required for lifting the right foot in phase III. In phase IV, the knee is bent to the back and placed on top of an uniaxial force sensor screwed to the support structure. Again, the additional sensor is only used for verification and not for controlling the robot. In phase V, the right hand is removed from the structure and the robot balances only on the left foot and the right knee. Note that in this experiment, the optimization assumes a point contact (see Sect. 3.4.3) for the right knee, generating only a unilateral force ${}^{\mathrm{KneeR}}f_{\mathrm{KneeR},z}$ perpendicular to the shin. (Note that the leading superscript indicates that the forces are specified in the knee frame $\mathcal{T}_{\mathrm{KneeR}}$ according to the notation introduced in Sect. 3.1.) Analogously, the right hand is also assumed to be a point contact generating only forces perpendicular to the palm.

The recorded data is shown in Fig. 7.8b: The hand contact is activated at $t = 7$ s, which causes the force at the right hand to rise from approximately zero (commanded force by the hand compliance) to the minimum contact force of $f^{\mathrm{min}}_{\mathrm{HandR},y} = 20$ N. Since the robot does not close the hand, the contact is treated as an unilateral contact. When the right foot is lifted, the load on the left leg increases from $f_{\mathrm{FootL},z} = 474$ N to 810 N, which corresponds to the total weight of the robot. The force at the right hand increases to values between $f_{\mathrm{HandR},y} = 25$ N and 50 N to counteract the torque that is induced about the x-axis of $\mathcal{W}$ by the distance between the CoM and the left foot. Note that in this scenario, the right hand is oriented such that the generated force acts parallel to the y-axis of the world frame $\mathcal{W}$ (see Fig. 7.8a). As soon as the right knee hits the support structure at $t = 20$ s, the impact creates a spike in the measured knee force. Adding the right knee to the set of support contacts reduces the contact force at the right hand again to its minimum of $f^{\mathrm{min}}_{\mathrm{HandR},y} = 20$ N. In this configuration, the robot can both utilize the right hand and the right knee to counteract the torque induced by the left foot about the x-axis of $\mathcal{W}$. But the constrained quadratic optimization (7.17)–(7.21) is configured to distribute the load preferably to the knee (${}^{\mathrm{KneeR}}f_{\mathrm{KneeR},z} = 120$ N) in order to reduce the stress on the hand. Of course, this also reduces the load on the left foot to $f_{\mathrm{FootL},z} = 705$ N, as the right knee contributes to supporting the weight of the robot. In summary, this experiment demonstrates that using the right knee for supporting the robot is more efficient than using the right hand because the knee can also contribute to carrying the weight of the robot, whereas the right hand can only be used to stabilize the CoM horizontally.

7.3.2.4 Balancing Solely on Knees

The last experiment presented in Henze et al. (2017) demonstrates the ability of the HMCB control framework to balance a humanoid robot solely on its knees, as shown in Fig. 7.9a. As both feet are in the air, the algorithm for estimating the state of the base/hip frame in Sect. 5.2 solely relies on the knee contacts and IMU measurements[1]. The world frame $\mathcal{W}$ is defined as shown in Fig. 7.9a, with the origin being located in the middle of both knees. The cQP (7.17)–(7.21) assumes the knees as rectangular contacts based on the model given in Sect. 3.4.2, each featuring a full six-dimensional contact wrench. In theory, the contacts would allow the controller to generate a six-dimension overall support wrench $\boldsymbol{F}^{\mathrm{res}}_{\mathrm{sup}}$ at the center of mass. But the null space projectors in (7.14) restrict the compliances $\boldsymbol{F}^{\mathrm{cpl}}_{\mathrm{c}}$, $\boldsymbol{F}^{\mathrm{cpl}}_{\mathrm{int}}$, and $\boldsymbol{\tau}^{\mathrm{cpl}}_{\mathrm{pose}}$ (see Table 7.2) to the subset of wrenches that can actually be generated with the joints of the robot. Note

[1] $\alpha_{\mathrm{IMU}} = 1$, $\alpha_{\mathrm{KneeR}} = \alpha_{\mathrm{KneeL}} = 0.5$, $\alpha_{\mathrm{FootR}} = \alpha_{\mathrm{FootL}} = 0$, and $\alpha_{\mathrm{HandR}} = \alpha_{\mathrm{HandL}} = 0$.

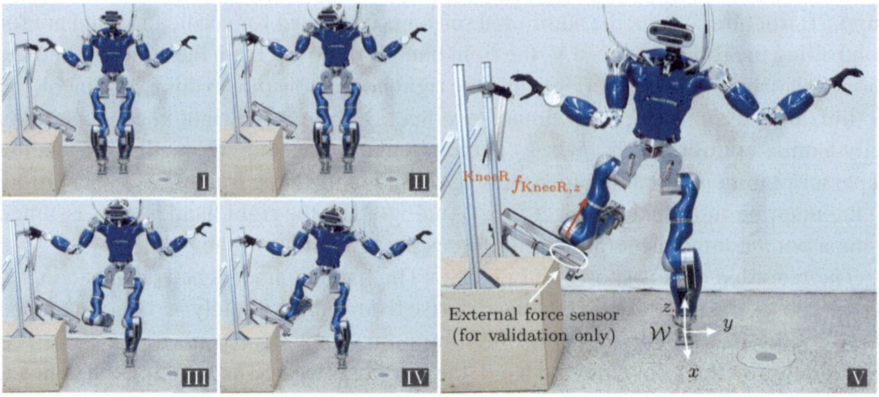

(a) Motion sequence during the experiment.

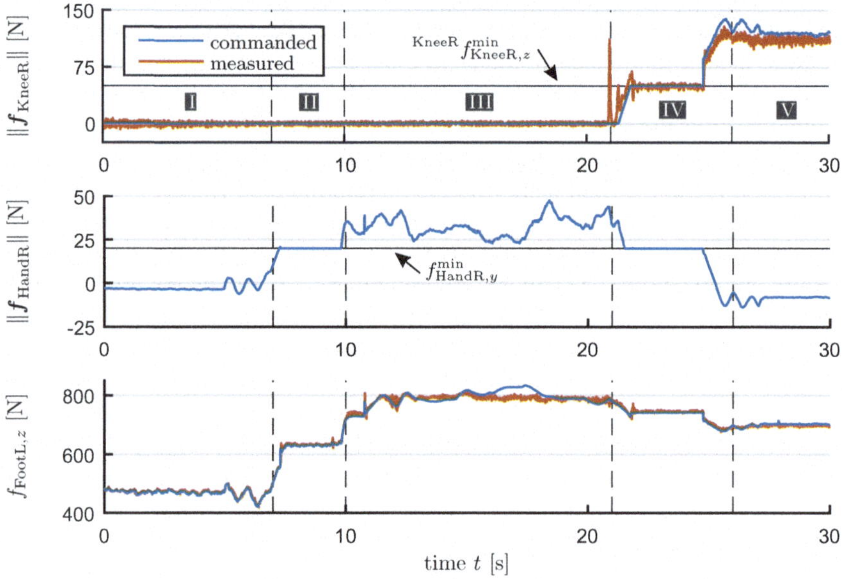

(b) Contact forces at the right hand, the right knee, and the left foot.

Fig. 7.8 The humanoid robot *TORO* robot lifts its right foot and places its right knee on the support structure

that *TORO* has a flat area of 3 cm by 18 cm at the top of the shin, which is used as a rectangular contact surface ($\mathcal{S}_{\text{KneeR}}$ and $\mathcal{S}_{\text{KneeL}}$) during the experiment. Starting from the initial configuration, the robot is disturbed by manually pulling and pushing the backpack, as shown in Fig. 7.9a. The results are given in Fig. 7.9b showing the motion of several prominent points in the saggital plane of the robot. Note that the right knee does not move due to the contact constraint $\mathbf{0} = \boldsymbol{v}_{\text{sup}} = \boldsymbol{J}_{\text{sup}}\, \boldsymbol{\nu}$. The right hand and the right foot manage to maintain their desired location throughout the experiment,

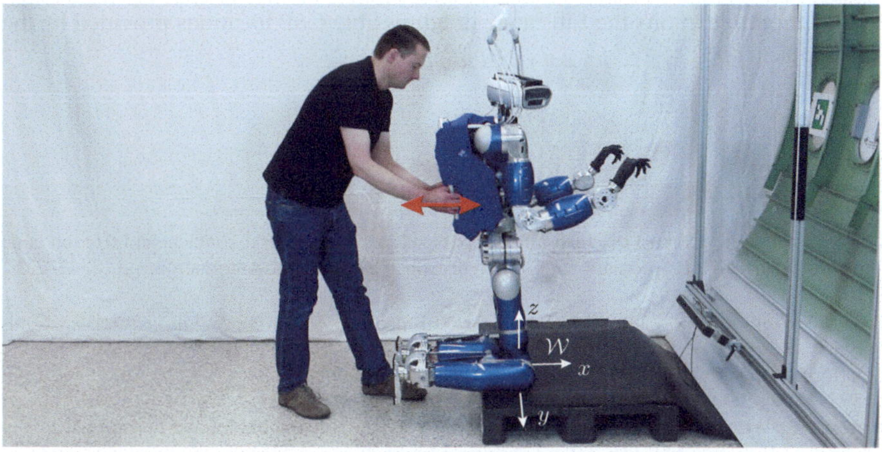

(a) Setup of the experiment. The red arrow represents the applied disturbance.

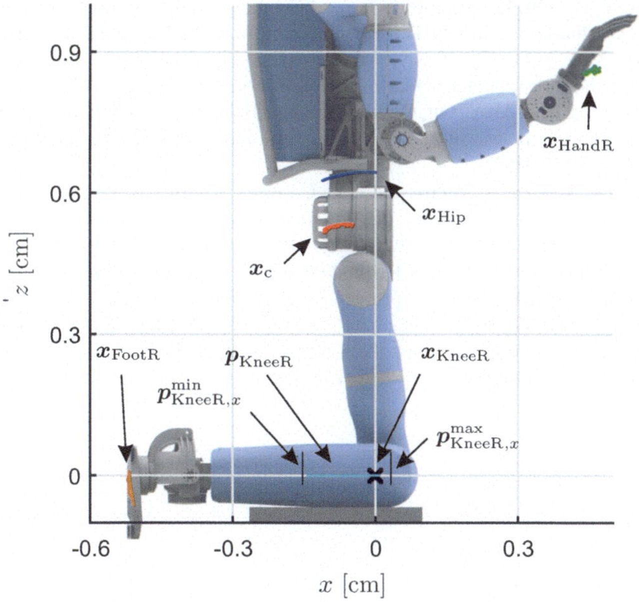

(b) Trajectory of several prominent points in the sagittal plane of the robot.

Fig. 7.9 The humanoid robot *TORO* balances solely on its knees

as the robot features enough DoFs to achieve both tasks. But due to the constrained kinematics, the hip and the CoM are restricted to a motion on a circular trajectory with the knee joints as central point, as shown in Fig. 7.9b. Note that the manually induced motion exploits the full range of admissible CoP locations provided by the knee contacts ($p^{\min}_{\mathrm{KneeR},x} \leq p_{\mathrm{KneeR},x} \leq p^{\max}_{\mathrm{KneeR},x}$).

References

Aghili, F., Su, C.Y.: Control of constrained robots subject to unilateral contacts and friction cone constraints. In: IEEE International Conference on Robotics and Automation, pp. 2347–2352 (2016)

Borst, C., Wimböck, T., Schmidt, F., Fuchs, M., Brunner, B., Zacharias, F., Giordano, P.R., Konietschke, R., Sepp, W., Fuchs, S., Rink, C., Albu-Schäffer, A., Hirzinger, G.: Rollin' Justin - mobile platform with variable base. In: IEEE International Conference on Robotics and Automation, pp. 1597–1598 (2009)

Dietrich, A.: Whole-Body Impedance Control of Wheeled Humanoid Robots, Springer Tracts in Advanced Robotics, vol. 116, 1st edn. Springer International Publishing, Switzerland (2016)

Dietrich, A., Ott, C., Park, J.: The hierarchical operational space formulation: stability analysis for the regulation case. IEEE Robot. Autom. Lett. **3**(2), 1120–1127 (2018)

Henze, B., Dietrich, A., Ott, C.: An approach to combine balancing with hierarchical whole-body control for legged humanoid robots. IEEE Robot. Autom. Lett. **1**(2), 700–707 (2016a)

Henze, B., Roa, M.A., Ott, C.: Passivity-based whole-body balancing for torque-controlled humanoid robots in multi-contact scenarios. Int. J. Robot. Res. **35**(12), 1522–1543 (2016b)

Henze, B., Dietrich, A., Roa, M.A., Ott, C.: Multi-contact balancing of humanoid robots in confined spaces: Utilizing knee contacts. In: IEEE/RSJ International Conference on Intelligent Robots and Systems, pp. 679–704 (2017)

Henze, B., Roa, M.A., Werner, A., Dietrich, A., Ott, C., Albu-Schäffer, A.: Experiments with human-inspired behaviors in a humanoid robot: quasi-static balancing using toe-off motion and stretched knees. In: IEEE International Conference on Robotics and Automation, pp. 2510–2516 (2019)

Ott, C.: Cartesian Impedance Control of Redundant and Flexible-Joint Robots, Springer Tracts in Advanced Robotics, vol. 49, 1st edn. Springer, Berlin (2008)

Werner, A., Henze, B., Rodriguez, D.A., Gabaret, J., Porges, O., Roa, M.A.: Multi-contact planning and control for a torque-controlled humanoid robot. In: IEEE/RSJ International Conference on Intelligent Robots and Systems, pp. 5708–5715 (2016)

Chapter 8
Balance Control Based on Reduced Dynamic Models

This section focuses on Model Predictive Control (MPC) for the balance control of humanoid robots. The basic idea of MPC is to choose the current control input such that the system shows an optimal behavior for a given time horizon, which ranges from the current to a future point in time. For this purpose, the future trajectory of the system is predicted and optimized based on the dynamic model of the plant. As the optimization often involves numerical methods, computational costs represent one of the limiting factors of MPC. As a consequence, many works on MPC in humanoid robotics resort to reduced dynamic models to decrease the numerical effort (Audren et al. 2014; Ibanez et al. 2012; Kajita et al. 2003; Wieber 2006). In contrast to the dynamic model presented in Sect. 3.2.2, which governs the whole body of the robot, reduced models usually focus on the center of mass dynamics only. Although reduced models only approximate the actual dynamics of the robot, MPC has the advantage of being able to react in advance to future events in the reference trajectory.

8.1 Interaction Aware Balancing via LIPM

We developed a balancing controller in Ott et al. (2013) based on the Linear Inverted Pendulum Model (LIPM) and Model Predictive Control. The architecture of the presented framework is schematically shown in Fig. 8.1. In order to allow for a modular design, the balancing controller only regulates the lower body of the robot, while the upper body can be subjected to another controller. For instance, we used a compliance controller for the upper body in Ott et al. (2013) to facilitate the kinesthetic teaching of a motion to the robot. The external forces and torques arising during the teaching and manipulation process are estimated by a disturbance observer and provided to the balance controller via a disturbance input $\boldsymbol{F}_{\text{int}}^{\text{obsv}}$. The balance controller out-

B. Henze, *Whole-Body Control for Multi-Contact Balancing of Humanoid Robots*, Springer Tracts in Advanced Robotics 143,
https://doi.org/10.1007/978-3-030-87212-0_8

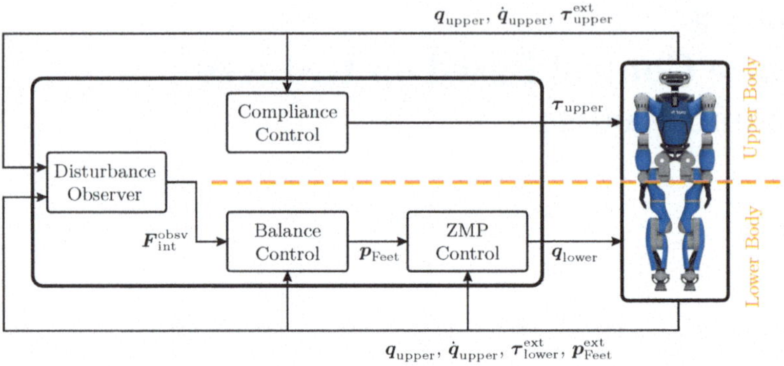

Fig. 8.1 Overview of the architecture of the interaction aware balancer

puts a desired ZMP position $\boldsymbol{p}_{\text{Feet}}$, which is implemented by a position-based ZMP controller, as in Krause et al. (2012). The presented approach is inspired by Ibanez et al. (2012), where a Cart-Table Model (Kajita et al. 2003) is used to derive the MPC controller.

The section is structured as follows: The model of the linear inverted pendulum is introduced in Sect. 8.1.1. The disturbance observer and the balance controller are detailed in Sects. 8.1.2 and 8.1.3. The ZMP controller is briefly reviewed in Sect. 8.1.4.

8.1.1 Linear Inverted Pendulum Model

While balancing and walking on a flat and horizontal floor, the dynamics of a humanoid robot is dominated by the horizontal translation of the center of mass. Based on this insight, Kajita et al. (2001) derived the Linear Inverted Pendulum Model (LIPM) from the general dynamic model (3.20) by isolating the CoM dynamics, incorporating the assumption of a constant height, and neglecting the angular momentum of the robot about the CoM. Alternatively, the model can also be derived by analyzing the angular momentum about the world frame $\mathcal{W}$. Without loss of generality, let us define $\mathcal{W}$ such that the origin is located on the floor surface underneath the setpoint of the CoM. The orientation of the world frame is chosen such that the x- and y-axis are parallel to the floor, as shown in Fig. 8.2. Furthermore, the LIPM requires the following assumptions:

- It is assumed that only the feet serve as support contacts. As the floor is flat and horizontal, one can compute a combined center of pressure $\boldsymbol{p}_{\text{Feet}} \in \mathbb{R}^3$ from both contact wrenches $\boldsymbol{F}_{\text{sup}}$, which lies on the surface of the floor.
- Furthermore, it is assumed that each one of the support wrenches stays well within the contact model given in Sect. 3.4, such that it can be omitted. Note that this

Fig. 8.2 Linear Inverted Pendulum Model (LIPM)

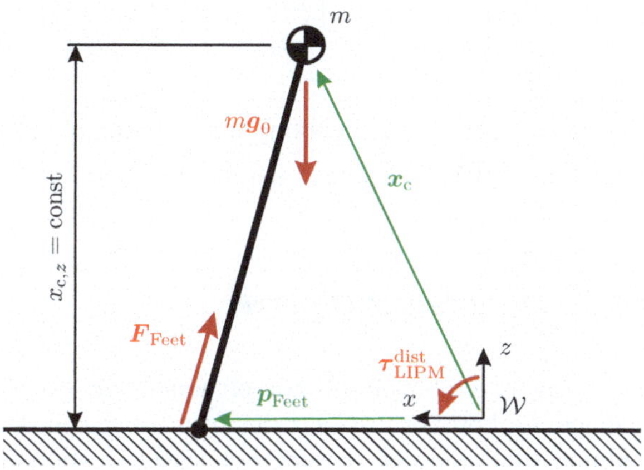

assumption is more restrictive than Assumption 3.1, which states that only the combined support wrench $\boldsymbol{F}^{\text{res}}_{\text{sup}}$ must be feasible. In contrast to the control approaches presented in Chaps. 6 and 7, the controller derived in this section cannot redistribute the load between the contacts if one of them is close to an overload.

The angular momentum $\boldsymbol{L}_c \in \mathbb{R}^3$ of the CoM about the world frame $\mathcal{W}$ is given by

$$\boldsymbol{L}_c = m\, x_{c,z} \begin{pmatrix} -\dot{x}_{c,y} \\ \dot{x}_{c,x} \\ 0 \end{pmatrix} \tag{8.1}$$

with $x_{c,z}$ denoting the height of the inverted pendulum. Note that the angular momentum of the robot about the CoM is neglected. Assuming a constant height, the conservation of angular momentum (Gross et al. 2014) states that

$$\dot{\boldsymbol{L}}_c = m\, x_{c,z} \begin{pmatrix} -\ddot{x}_{c,y} \\ \ddot{x}_{c,x} \\ 0 \end{pmatrix} = \boldsymbol{S}(\boldsymbol{p}_{\text{Feet}})\boldsymbol{f}^{\text{res}}_{\text{sup}} + \boldsymbol{S}(\boldsymbol{x}_c) \begin{pmatrix} 0 \\ 0 \\ -m \left\|\boldsymbol{g}_0\right\| \end{pmatrix} + \boldsymbol{\tau}^{\text{LIPM}}_{\text{dist}} \tag{8.2}$$

with the right-hand side specifying the sum of all torques about $\mathcal{W}$. The cross-product matrix $\boldsymbol{S}(\bullet)$ is given in Definition 3.1. Note that the external forces and torques acting on the robot are split into the overall support force $\boldsymbol{f}^{\text{res}}_{\text{sup}}$ and into a torque $\boldsymbol{\tau}^{\text{LIPM}}_{\text{dist}} = (\, \tau^{\text{LIPM}}_{\text{dist},x}\ \tau^{\text{LIPM}}_{\text{dist},y}\ 0\,)^T$ about the origin of $\mathcal{W}$ comprising the external disturbances. The above equation can be simplified into

$$m\, x_{c,z} \begin{pmatrix} -\ddot{x}_{c,y} \\ \ddot{x}_{c,x} \end{pmatrix} = m \left\|\boldsymbol{g}_0\right\| \begin{pmatrix} x_{c,y} - p_{\text{Feet},y} \\ -x_{c,x} + p_{\text{Feet},x} \end{pmatrix} + \begin{pmatrix} \tau^{\text{LIPM}}_{\text{dist},x} \\ \tau^{\text{LIPM}}_{\text{dist},y} \end{pmatrix} \tag{8.3}$$

by omitting the third line and considering that $f^{\text{res}}_{\text{sup},z} = m \left\|\boldsymbol{g}_0\right\|$. Reordering finally leads to the Linear Inverted Pendulum Model (LIPM)

$$\begin{pmatrix} \ddot{x}_{c,x} \\ \ddot{x}_{c,y} \end{pmatrix} = \omega_0^2 \left\{ \begin{pmatrix} p_{\text{Feet},x} \\ p_{\text{Feet},y} \end{pmatrix} - \begin{pmatrix} x_{c,x} \\ x_{c,y} \end{pmatrix} \right\} + \frac{1}{m\, x_{c,z}} \begin{pmatrix} \tau_{\text{dist},y}^{\text{LIPM}} \\ -\tau_{\text{dist},x}^{\text{LIPM}} \end{pmatrix} \tag{8.4}$$

with $\omega_0 = \sqrt{\left\| \boldsymbol{g}_0 \right\| / x_{c,z}}$ denoting the eigenfrequency of the LIPM. Note that (8.4) extends the LIPM by Kajita et al. (2001) with the additional disturbance input $\boldsymbol{\tau}_{\text{dist}}^{\text{LIPM}}$.

8.1.2 *Disturbance Observer*

There are several methods for obtaining the external disturbances acting on a humanoid robot: For instance, the disturbance can be explicitly measured by using additional force-torque sensors if the exact location of the disturbance is known. If the disturbance is caused by an upper body controller as in Ott et al. (2013), then it could also be estimated from the forces and torques commanded to the upper body (e. g. from $\boldsymbol{F}_{\text{int}}^{\text{cpl}}$ in the case of a Cartesian compliance). However, to be independent of additional sensors or information on the upper body controller, in Ott et al. (2013) we suggested employing a momentum-based disturbance observer as proposed by De Luca et al. (2006). The outline of the observer is given in Fig. 8.3.

The generalized momentum of the complete robot is given by $\boldsymbol{\chi} = \boldsymbol{M}\boldsymbol{\nu} \in \mathbb{R}^{6+n}$ considering all $6+n$ DoF of the robot. The time derivative

$$\begin{aligned} \dot{\boldsymbol{\chi}} &= \boldsymbol{C}^T\boldsymbol{\nu} - \boldsymbol{g} + \begin{pmatrix} \boldsymbol{0} \\ \boldsymbol{\tau} \end{pmatrix} + \boldsymbol{\tau}_{\text{ext}} \\ &= \boldsymbol{C}^T\boldsymbol{\nu} - \boldsymbol{g} + \begin{pmatrix} \boldsymbol{0} \\ \boldsymbol{\tau} \end{pmatrix} + \boldsymbol{J}_{\text{sup}}^T \boldsymbol{F}_{\text{sup}}^{\text{ext}} + \boldsymbol{\tau}_{\text{dist}} \end{aligned} \tag{8.5}$$

can be computed by exploiting the robot dynamics (3.20) and the passivity property $\dot{\boldsymbol{M}} = \boldsymbol{C} + \boldsymbol{C}^T$ (Ott 2008). Note that the generalized vector of external forces and torques $\boldsymbol{\tau}_{\text{ext}} = \boldsymbol{J}_{\text{sup}}^T \boldsymbol{F}_{\text{sup}}^{\text{ext}} + \boldsymbol{\tau}_{\text{dist}}$ is split into the wrench at the support end-

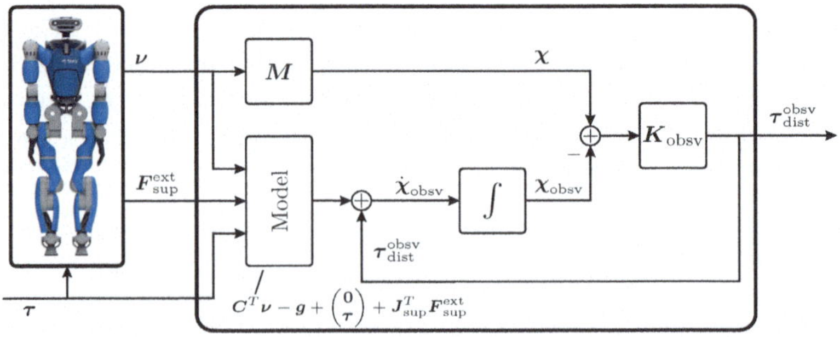

Fig. 8.3 Architecture of the disturbance observer

effectors $\boldsymbol{F}_{\text{sup}}^{\text{ext}}$ and into the vector of external disturbances $\boldsymbol{\tau}_{\text{dist}} \in \mathbb{R}^{6+n}$ similar to (3.24). It is assumed that the quantities $\boldsymbol{\tau}$, $\boldsymbol{F}_{\text{sup}}^{\text{ext}}$, and $\boldsymbol{\nu}$ can be directly measured or derived from measurements in contrast to $\boldsymbol{\tau}_{\text{dist}}$, which is to be reconstructed by the observer. Note that in the case of the humanoid robot *TORO*, $\boldsymbol{F}_{\text{sup}}^{\text{ext}}$ can be obtained from the force-torque sensors in the feet and $\boldsymbol{\tau}$ from the torque sensors in each joint (see Chap. 5). Based on an estimated disturbance $\boldsymbol{\tau}_{\text{dist}}^{\text{obsv}} \in \mathbb{R}^{6+n}$, we can compute the time derivative

$$\dot{\boldsymbol{\chi}}_{\text{obsv}} = \boldsymbol{C}^T \boldsymbol{\nu} - \boldsymbol{g} + \begin{pmatrix} \boldsymbol{0} \\ \boldsymbol{\tau} \end{pmatrix} + \boldsymbol{J}_{\text{sup}}^T \boldsymbol{F}_{\text{sup}}^{\text{ext}} + \boldsymbol{\tau}_{\text{dist}}^{\text{obsv}} \tag{8.6}$$

of the estimated momentum $\dot{\boldsymbol{\chi}}_{\text{obsv}} \in \mathbb{R}^{6+n}$. The derivative $\dot{\boldsymbol{\chi}}_{\text{obsv}}$ can be integrated over time and compared to the actual generalized momentum $\boldsymbol{\chi} = \boldsymbol{M}\boldsymbol{\nu}$, which is based on a measurement of $\boldsymbol{\nu}$. The resulting error is fed back using the diagonal and positive definite gain matrix $\boldsymbol{K}_{\text{obsv}} \in \mathbb{R}^{(6+n)\times(6+n)}$ according to

$$\begin{aligned}
\boldsymbol{\tau}_{\text{dist}}^{\text{obsv}} &= \boldsymbol{K}_{\text{obsv}} \left\{ \boldsymbol{\chi} - \boldsymbol{\chi}_{\text{obsv}} \right\} \\
&= \boldsymbol{K}_{\text{obsv}} \left\{ \boldsymbol{M}\boldsymbol{\nu} - \int_0^t \dot{\boldsymbol{\chi}}_{\text{obsv}}(t)\, dt - \boldsymbol{\chi}_{\text{obsv}}(0) \right\} \\
&= \boldsymbol{K}_{\text{obsv}} \left\{ \boldsymbol{M}\boldsymbol{\nu} - \int_0^t \left(\boldsymbol{C}^T \boldsymbol{\nu} - \boldsymbol{g} + \begin{pmatrix} \boldsymbol{0} \\ \boldsymbol{\tau} \end{pmatrix} + \boldsymbol{J}_{\text{sup}}^T \boldsymbol{F}_{\text{sup}}^{\text{ext}} + \boldsymbol{\tau}_{\text{dist}}^{\text{obsv}} \right) dt - \boldsymbol{\chi}_{\text{obsv}}(0) \right\}.
\end{aligned} \tag{8.7}$$

By inserting (8.5) and (8.6) into (8.7), it can be verified that the observer shows a stable first-order dynamics $\dot{\boldsymbol{\tau}}_{\text{dist}}^{\text{obsv}} = \boldsymbol{K}_{\text{obsv}} \left(\boldsymbol{\tau}_{\text{dist}} - \boldsymbol{\tau}_{\text{dist}}^{\text{obsv}} \right)$. Thus, $\boldsymbol{\tau}_{\text{dist}}^{\text{obsv}}$ represents a filtered estimate for the generalized vector $\boldsymbol{\tau}_{\text{dist}}$ of external disturbances (De Luca et al. 2006).

If we assume that the external disturbances exclusively act on the interaction end-effectors of the robot (in many cases the hands), then

$$\boldsymbol{\tau}_{\text{dist}}^{\text{obsv}} = \boldsymbol{J}_{\text{int}}^T \boldsymbol{F}_{\text{int}}^{\text{obsv}} \tag{8.8}$$

relates $\boldsymbol{\tau}_{\text{dist}}^{\text{obsv}}$ to the estimated interaction wrenches $\boldsymbol{F}_{\text{int}}^{\text{obsv}}$ (see also (3.24)). Thus, the latter can be obtained via

$$\boldsymbol{F}_{\text{int}}^{\text{obsv}} = \boldsymbol{J}_{\text{int}}^{\boldsymbol{Q}_{\text{obsv}}+,T} \boldsymbol{\tau}_{\text{ext}}^{\text{obsv}} \tag{8.9}$$

using the weighted pseudoinverse $\boldsymbol{J}_{\text{int}}^{\boldsymbol{Q}_{\text{obsv}}+} = \boldsymbol{Q}_{\text{obsv}}^{-1} \boldsymbol{J}_{\text{int}}^T \left(\boldsymbol{J}_{\text{int}} \boldsymbol{Q}_{\text{obsv}} \boldsymbol{J}_{\text{int}}^T \right)^{-1}$.

8.1.3 Controller Derivation

In order to employ Model Predictive Control, we discretized the LIPM (8.4) in Ott et al. (2013) using the Euler polygonal method (Bronshtein et al. 2015) leading to

$$\underbrace{\begin{pmatrix} x_{c,x}[l+1] \\ x_{c,y}[l+1] \\ \dot{x}_{c,x}[l+1] \\ \dot{x}_{c,y}[l+1] \end{pmatrix}}_{\hat{\xi}[l+1]} = \underbrace{\begin{bmatrix} 1 & 0 & T & 0 \\ 0 & 1 & 0 & T \\ \omega_0^2 T & 0 & 1 & 0 \\ 0 & \omega_0^2 T & 0 & 1 \end{bmatrix}}_{\hat{A}[l]} \underbrace{\begin{pmatrix} x_{c,x}[l] \\ x_{c,y}[l] \\ \dot{x}_{c,x}[l] \\ \dot{x}_{c,y}[l] \end{pmatrix}}_{\hat{\xi}[l]} + \underbrace{\begin{bmatrix} 0 & 0 \\ 0 & 0 \\ -\omega_0^2 T & 0 \\ 0 & -\omega_0^2 T \end{bmatrix}}_{\hat{B}[l]} \underbrace{\begin{pmatrix} p_{\text{Feet},x}[l] \\ p_{\text{Feet},y}[l] \end{pmatrix}}_{\hat{u}[l]}$$
$$+ \underbrace{\begin{bmatrix} 0 & 0 \\ 0 & 0 \\ 0 & T/(m\, x_{c,z}) \\ -T/(m\, x_{c,z}) & 0 \end{bmatrix}}_{\hat{E}[l]} \underbrace{\begin{pmatrix} \tau_{\text{dist},x}^{\text{LIPM}} \\ \tau_{\text{dist},y}^{\text{LIPM}} \end{pmatrix}}_{\hat{e}[l]} \tag{8.10}$$

based on the sampling rate $T \in \mathbb{R}$. Note that all quantities are given as a function $\bullet\,[l]$ of the current time step $l \in \mathbb{N}_0$. The evolution of the system state $\hat{\xi}\,[l]$ is characterized by the matrix $\hat{A}\,[l]$. The control input $\hat{u}\,[l]$ and the disturbance input $\hat{e}\,[l]$ are mapped via the matrices $\hat{B}\,[l]$ and $\hat{E}\,[l]$ to $\hat{\xi}\,[l+1]$.

In order to predict the behavior of the LIPM, the system state $\hat{\xi}\,[l]$ must be predicted for the next $N \in \mathbb{N}$ time steps. As the matrices $\hat{A}\,[l]$, $\hat{B}\,[l]$, and $\hat{E}\,[l]$ are constant, the prediction can be obtained by iteratively inserting (8.10) into itself, which leads to

$$\underbrace{\begin{pmatrix} \hat{\xi}[l+1] \\ \vdots \\ \hat{\xi}[l+N] \end{pmatrix}}_{\hat{\xi}} = \underbrace{\begin{bmatrix} \hat{A}[l] \\ \vdots \\ \hat{A}[l]^N \end{bmatrix}}_{\hat{A}} \hat{\xi}[l] + \underbrace{\begin{bmatrix} \hat{B}[l] & \dots & \mathbf{0} \\ \vdots & \ddots & \vdots \\ \hat{A}[l]^{N-1}\hat{B}[l] & \dots & \hat{B}[l] \end{bmatrix}}_{\hat{B}} \underbrace{\begin{pmatrix} \hat{u}[l] \\ \vdots \\ \hat{u}[l+N-1] \end{pmatrix}}_{\hat{u}}$$
$$+ \underbrace{\begin{bmatrix} \hat{E}[l] & \dots & \mathbf{0} \\ \vdots & \ddots & \vdots \\ \hat{A}[l]^{N-1}\hat{E}[l] & \dots & \hat{E}[l] \end{bmatrix}}_{\hat{E}} \underbrace{\begin{pmatrix} \hat{e}[l] \\ \vdots \\ \hat{e}[l+N-1] \end{pmatrix}}_{\hat{e}}. \tag{8.11}$$

The goal of the controller is to minimize the quadratic cost function

$$\hat{u}^{\text{opt}} = \underset{\hat{u}}{\operatorname{argmin}} \frac{1}{2}\hat{\xi}^T Q_{\hat{\xi}}\,\hat{\xi} + \frac{1}{2}\hat{u}^T Q_{\hat{u}}\,\hat{u} + \frac{1}{2}\delta_{\hat{e}}^T Q_{\hat{e}}\,\delta_{\hat{e}} \tag{8.12}$$

with respect to the extrapolated dynamics (8.11) and the constraints

$$\delta_{\hat{e}} = \hat{B}\,\hat{u} + \hat{E}\,\hat{e}, \tag{8.13}$$
$$\hat{\xi}\,[l+N] = \hat{\xi}_{\text{end}}. \tag{8.14}$$

The weighting matrices $Q_{\hat{\xi}} \in \mathbb{R}^{4N\times 4N}$, $Q_{\hat{u}} \in \mathbb{R}^{2N\times 2N}$, and $Q_{\hat{e}} \in \mathbb{R}^{2N\times 2N}$ are symmetric and positive definite. An error of the state vector $\hat{\xi}$ is weighted with $Q_{\hat{\xi}}$, while the control input $\hat{\boldsymbol{u}}$ is penalized with $\boldsymbol{Q}_{\hat{u}}$. The terminal constraint (8.14) enforces the stability of the closed-loop system (see Maciejowski (2002)).

The disturbance rejection of the controller can be parameterized via the weighting matrix $\boldsymbol{Q}_{\hat{e}}$ and the soft constraint (8.13). The latter demands a static equilibrium between the position $\boldsymbol{p}_{\text{Feet}}$ of the CoP and the disturbance $\boldsymbol{\tau}_{\text{dist}}^{\text{LIPM}}$ over the full length of the prediction. For instance, let us consider the case of an external force that acts at the CoM pointing to the back of the robot. To compensate for the resulting torque $\boldsymbol{\tau}_{\text{dist}}^{\text{LIPM}}$, the controller will move the CoP to the back, as shown in Fig. 8.2. Note that only the disturbance observer from Sect. 8.1.2 only provides $\boldsymbol{\tau}_{\text{dist}}^{\text{LIPM}}$ for the current point in time, whereas future disturbances are unknown. Thus, a disturbance model is used to predict future disturbances by assuming that the disturbance is constant over the complete time span of the prediction ($\hat{\boldsymbol{e}}\,[l+N] = \cdots = \hat{\boldsymbol{e}}\,[l]$).

Note that the vector $\hat{\boldsymbol{u}}^{\text{opt}}$ contains the feedforward control input from the current point in time l until the end of the prediction $l+N$. In order to obtain the control law, the feedback loop must be closed by extracting the current control input $\hat{\boldsymbol{u}}^{\text{opt}}\,[l]$ and recomputing it at every time step. Note that the optimization problem (8.12)–(8.14) can be solved analytically due to the absence of inequality constraints, which results in a control law of the form

$$\hat{\boldsymbol{u}}^{\text{opt}}\,[l] = \boldsymbol{K}_{\hat{\xi}}\,\hat{\boldsymbol{\xi}}\,[l] + \boldsymbol{K}_{\text{end}}\,\hat{\boldsymbol{\xi}}_{\text{end}} + \boldsymbol{K}_{\hat{e}}\,\hat{\boldsymbol{e}}\,[l]\,. \tag{8.15}$$

As the gain matrices $\boldsymbol{K}_{\hat{\xi}} \in \mathbb{R}^{2\times 4}$, $\boldsymbol{K}_{\text{end}} \in \mathbb{R}^{2\times 4}$ and $\boldsymbol{K}_{\hat{e}} \in \mathbb{R}^{2\times 2}$ are constant, the optimization can be solved offline.

The asymptotic stability of the resulting closed-loop dynamics can be shown by applying (Maciejowski 2002, Theorem 6.1), which requires a terminal constraint for the optimization problem and a feasible solution. Since there always exists a feasible solution for (8.12)–(8.14), the presented controller is asymptotically stable.

If we assume that the external disturbances act solely on the interaction end-effectors (here the hands), then

$$\hat{\boldsymbol{e}}\,[l] = \begin{pmatrix} \tau_{\text{dist},x}^{\text{LIPM}} \\ \tau_{\text{dist},y}^{\text{LIPM}} \end{pmatrix} = \text{diag}(1,1,0,0,0,0)\ \boldsymbol{Ad}(\boldsymbol{I},\boldsymbol{x}_{\text{c}})^T\ \boldsymbol{Ad}_{\text{int}}^T\ \boldsymbol{F}_{\text{int}}^{\text{obsv}} \tag{8.16}$$

can be computed from $\boldsymbol{F}_{\text{int}}^{\text{obsv}}$ in (8.9). Here, the transpose of $\boldsymbol{Ad}_{\text{int}}\,\boldsymbol{Ad}(\boldsymbol{I},\boldsymbol{x}_{\text{c}})$ is used to map $\boldsymbol{F}_{\text{int}}^{\text{obsv}}$ via the CoM frame C to the world frame $\mathcal{W}$. The diagonal selection matrix is used to isolate the torque about the x- and y-axis of $\mathcal{W}$.

8.1.4 Position-Based ZMP Controller

The presented ZMP controller by Krause et al. (2012) is used to convert the desired ZMP position $\boldsymbol{p}_{\text{Feet}}$ resulting from the balance controller in Sect. 8.1.3 into a desired joint position commanded to the lower body of the robot (see Fig. 8.1). The actual

ZMP position $\boldsymbol{p}_{\text{Feet}}^{\text{ext}}$ can be obtained from the six-dimensional force-torque sensors in the feet of *TORO* (see Chap. 5). Under the assumption of a linear inverted pendulum model (8.4) for the center of mass dynamics, the ZMP can be related to an equivalent force vector $(\, f_{c,x}^{\text{ext}} \; f_{c,y}^{\text{ext}}\,)$ on the CoM:

$$\begin{pmatrix} f_{c,x}^{\text{ext}} \\ f_{c,y}^{\text{ext}} \end{pmatrix} = m \begin{pmatrix} \ddot{x}_{c,x} \\ \ddot{x}_{c,y} \end{pmatrix} = m\,\omega_0^2 \left\{ \begin{pmatrix} p_{\text{Feet},x}^{\text{ext}} \\ p_{\text{Feet},y}^{\text{ext}} \end{pmatrix} - \begin{pmatrix} x_{c,x} \\ x_{c,y} \end{pmatrix} \right\} + \frac{1}{x_{c,z}} \begin{pmatrix} \tau_{\text{dist},y}^{\text{LIPM}} \\ -\tau_{\text{dist},x}^{\text{LIPM}} \end{pmatrix}. \quad (8.17)$$

It is well-established that a position-based force set point regulator can be implemented by an integral control action (Roy and Whitcomb 2002)

$$\begin{pmatrix} x_{c,d,x} \\ x_{c,d,y} \end{pmatrix} = \boldsymbol{K}_{\text{ZMP}} \int_0^t \left\{ \begin{pmatrix} f_{c,d,x} \\ f_{c,d,y} \end{pmatrix} - \begin{pmatrix} f_{c,x}^{\text{ext}} \\ f_{c,y}^{\text{ext}} \end{pmatrix} \right\} dt + \begin{pmatrix} x_{c,d,x}(t=0) \\ x_{c,d,y}(t=0) \end{pmatrix} \quad (8.18)$$

with $\boldsymbol{K}_{\text{ZMP}} \in \mathbb{R}^{2\times 2}$ denoting a positive definite gain matrix and $(\, f_{c,d,x} \; f_{c,d,y}\,)^T$ a vector of desired horizontal forces on the CoM. Inserting (8.17) for the actual and the desired force yields

$$\begin{pmatrix} x_{c,d,x} \\ x_{c,d,y} \end{pmatrix} = m\,\omega_0^2\, \boldsymbol{K}_{\text{ZMP}} \int_0^t \left\{ \begin{pmatrix} p_{\text{Feet},x} \\ p_{\text{Feet},y} \end{pmatrix} - \begin{pmatrix} p_{\text{Feet},x}^{\text{ext}} \\ p_{\text{Feet},y^{\text{ext}}} \end{pmatrix} \right\} dt + \begin{pmatrix} x_{c,d,x}(t=0) \\ x_{c,d,y}(t=0) \end{pmatrix}, \quad (8.19)$$

which can be converted via inverse kinematics into a command for the desired joint position $\boldsymbol{q}_{\text{lower}}$ of the lower body.

8.1.5 Experimental Evaluation

To evaluate the presented balancing controller, we reported on an experiment in Ott et al. (2013) with the humanoid robot *TORO* (see Sect. 5). The balancing controller was in charge of the lower body, while the upper body was regulated by a Cartesian compliance controller, which stabilized the position and orientation of the hands with respect to the hip of the robot. To obtain a reproducible force generation, the robot was placed in front of a table (see Fig. 8.4a), which served as a stationary obstacle for the arms. By commanding both hands 9 cm to the front (relative to the hip), the robot was programmed to push against the table, hold the commanded position for 5 s, and then move back again. This experiment was conducted once for the balancing controller, taking into account the external forces resulting from the contact with the table ($\boldsymbol{K}_{\hat{e}} \neq \boldsymbol{0}$) and once without accounting for the external forces ($\boldsymbol{K}_{\hat{e}} = \boldsymbol{0}$).

The results are shown in Fig. 8.4b, c: The excursion of the combined CoP $\boldsymbol{p}_{\text{Feet}}$ is higher for $\boldsymbol{K}_{\hat{e}} \neq \boldsymbol{0}$ than for $\boldsymbol{K}_{\hat{e}} = \boldsymbol{0}$, which is consistent with the observation that the contact force is higher as well. Note that the force resulting from the position difference of CoP and CoM must be equal to the contact force. The most important fact is that the CoM moves about 1.5 cm to the front if $\boldsymbol{K}_{\hat{e}} \neq \boldsymbol{0}$ in contrast to 3.0 cm to the back if $\boldsymbol{K}_{\hat{e}} = \boldsymbol{0}$. This means that by taking the disturbances into account, the influence on the CoM can be reduced significantly.

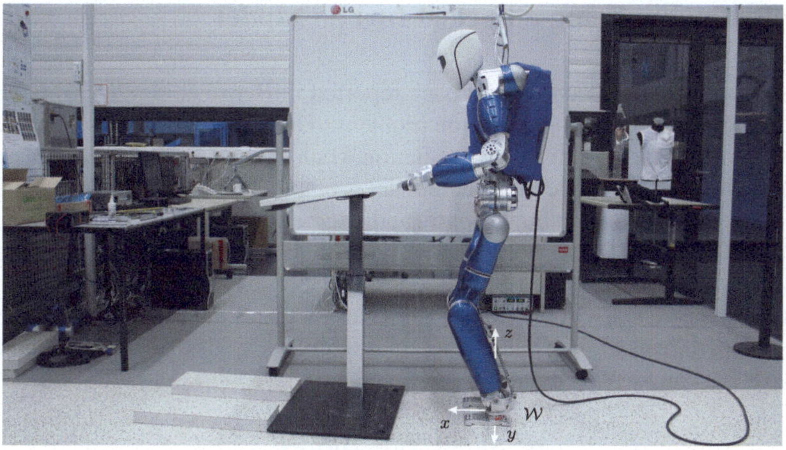

(a) Setup of the experiment.

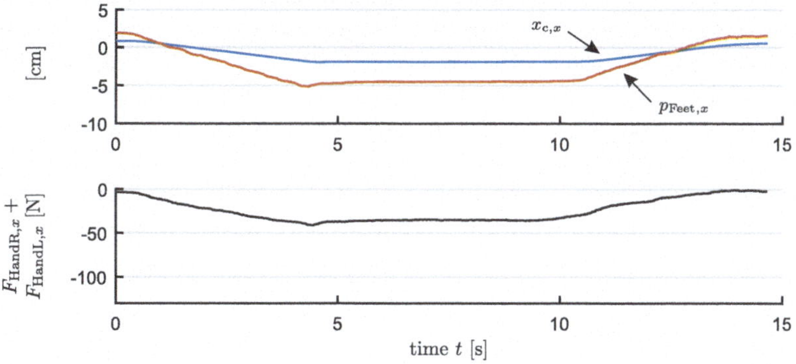

(b) Control error with the consideration of external disturbances ($\boldsymbol{K}_{\hat{\boldsymbol{e}}} \neq \boldsymbol{0}$).

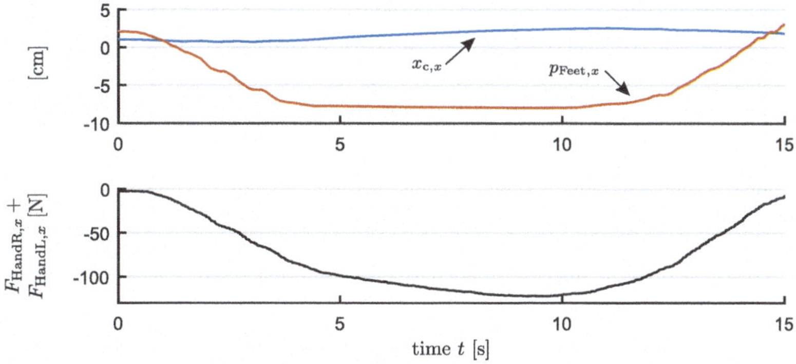

(c) Control error without the consideration of external disturbances ($\boldsymbol{K}_{\hat{\boldsymbol{e}}} = \boldsymbol{0}$).

Fig. 8.4 Comparison of the control error with and without the consideration of external disturbances

8.2 Balancing Using a Locked Inertia Model

This section presents a control approach reported in Henze et al. (2014a) based on MPC, which requires the optimization problem to be solved online. Using the complete dynamic model (3.20) for the prediction would consume a lot of computational power, therefore a simpler conceptual model is employed to speed up the prediction process. Let us assume that both support and interaction end-effectors maintain their posture with respect to the world frame $\mathcal{W}$. The dynamics of the robot is approximated via the Locked Inertia Model (Murray 1997; Nenchev et al. 2019), which only considers the DoFs of the CoM frame C and neglects the DoFs in joint space. In other words, the robot is considered one single rigid body. Thus, the general dynamic model (3.20) can be reduced to

$$\boldsymbol{M}_{\mathrm{cc}}\, \dot{\boldsymbol{v}}_{\mathrm{c}} = m\boldsymbol{g}_0 + \boldsymbol{Ad}_{\mathrm{sup}}^T\, \boldsymbol{F}_{\mathrm{sup}} \tag{8.20}$$

by isolating the DoFs of the CoM frame C and omitting negligible terms, that is the Coriolis/centrifugal effects and $\boldsymbol{M}_{\mathrm{cq}}$. Furthermore, it is assumed that all task frames $\mathcal{T}_i$ are located at the end-effectors as in Chap. 6. Besides, all end-effectors are used to support the robot, which leads to $\Psi_{\mathrm{sup}} = \Psi$ and $\Psi_{\mathrm{int}} = 0$. The associated support wrenches $\boldsymbol{F}_{\mathrm{sup}} \in \mathbb{R}^{6\Psi_{\mathrm{sup}}}$ are used as control input to stabilize the CoM frame C.

An overview of the architecture of the presented controller is given in Fig. 8.5. The MPC balancer can be combined with the Null space controller presented in Sect. 6.2 to deal with singular configurations and/or redundant kinematics.

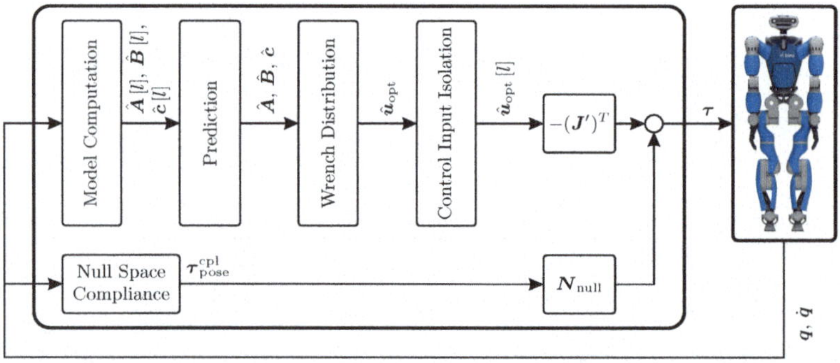

Fig. 8.5 Overview of the architecture of the balancing controller based on the locked inertia model

8.2.1 Controller Derivation

In order to prepare the discretization of the reduced model (8.20), let us describe the configuration of the CoM frame C with a vector $\boldsymbol{\xi}_c = (\boldsymbol{x}_c^T \ \boldsymbol{\phi}_c^T)^T$, which combines the CoM position $\boldsymbol{x}_c$ with $\boldsymbol{\phi}_c \in \mathbb{R}^3$ describing the orientation of C in Euler angles. Note that Euler angles represent a minimal parameterization, although it is not singularity free (Shuster 1993). The exact location of the singularity in $SO(3)$ depends on the order and axis of the basic rotations forming the three Euler angles. We decided in Henze et al. (2014a) to use the consecutive roll-pitch-yaw notation of the Euler angles, which shows a singularity for a pitch angle of $\pm 90°$. However, any other notation is also admissible as long as there is sufficient distance between the expected CoM orientation $\boldsymbol{\phi}_c$ and the singularity. As $\boldsymbol{\phi}_c(\boldsymbol{R}_c)$ is a function of the rotation matrix $\boldsymbol{R}_c$, a linear mapping $\dot{\boldsymbol{\phi}}_c = \boldsymbol{\Phi}'_c(\boldsymbol{\phi}_c)\,\boldsymbol{\omega}_c$ with $\boldsymbol{\Phi}'_c(\boldsymbol{\phi}_c) \in \mathbb{R}^{3\times 3}$ can be found linking $\dot{\boldsymbol{\phi}}_c$ with the angular velocity $\boldsymbol{\omega}_c$ (Shuster 1993). Let us generalize the mapping in the form of $\dot{\boldsymbol{\xi}}_c = \boldsymbol{\Phi}(\boldsymbol{\xi}_c)\,\boldsymbol{v}_c$ connecting $\dot{\boldsymbol{\xi}}_c$ with the six-dimensional Cartesian velocity $\boldsymbol{v}_c$. Based on this notation, the continuous model (8.20) can be discretized in time using the Euler polygonal method (Bronshtein et al. 2015) with $T \in \mathbb{R}$ denoting the sampling rate:

$$
\underbrace{\begin{pmatrix} \boldsymbol{\xi}_c[l+1] \\ \boldsymbol{v}_c[l+1] \end{pmatrix}}_{\hat{\boldsymbol{\xi}}[l+1]} = \underbrace{\begin{bmatrix} \boldsymbol{I} & T\,\boldsymbol{\Phi}_c[l] \\ \boldsymbol{0} & \boldsymbol{I} \end{bmatrix}}_{\hat{\boldsymbol{A}}[l]} \underbrace{\begin{pmatrix} \boldsymbol{\xi}_c[l] \\ \boldsymbol{v}_c[l] \end{pmatrix}}_{\hat{\boldsymbol{\xi}}[l]} + \underbrace{\begin{bmatrix} \boldsymbol{0} \\ T\,\boldsymbol{M}_{cc}^{-1}[l]\,\boldsymbol{Ad}^T[l] \end{bmatrix}}_{\hat{\boldsymbol{B}}[l]} \underbrace{\boldsymbol{F}_{sup}[l]}_{\hat{\boldsymbol{u}}[l]} + \underbrace{\begin{pmatrix} \boldsymbol{0} \\ T\,\boldsymbol{M}_{cc}^{-1}[l]\,\boldsymbol{g}_0 \end{pmatrix}}_{\hat{\boldsymbol{c}}[l]} \tag{8.21}
$$

Again, $\bullet\,[l]$ indicates that all quantities are a function of the current time step $l \in \mathbb{N}_0$. To provide the controller with a prediction for the next $N \in \mathbb{N}_0$ time steps, the state vector is extrapolated to

$$
\underbrace{\begin{pmatrix} \hat{\boldsymbol{\xi}}[l+1] \\ \vdots \\ \hat{\boldsymbol{\xi}}[l+N] \end{pmatrix}}_{\hat{\boldsymbol{\xi}}} = \underbrace{\begin{bmatrix} \hat{\boldsymbol{A}}[l] \\ \vdots \\ \hat{\boldsymbol{A}}[l]^N \end{bmatrix}}_{\hat{\boldsymbol{A}}} \hat{\boldsymbol{\xi}}[l] + \underbrace{\begin{bmatrix} \hat{\boldsymbol{B}}[l] & \dots & \boldsymbol{0} \\ \vdots & \ddots & \vdots \\ \hat{\boldsymbol{A}}[l]^{N-1}\,\hat{\boldsymbol{B}}[l] & \dots & \hat{\boldsymbol{B}}[l] \end{bmatrix}}_{\hat{\boldsymbol{B}}} \underbrace{\begin{pmatrix} \hat{\boldsymbol{u}}[l] \\ \vdots \\ \hat{\boldsymbol{u}}[l+N-1] \end{pmatrix}}_{\hat{\boldsymbol{u}}} + \underbrace{\begin{pmatrix} \hat{\boldsymbol{c}}[l] \\ \vdots \\ \hat{\boldsymbol{A}}[l]^{N-1}\,\hat{\boldsymbol{c}}[l] \end{pmatrix}}_{\hat{\boldsymbol{c}}}, \tag{8.22}
$$

which includes the assumption that $\boldsymbol{\Phi}_{\mathrm{c}}[l]$, $\boldsymbol{M}_{\mathrm{cc}}[l]$, and $\boldsymbol{Ad}[l]$ are constant during the time span of the prediction. This assumption can be justified by a sufficiently short prediction horizon with respect to the motion of the robot.

The MPC algorithm minimizes the difference between the predicted state $\hat{\boldsymbol{\xi}}$ and a desired state vector $\hat{\boldsymbol{\xi}}_d \in \mathbb{R}^{6N}$ as well as the difference between $\hat{\boldsymbol{u}}$ and a desired future input $\hat{\boldsymbol{u}}_d \in \mathbb{R}^{6N}$ by solving the following constrained quadratic optimization problem (cQP):

$$\hat{\boldsymbol{u}}^{\mathrm{opt}} = \underset{\hat{\boldsymbol{u}}}{\mathrm{argmin}} \left(\frac{1}{2} \boldsymbol{\delta}_{\hat{\xi}}^T \boldsymbol{Q}_{\hat{\xi}} \boldsymbol{\delta}_{\hat{\xi}} + \frac{1}{2} \boldsymbol{\delta}_{\hat{u}}^T \boldsymbol{Q}_{\hat{u}} \boldsymbol{\delta}_{\hat{u}} \right) \tag{8.23}$$

minimizing the residua

$$\boldsymbol{\delta}_{\hat{\xi}} = \hat{\boldsymbol{\xi}} - \hat{\boldsymbol{\xi}}_d \tag{8.24}$$

$$\boldsymbol{\delta}_{\hat{u}} = \hat{\boldsymbol{u}} - \hat{\boldsymbol{u}}_d \tag{8.25}$$

with respect to the extrapolated dynamics (8.22) and to the inequality constraints

$$\boldsymbol{A}_i[l]\ \boldsymbol{F}_i[l] \leq \boldsymbol{b}_i[l] \qquad \forall i = 1 \ldots \psi_{\mathrm{sup}}, \forall l = 1 \ldots N, \tag{8.26}$$

$$\boldsymbol{B}_\tau[l]\ \boldsymbol{F}[l] \leq \boldsymbol{\tau}^{\max} \qquad \forall l = 1 \ldots N. \tag{8.27}$$

The residua for the state vector $\boldsymbol{\delta}_{\hat{\xi}}$ and for the control input $\boldsymbol{\delta}_{\hat{u}}$ are penalized via the symmetric and positive definite weighting matrices $\boldsymbol{Q}_{\hat{\xi}} \in \mathbb{R}^{6N \times 6N}$ and $\boldsymbol{Q}_{\hat{u}} \in \mathbb{R}^{6N \times 6N}$, respectively. Note that the control input $\hat{\boldsymbol{u}} \in \mathbb{R}^{6N}$ comprises all end-effector wrenches $\boldsymbol{F}[l]$ for the complete time span of the prediction $(l \ldots l+N)$. The vector $\hat{\boldsymbol{u}}$ and the weighting matrix $\boldsymbol{Q}_{\hat{u}}$ can be used analogously to Sect. 6.1 to achieve the following tasks:

- Providing a default wrench distribution $\boldsymbol{F}_{\mathrm{sup}}^{\mathrm{def}}[l]\ \forall l \ldots l+N$ to the support end-effectors to regularize the wrench distribution.
- Ensuring that the wrenches of the support end-effectors match the Cartesian compliances $\boldsymbol{F}_{\mathrm{int}}^{\mathrm{cpl}}[l]\ \forall l \ldots l+N$. The latter requires $\boldsymbol{F}_{\mathrm{int}}^{\mathrm{cpl}}$ not only for the current but also for the future time steps, which can be remedied by assuming that $\boldsymbol{F}_{\mathrm{int}}^{\mathrm{cpl}}$ is constant during the time span of the prediction.

The inequality constraints (8.26) and (8.27) ensure the contact model for all support contacts and the actuator limitations during the complete time span of the prediction.

Note that the vector $\hat{\boldsymbol{u}}^{\mathrm{opt}}$ contains the feedforward control input from the current point in time l until the end of the prediction $l+N$. In order to close the feedback loop, the current control input $\hat{\boldsymbol{u}}^{\mathrm{opt}}[l]$ is extracted from $\hat{\boldsymbol{u}}^{\mathrm{opt}}$ and applied to the joints of the robot via

$$\boldsymbol{\tau}[l] = -(\boldsymbol{J}'[l])^T\ \hat{\boldsymbol{u}}^{\mathrm{opt}}[l] \tag{8.28}$$

analogously to (6.17).

8.2.2 Evaluation in Simulation

In Henze et al. (2014a), we reported on simulation results involving the humanoid robot TORO (see Chap. 5) to evaluate the presented controller. As shown in Fig. 8.6a, the scenario comprises the robot TORO standing in front of a wall using the hands and the feet as supporting contacts. The hands are considered point contacts, whereas the feet are able to transmit a complete six-dimensional wrench (see Sect. 3.4). The MPC controller comprises $N = 10$ time steps, which are $T = 50\,\text{ms}$ apart from each other, such that the prediction covers a time span of 0.5 s in total.

Two instantaneous jumps in the desired position and orientation of the CoM frame C are commanded to the robot to test the response of the system. The results

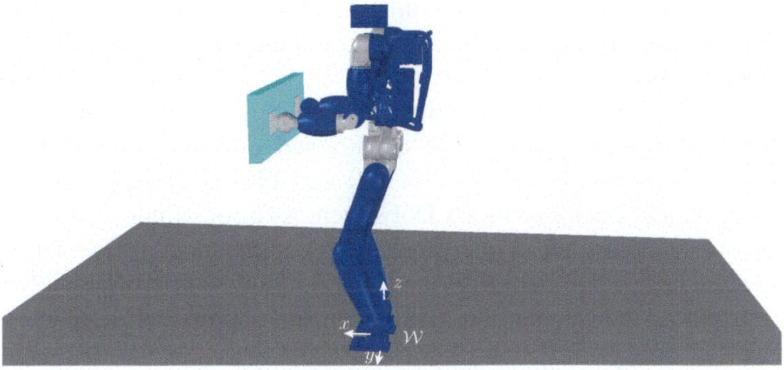

(a) Setup of the simulation.

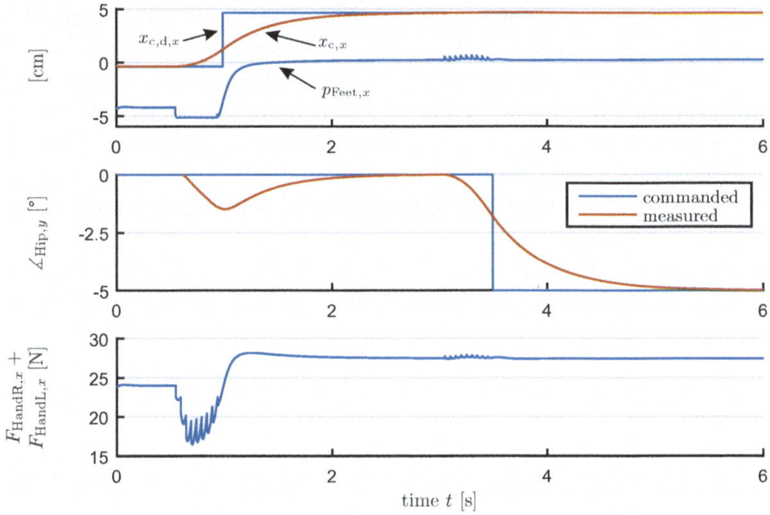

(b) Motion of the robot and hand forces during the simulation.

Fig. 8.6 Simulated response to a step in the desired CoM position and hip orientation

are shown in Fig. 8.6b. The first step occurs at $t = 1$ s, setting the desired CoM position 5 cm to the front. The controller reacts to the jump in $\hat{\boldsymbol{\xi}}_d$ by continuously moving the actual CoM to the new setpoint. What is interesting here is that the controller starts to react 0.5 s before the actual jump in $\hat{\boldsymbol{\xi}}_d$ takes place, which corresponds exactly to the length of the prediction. In order to accelerate, the controller moves the combined CoP of the feet $\boldsymbol{p}_{\text{Feet}}$ first to the front and then to the back to decelerate again. In order to generate the acceleration, the controller also lowers the contact forces in both hands to approximately 9 N each, although $\hat{\boldsymbol{u}}_d$ specifies a default contact force of 15 N. The steady state is reached at $t = 2.5$ s. The second jump occurs in the signal for the desired pitch angle of the CoM/hip at $t = 3.5$ s. Again, the controller reacts to the change in the desired position exactly 0.5 s prior to the actual jump in $\hat{\boldsymbol{\xi}}_d$. The steady state is reached 2.5 s afterwards. Note that the "oscillations" that can be seen in the plots of the hand forces are caused by the fact that the controller is executed at a rate of 1 ms, whereas the points in time used for the prediction are 50 ms apart from each other. This means that the vector $\hat{\boldsymbol{\xi}}_d$ containing the commanded step inputs changes every 50 ms.

The reason for validating the controller in simulation and not on the actual robot is the required computing power. The constrained quadratic optimization problem (8.23)–(8.27) comprises $N \cdot \Psi \cdot 6 = 240$ variables considering $N = 10$ time steps and $\Psi = 4$ end-effectors with six DoFs each. As a consequence, it was not possible to execute the controller in real-time on the computers of *TORO*. One method for reducing the computational effort could be the polyhedron of feasible support wrenches introduced in Sect. 3.6. The MPC algorithm optimizes the trajectory for the center of mass frame C such that the required contact forces are feasible with respect to the contact model (8.26). But in order to validate the contact model, it is not necessary to explicitly compute the wrench distribution. Instead, it is sufficient to check whether the required overall support wrench $\boldsymbol{F}_{\text{sup}}^{\text{res}} = \boldsymbol{Ad}_{\text{sup}}^{T} \boldsymbol{F}_{\text{sup}}$ lies within the polygon of feasible support wrenches. Thus, it is only necessary to compute the explicit wrench distribution $\hat{\boldsymbol{u}}^{\text{opt}}[l]$ for the current time step in the prediction in order to obtain the control torques $\boldsymbol{\tau}[l]$ via (8.28). For all other/future time steps, it is sufficient to check the feasibility of the trajectory via the resulting support wrench. Thus, the size of the constrained quadratic optimization problem can be reduced from $N \cdot \Psi \cdot 6 = 240$ to $1 \cdot \Psi \cdot 6 + (N - 1) \cdot 6 = 60$ variables. Note that computing the polygon of feasible wrenches also requires some computational effort. However, the polyhedron is constant for a given contact configuration and must be computed only once, which can also be done offline if the contact configuration is known in advance.

References

Audren, H., Vaillant, J., Kheddar, A., Escande, A., Kaneko, K., Yoshida, E.: Model preview control in multi-contact motion-application to a humanoid robot. In: IEEE/RSJ International Conference on Intelligent Robots and Systems, pp. 4030–4035 (2014)

Bronshtein, I.N., Semendyayev, K.A., Musiol, G., Mühlig, H.: Handbook of Mathematics, 6th edn. Springer, Berlin (2015)

De Luca, A., Albu-Schäffer, A., Haddadin, S., Hirzinger, G.: Collision detection and safe reaction with the DLR-III lightweight manipulator arm. In: IEEE/RSJ International Conference on Intelligent Robots and Systems, pp. 1623–1630 (2006)

Gross, D., Hauger, W., Schröder, J., Wall, W.A., Govindjee, S.: Engineering Mechanics 3, 2nd edn. Springer, Berlin (2014)

Henze, B., Ott, C., Roa, M.A.: Posture and balance control for humanoid robots in multi-contact scenarios based on model predictive control. In: IEEE/RSJ International Conference on Intelligent Robots and Systems, pp. 3253–3258 (2014a)

Ibanez, A., Bidaud, P., Padois, V.: Unified preview control for humanoid postural stability and upper-limb interaction adaptation. In: IEEE/RSJ International Conference on Intelligent Robots and Systems, pp. 1801–1808 (2012)

Kajita, S., Kanehiro, F., Kando, K., Yokoi, K., Hirukawa, H.: The 3D linear inverted pendulum mode: a simple modeling for a biped walking pattern generation. In: IEEE/RSJ International Conference on Intelligent Robots and Systems, pp. 239–246 (2001)

Kajita, S., Kanehiro, F., Kaneko, K., Fujiwara, K., Harada, K., Yokoi, K., Hirukawa, H.: Biped walking pattern generation by using preview control of zero-moment point. In: IEEE International Conference on Robotics and Automation, pp. 1620–1626 (2003)

Krause, M., Englsberger, J., Wieber, P.B., Ott, C.: Stabilization of the capture point dynamics for bipedal walking based on model predictive control. In: IFAC Symposium on Robot Control, pp. 165–171 (2012)

Maciejowski, J.M.: Predictive Control with Constraints, 1st edn. Prentice Hall, Edingburgh Gate (2002)

Murray, R.M.: Nonlinear control of mechanical systems: a Lagrangian perspective. Annu. Rev. Control **21**, 31–42 (1997)

Nenchev, D.N., Konno, A., Tsujita, T.: Humanoid Robots: Modeling and Control. Butterworth-Heinemann, Oxford (2019)

Ott, C.: Cartesian Impedance Control of Redundant and Flexible-Joint Robots, Springer Tracts in Advanced Robotics, vol. 49, 1st edn. Springer, Berlin (2008)

Ott, C., Henze, B., Lee, D.: Kinesthetic teaching of humanoid motion based on whole-body compliance control with interaction-aware balancing. In: IEEE/RSJ International Conference on Intelligent Robots and Systems, pp. 4615–4621 (2013)

Roy, J., Whitcomb, L.L.: Adaptive force control of position/velocity controlled robots: theory and experiment. IEEE Trans. Robot. Autom. **18**(2), 121–137 (2002)

Shuster, M.D.: A survey of attitude representations. J. Astronaut. Sci. **41**(4), 439–517 (1993)

Wieber, P.B.: Trajectory free linear model predictive control for stable walking in the presence of strong perturbations. In: IEEE-RAS International Conference on Humanoid Robots, pp. 137–142 (2006)

Chapter 9
Applications

This chapter reports on several applications in which the control framework from Chaps. 6 and 7 was used to facilitate a whole-body motion of the humanoid robot *TORO* (see Chap. 5) while balancing. The applications demonstrate that the whole-body controllers can be integrated into the software ecosystem of a humanoid robot, which features a multitude of components besides the balancing controller, such as vision and planning. The variety of applications emphasizes the versatility and the robustness of the developed whole-body control approaches.

The chapter is structured as follows: Sect. 9.1 presents two scenarios, where the balancing controller was combined with planning and vision algorithms to facilitate autonomous grasping or stair climbing. In Sect. 9.2, the whole-body balancer is combined with different types of input devices to teleoperate the humanoid robot *TORO*. Section 9.3 shows *TORO* in a real-world scenario from the field of aircraft manufacturing.

9.1 Combining Planning with Whole-Body Balancing

To employ humanoid robots for disaster scenarios or in the field of industrial manufacturing, it is essential that the robots show a certain level of autonomy, which can be achieved by combining whole-body control with planning and perception. This section presents the works of Sundaram et al. (2018) and Werner et al. (2016), where the MCB controller from Sect. 6.1 was combined with planning and perception algorithms for autonomous grasping and multi-contact stair climbing, respectively. The MCB controller was extended in Henze et al. (2019) to increase the achievable stair height from 5 cm to 18 cm, which corresponds to an ordinary staircase in Germany (DIN e.V. 2016).

B. Henze, *Whole-Body Control for Multi-Contact Balancing of Humanoid Robots*,
Springer Tracts in Advanced Robotics 143,
https://doi.org/10.1007/978-3-030-87212-0_9

9.1.1 Autonomous Grasping

Sundaram et al. (2018) presented an integrated pipeline for generating autonomous grasping behaviors with humanoid robots. The pipeline comprises several components for processing visual information, planning, and control, as shown in the system overview in Fig. 9.1. The presented scenario involves the humanoid robot *TORO* autonomously approaching and picking up an object from a table (see Fig. 9.2a). For this purpose, the robot first localizes the table via its on-board cameras using *AprilTags* (Olson 2011) attached to the table. The subsequent approach utilizes visual odometry and *ORB-SLAM2* (Mur-Artal and Tardós 2017) for localization and a ZMP-based walking controller (Englsberger et al. 2011) for locomotion. As soon as the final position with respect to the table is reached, the framework enters the manipulation phase by visually recognizing and classifying the objects on the tabletop as a set of primitive geometries (Büttner et al. 2016). The latter is used by the grasp planner to synthesize a set of parametric grasps, which are filtered and evaluated based on a reachability analysis by Porges et al. (2014). If no feasible grasps can be found, then the framework performs a repositioning of the base by providing an adjusted goal to the walking controller. The adjusted goal is selected from a set of predefined base positions, which are ranked based on the reachability of the computed grasps at these positions. Otherwise, a sorted list of grasps is handed over to trajectory planning, which tries to compute a feasible, quasi-static trajectory starting with the preferred grasp from the list. If no feasible trajectory can be found for any of the grasps, then the framework can again trigger a repositioning of the base. To reduce the complexity of the trajectory planning, the algorithm considers only a subsystem of the robot kinematics. For this application, *TORO* always uses the right hand to grasp the object, which is why the reduced kinematic chain comprises all six DoFs of the right arm and the joint located between the torso and the pelvis (see Chap. 5). Here, the pelvis serves as a fixed base of the kinematic model. The left arm and the legs are irrelevant for the grasping process and are thereby omitted. Thus, the planning problem is reduced to an ordinary 7-DoF robot with a serial kinematic chain and a fixed base.

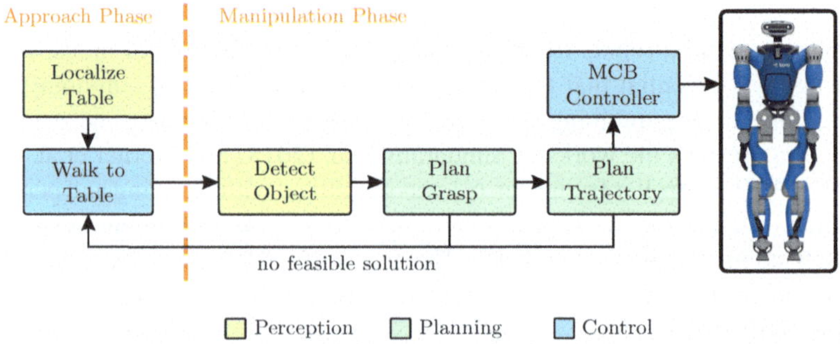

Fig. 9.1 Overview of the integrated pipeline for autonomous grasping with humanoid robots

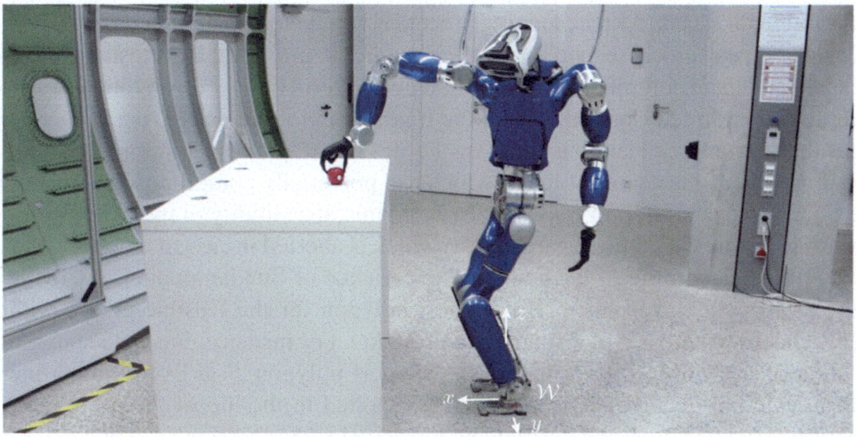

(a) Grasp execution using the MCB controller.

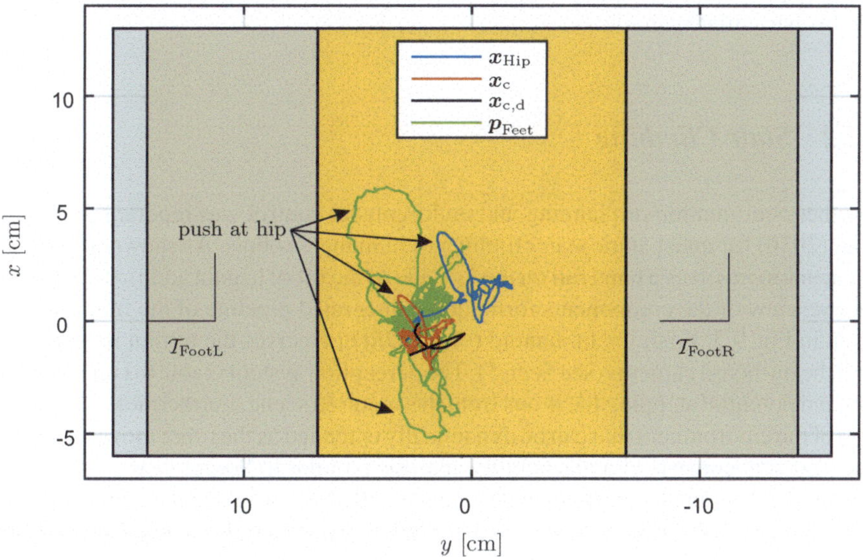

(b) Trajectory of the CoM and the hip relative to the support polygon $\mathcal{S}_{\mathrm{SP}}$ (yellow). The contact areas of the feet $\mathcal{S}_{\mathrm{FootR}}$ and $\mathcal{S}_{\mathrm{FootL}}$ are given in gray.

Fig. 9.2 The humanoid robot *TORO* autonomously picking up an object from a table

After the planning is completed, the desired trajectory is executed on the robot using the whole-body controller (MCB) presented in Sect. 6.1, which coordinates the motion of the right hand with the rest of the body to hold the balance. The communication between planner and controller is established via the interface described in Sect. 6.6. The feet are configured as supporting contacts, while the hands are assigned to the interaction task. As the planning process is conducted in joint space, the desired trajectory is post-processed via forward kinematics to provide the Cartesian goals for the CoM and hand compliances of the MCB controller. The accompanying null space compliance is directly provided with the trajectory in joint space. As the planning algorithm assumes a stationary pelvis, the desired trajectory comprises an intentional motion of the center of mass, which can potentially compromise balancing if the CoM leaves the support polygon. To avoid this, the validity of the CoM position is verified before executing the trajectory, which is aborted in case of an invalid CoM location. But according to the empirical experience of Sundaram et al. (2018), the CoM always stays well inside the support polygon for the feasible grasps within the reachable workspace of a single arm of *TORO*. For instance, Fig. 9.2b shows the location of the CoM with respect to the support polygon $\mathcal{S}_{\text{SP}}$. Even though there was an external disturbance (manual push) applied to the hip of the robot during the trajectory execution, the desired and the actual CoM location ($\boldsymbol{x}_{\text{c,d}}$ and $\boldsymbol{x}_{\text{c}}$) kept a sufficient distance to the borders of the support polygon. The same holds for the CoP $\boldsymbol{p}_{\text{Feet}}$ of the feet. Thus, using a fixed pelvis for planning is a valid assumption for this particular scenario.

9.1.2 *Stair Climbing*

Another combination of planning and multi-contact control was reported in Werner et al. (2016) for quasi-static stair climbing of humanoid robots. As shown in Fig. 9.4, the scenario involves a hand rail on the left side of the robot to gain additional support. An overview of the components forming the integrated pipeline of the framework is given in Fig. 9.3. First, the humanoid robot *TORO* perceives the terrain in front of it with the on-board cameras (see Sect. 5). The perception system is able to isolate obstacles and potential stepping locations from the acquired scene. Furthermore, it builds a map of the environment that can be dynamically extended as the robot moves forward. The next step consists of a reachability analysis (similar to Porges et al. (2014)) that isolates the stepping locations that are kinematically feasible given the current contact configuration, joint limitations, and static balancing. Once the reachable locations

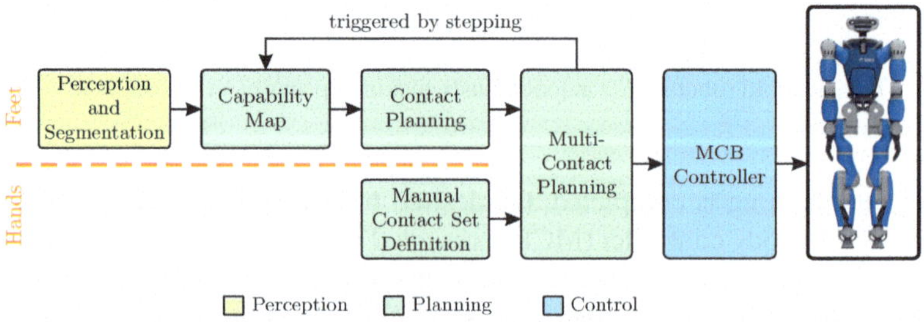

Fig. 9.3 Overview of the integrated pipeline for stair climbing

have been identified, the module for contact planning computes an optimal stepping sequence using a graph-based approach. Note that perception and contact planning only provide the contact sequence for the feet, while the hand contacts are manually defined for reasons of simplicity. The next component in the pipeline is a path planning algorithm based on *CBiRRT* (Berenson et al. 2009), which is able to handle the closed kinematic chain arising from the multi-contact scenario. The algorithm computes a quasi-static trajectory for bringing the robot from one contact configuration to the next one accounting for joint limitations, self-collisions, environment-collisions, kinematic singularities, and static balance. After the robot has taken a step, the multi-contact planner can trigger an update of the capability map and the contact planning to account for the new contact configuration and the terrain ahead. Finally, the quasi-static trajectory is handed over to the MCB controller presented in Sect. 6.1 to allow for a robust and compliant execution of the whole-body trajectory. The communication between the multi-contact planner and the controller is established via the interface described in Sect. 6.6. Note that stair climbing involves the attaching and detaching of contacts, which means that the task assignment of the end-effectors (support or interaction) changes during the motion sequence. Thus, the interface from Sect. 6.6 is extended with means of specifying and updating the task assignment online based on the procedure for contact switching described in Sect. 6.7.3.

One of the limiting factors during quasi-static stair climbing is the maximum torque that can be delivered by the knee joints. As shown in Fig. 9.5a, the length of the rear leg limits the height of the hip, which causes a significant flexion of the front leg. The flexion in turn correlates with the lever arm between the knee joint and the contact force of the front leg. Moreover, the robot has to shift its entire weight to the front leg in order to detach the rear foot. Thus, the combination of high load and high lever arm causes a significant torque in the knee joint of the front leg. For this reason, the humanoid robot *TORO* was only able to master stairs with a height of 5 cm during the experiments reported in Werner et al. (2016). The corresponding motion sequence is shown in Fig. 9.4. In order to overcome this limitation, we extended the MCB controller in Henze et al. (2019) to allow for toe-off motions, as shown in Fig. 9.6.

The advantage of performing a toe-off motion with the rear foot is that it extends the effective length of the rear leg, and therefore the kinematic capability of the robot. As shown in Fig. 9.5b, this extension allows the robot to increase the height of the hip, which reduces the knee flexion of the front leg. This in turn results in a shorter lever arm and consequently in a lower torque in the knee. In order to analyze this effect, a simulation study was conducted in Henze et al. (2019) with *TORO* climbing stairs of 28 cm width and without toe-off motion. Without toe-off, the required torque in the knee reached 96% of the maximum torque possible at this joint, which is too close to the limit to be executed on the actual robot, considering joint friction and measurement noise. By performing the toe-off motion, the knee torque can be reduced to 76% of the maximum torque, which leaves enough margin for executing the motion on the actual robot. Therefore, climbing stairs of this particular height (28 cm) is only possible by using a toe-off motion to extend the length of the rear leg.

In order to implement toe-off motions, we modified the MCB controller by shifting and rotating the end-effector frames of the feet as shown in Fig. 9.6b. During

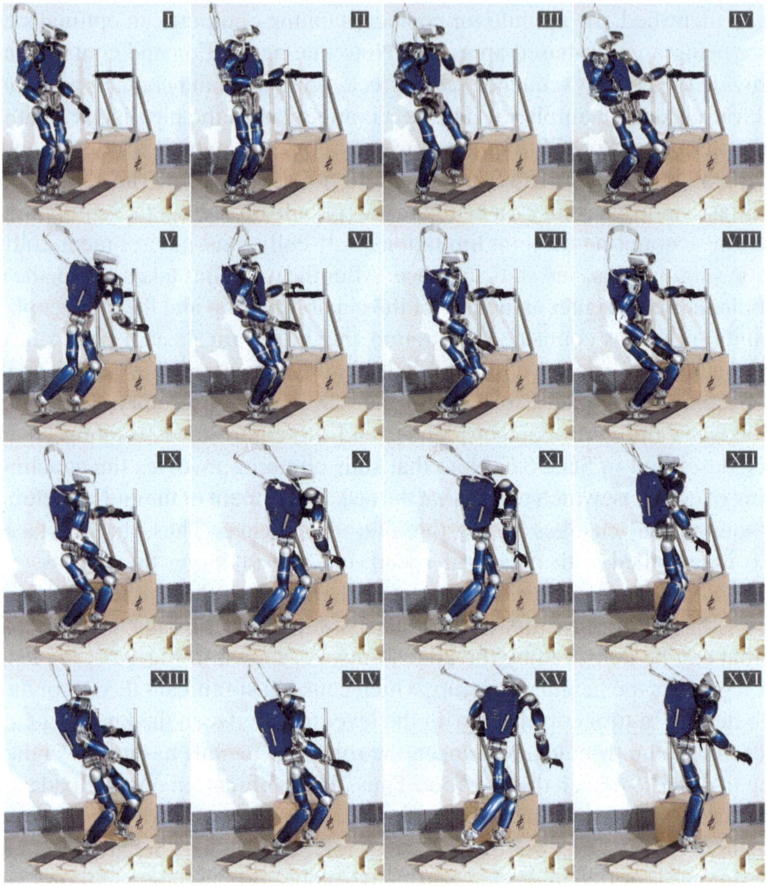

Fig. 9.4 The humanoid robot *TORO* climbing stairs with a height of 5 cm

Fig. 9.5 Impact of a toe-off motion on the joint torque of the front knee

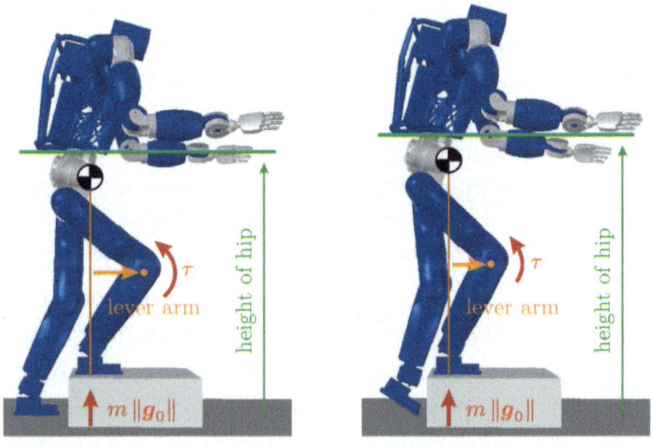

(a) Without toe-off motion. (b) With toe-off motion.

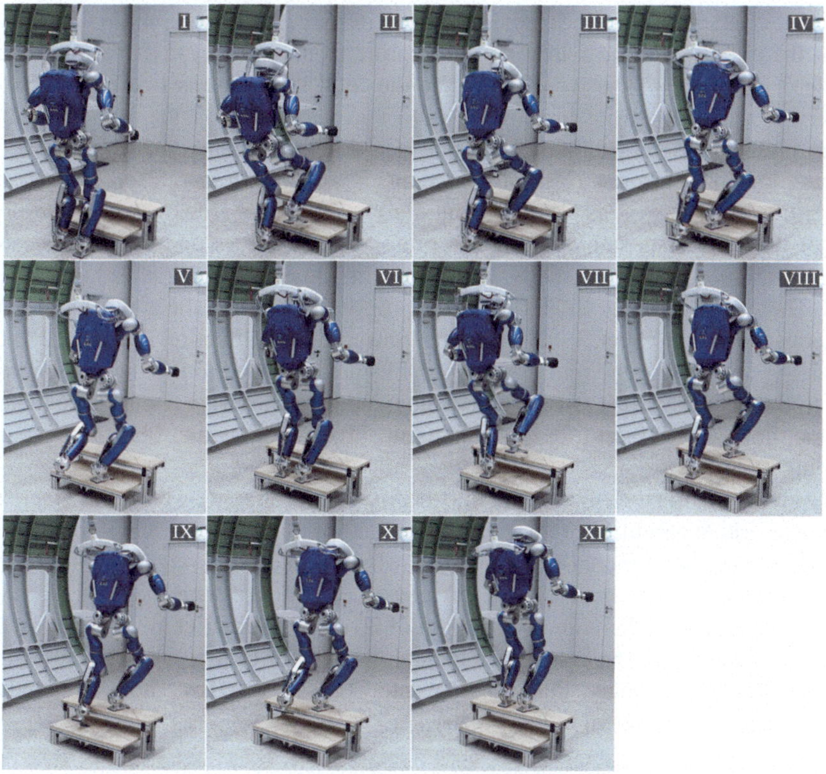

(a) Motion sequence. The toe-off motion is performed during phase IV and IX.

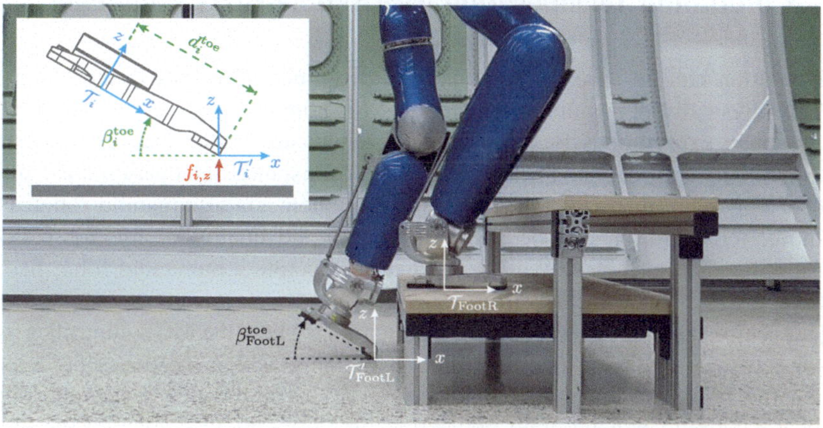

(b) Toe-off motion with the rear foot. The end-effector frame is moved by an offset d_i^{toe} and inclined by an angle β_i^{toe}.

Fig. 9.6 The humanoid robot *TORO* climbing stairs with a height of 18 cm

flat contact, the end-effector frame is located in its default configuration $\mathcal{T}_i$ underneath the ankle, with the x-axis pointing to the front and the z-axis being perpendicular to the sole of the foot. To incline the foot, the frame is first moved by an offset $d_i^{\text{toe}} \in \mathbb{R}$ to the front edge of the contact area. Afterwards, the frame is rotated by an angle $\beta_i^{\text{toe}} \in \mathbb{R}$, such that the z-axis is no longer perpendicular to the sole but to the floor. In theory, the resulting line contact is able to transmit forces in x-, y-, and z-direction, as well as torques about the x- and z-axis. However, it is questionable if the contact is stable enough to provide those forces and torques, because an inclination of the foot usually takes place in situations with rather low vertical load. Therefore, in Henze et al. (2019) we proposed modeling the contact as a point contact according to Sect. 3.4.3. Thus, the translational z-direction is assigned to the balancing task, while the remaining five DoFs are allocated to the interaction task. The five remaining DoFs are therefore stabilized by a Cartesian compliance including the inclination of the foot, which significantly increases the robustness against slipping and tilting.

By incorporating toe-off motions into the MCB controller, we were able to increase the kinematic capabilities of the robot by extending the effective length of the rear leg. This leads to an increased hip height, which reduces the required joint torque in the front knee. By including toe-off phases, we moved from negotiating stairs with a step height of 5 cm and step length of 20 cm in Werner et al. (2016) to stairs with a step height of 18 cm and step length of 28 cm in Henze et al. (2019), which corresponds to an ordinary staircase in Germany (DIN e.V. 2016). The corresponding motion sequence is shown in Fig. 9.6a.

9.2 Combining Teleoperation with Whole-Body Balancing

Teleoperation can be used to remotely control humanoid robots in environments that are physically demanding or dangerous for humans, such as disaster scenarios or space exploration. Besides this, the healthcare sector also provides several interesting applications for teleoperation. For instance, teleoperation can allow medical personnel to remotely interact with a patient via a humanoid robot. Furthermore, teleoperation can be used to increase the quality of life for patients with mobility impairments by allowing them to operate a humanoid robot like an avatar.

This section presents two exemplary applications where the MCB controller from Sect. 6.1 was combined with two different human-machine interfaces for teleoperating the humanoid robot *TORO*. Section 9.2.1 demonstrates a wearable input device for helping patients with upper limb impairments to command the hands of *TORO*. Another input device for controlling the arms is presented in Sect. 9.2.2, which provides haptic feedback to the user to inform her/him about the consequences of the commanded actions. The setup is enriched with an automatic CoM task for moving the CoM in conjunction with the hands to increase the workspace of the robot.

9.2.1 Bimanual Teleoperation Using a Wearable, Ultralight Input Device

Gijsberts et al. (2014) demonstrated the use of a myoelectric human-machine interface for teleoperating the humanoid robot *TORO*. Due to the lack of a whole-body controller at that time, there was no coordination between the lower and the upper body of *TORO*. Furthermore, the interface only comprised the right arm of the robot. We extended the framework in Porges et al. (2019) by combining the input device with the whole-body controller from Sect. 6.1 (MCB). This allows the operator to command both arms simultaneously, while the MCB controller ensures the balance of the robot in a compliant and robust way. The improvements on the human-machine interface (HMI) mainly address the wearability and usability of the input device.

As shown in Fig. 9.7, the interface comprises two *Myo* gesture control armbands by *Thalmic Labs* to predict the finger actuation of the human operator via surface electromyography (sEMG). The sEMG data is processed on a Laptop using Incremental Ridge Regression with Random Fourier Features in order to provide a command for the finger configurations of *TORO*. The Cartesian poses of the operator's hands are measured via two *Vive* motion trackers by *HTC*. The tracking data is processed on the same Laptop using the cross-platform game engine *Unity* developed by *Unity Technologies*. The measured Cartesian poses are provided to the robot in the form of the setpoints for the hand compliances. For the experiment, the master system provides a combined data stream at 50 Hz, which is transmitted via *Wi-Fi* to the humanoid robot *TORO*. The data stream is processed by the MCB controller, which is executed at a rate of 1 kHz on the real-time computer of *TORO* (see Chap. 5). The interface is completely wireless and light-weight (about 180 g per arm), which gives the operator the freedom to move around.

The capabilities of the framework are demonstrated in Porges et al. (2019) via several complex daily-living tasks, which require a robust and dexterous two-handed manipulation (see Fig. 9.8). The tasks include unscrewing a lid from a bottle, pouring fluid from a bottle into a cooking pot, and holding a wireless phone while dialing a number. The video of the experiment can also be found on the *YouTube*-channel (RMC 2019) of the *Institute of Robotics and Mechatronics*.

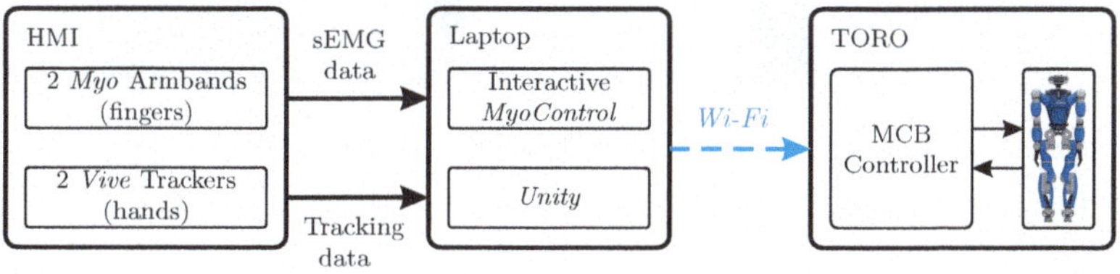

Fig. 9.7 Overview of the framework for bimanual teleoperation of humanoid robots

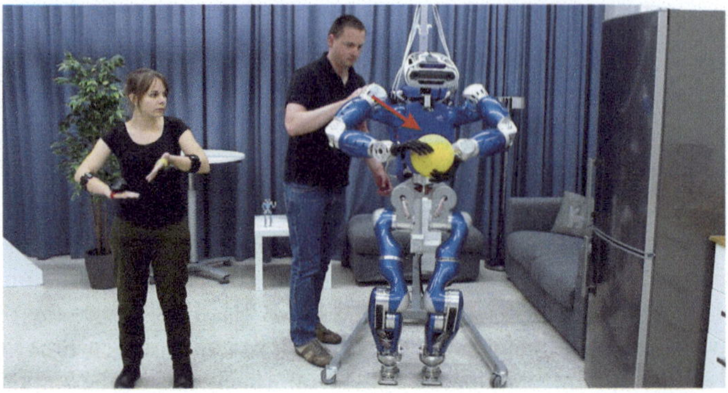

(a) Bimanual handling of a ball while balancing compliantly in the presence of external disturbances (red arrow).

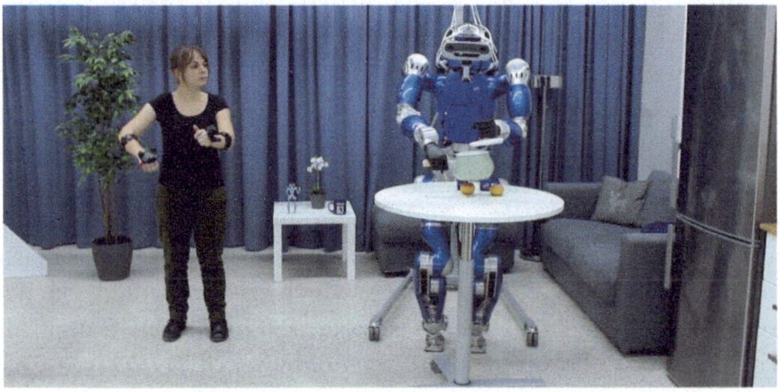

(b) Removing a lid from a cooking pot.

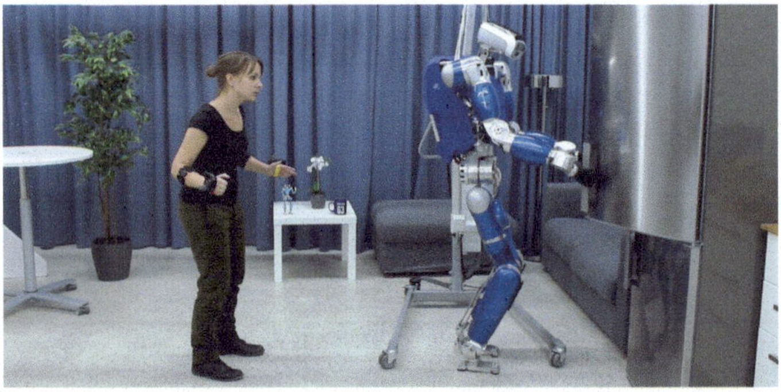

(c) Opening a fridge.

Fig. 9.8 Bimanual Teleoperation of *TORO* using a wearable, ultralight input device

9.2.2 *Teleoperation Using a Task-Relevant Haptic Interface*

In Abi-Farraj et al. (2018), we presented another interface for teleoperating humanoid robots, which can inform the operator about the consequences of the commanded actions via haptic feedback. The interface is enhanced with an automatic task for the CoM to increase the workspace of the humanoid robot. A schematic overview of the interface is given in Fig. 9.9. The framework uses *HUG* (Sagardia et al. 2016) as an input device on the master side. *HUG* is a haptic user interface developed by *DLR* that consists of two *DLR-KUKA* Lightweight robot arms mounted on a column, as shown in Fig. 9.10. Each arm features seven DoFs, a workspace of 1.1 m, and a nominal payload of 20 kg. The humanoid robot *TORO* (see Chap. 5) is on the slave side, and it is controlled by the MCB balancer presented in Sect. 6.1. Both sides are connected via the interface detailed in Sect. 6.6.

The operator can move the hands of the slave by moving the end-effectors of the master device. The Cartesian end-effector velocities are measured on the master

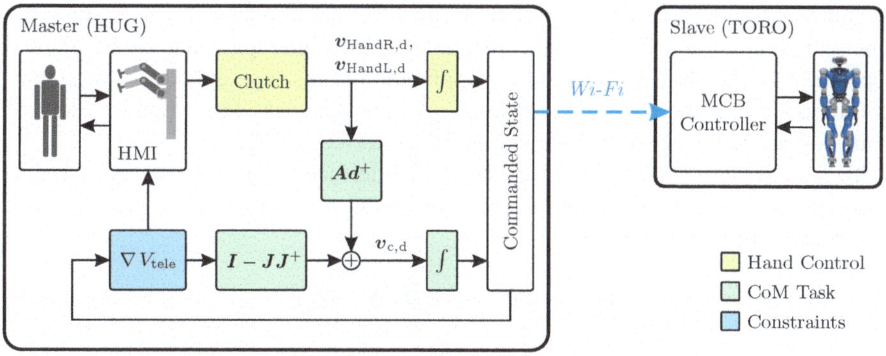

Fig. 9.9 Overview of the task-relevant haptic interface

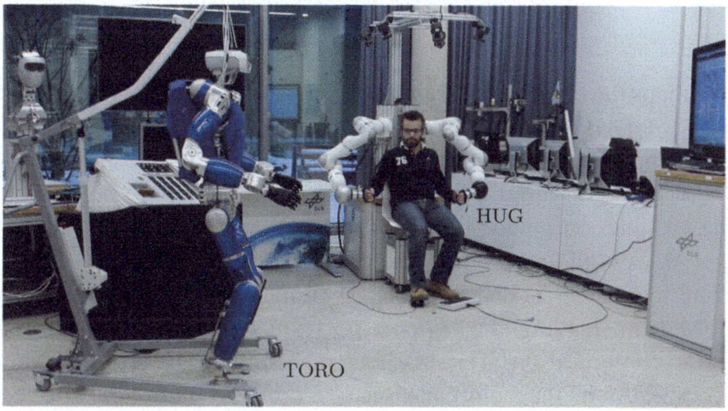

Fig. 9.10 Teleoperating the humanoid robot *TORO* via the haptic user interface *HUG*

side, integrated into a desired pose for the left and the right hand, and transmitted via *Wi-Fi* to the slave. Note that the software implementation comprises a "clutching mechanism", which allows the operator to disengage from the slave side and move the arms of the input device (HUG) into a more convenient configuration without actually moving the slave (*TORO*).

To prevent the operator from bringing the slave robot into an undesirable configuration, the input device (*HUG*) provides haptic feedback to the user. In particular, a positive definite potential function

$$V_{\text{tele}} = \sum_k V_{\text{SP}}^{\text{tele}}(d_k) + V_{\text{HandR}}^{\text{tele}}(d_{\text{HandR}}) + V_{\text{HandL}}^{\text{tele}}(d_{\text{HandL}}) \tag{9.1}$$

is defined in Abi-Farraj et al. (2018). The first term encodes the distance $d_k \in \mathbb{R}$ between the CoM (projected onto the floor) and the edges of the support polygon $\mathcal{S}_{\text{SP}}$ in order to inform the user if the CoM approaches the border of $\mathcal{S}_{\text{SP}}$. The potentials $V_{\text{HandR}}^{\text{tele}}(d_{\text{HandR}})$ and $V_{\text{HandR}}^{\text{tele}}(d_{\text{HandL}})$ evaluate the distance between the torso and the right or the left hand, respectively. If the user tries to stretch the arms beyond a given maximum distance, the potentials are activated to inform the user about the kinematic limitation. Note that the constraints are a design aspect that can be adapted depending on the slave system and the application. Thus, V_{tele} encodes several constraints regarding the robot's kinematic and static balance. The wrenches derived from V_{tele} are fed back to the user via the input device *HUG*.

To increase the workspace of the slave, the interface features a task for automatically moving the center of mass frame C of *TORO*. As shown in Fig. 9.9, the CoM task consists of two elements:

$$\boldsymbol{v}_{\text{c,d}} = \boldsymbol{Ad}^+ \ \boldsymbol{v}_d + \boldsymbol{J}_{\text{c}} \, \boldsymbol{N}_{\text{tele}} \, \boldsymbol{\nu}_{\text{tele}} \tag{9.2}$$

with

$$\boldsymbol{v}_d = \begin{pmatrix} \boldsymbol{v}_{\text{FootR,d}} \\ \boldsymbol{v}_{\text{FootL,d}} \\ \boldsymbol{v}_{\text{HandR,d}} \\ \boldsymbol{v}_{\text{HandL,d}} \end{pmatrix} = \begin{pmatrix} \boldsymbol{0} \\ \boldsymbol{0} \\ \boldsymbol{v}_{\text{HandR,d}} \\ \boldsymbol{v}_{\text{HandL,d}} \end{pmatrix}. \tag{9.3}$$

The first term maps the commanded hand velocities to the center of mass via the pseudoinverse $\boldsymbol{Ad}^+ = (\boldsymbol{Ad}^T \boldsymbol{Ad})^{-1} \boldsymbol{Ad}^T$ of the adjoint matrix $\boldsymbol{Ad}$ as defined in (3.23). This choice can be motivated with (3.23), which maps $\boldsymbol{v}_{\text{c}}$ to the end-effector velocities $\boldsymbol{v}$ via $\boldsymbol{v} = \boldsymbol{Ad} \ \boldsymbol{v}_{\text{c}}$ assuming $\dot{\boldsymbol{q}} = \boldsymbol{0}$. Thus, $\boldsymbol{v}_{\text{c,d}}$ in (9.2) can be interpreted as a "mean value" of the commanded end-effector velocities $\boldsymbol{v}_d$. Note that $\boldsymbol{v}_{\text{FootR,d}} = \boldsymbol{0}$ and $\boldsymbol{v}_{\text{FootL,d}} = \boldsymbol{0}$, as both feet are in contact with the floor.

The second term in (9.2) drives the CoM frame C via the potential V_{tele} to prevent the CoM from leaving the support polygon $\mathcal{S}_{\text{SP}}$ or the arms from moving into a singular configuration. The gradient of V_{tele} is used to compute a desired velocity $\boldsymbol{\nu}_{\text{tele}} \in \mathbb{R}^{6+n}$ in the configuration space of the humanoid robot. The null space projector $\boldsymbol{N}_{\text{tele}} = (\boldsymbol{I} - \boldsymbol{J}\boldsymbol{J}^+) \in \mathbb{R}^{(6+n)\times(6+n)}$ maps $\boldsymbol{\nu}_{\text{tele}}$ to the null space of the end-effector task. The result is then mapped via the center of mass Jacobian matrix $\boldsymbol{J}_{\mathrm{c}} = [\,\boldsymbol{I}\;\boldsymbol{0}\,] \in \mathbb{R}^{6\times(6+n)}$ to the desired CoM velocity.

As soon as the slave robot is brought close to a singular arm configuration or to the boundary of the support polygon, the operator is informed about these constraints via the implemented force feedback. Note that the forces provided to the user increase the closer the slave robot gets to the constraints. As a last resort, the "clutching mechanism" is used to automatically sever the connection between master and slave such that the operator cannot move the slave beyond the boundaries.

9.3 Real-World Applications: Employing Humanoid Robots in Aircraft Manufacturing

The objective of the European research project *COMANOID*[1] (Comanoid 2019; Kheddar et al. 2019) was to evaluate the potential of employing humanoid robots in aircraft manufacturing. The consortium consisted of several research institutions from France, Italy, and Germany, including the *German Aerospace Center* (*DLR*). The industrial partner was represented by *Airbus Group Innovations*, which provided the use case and the testing facilities for the final demonstration.

Although several automation solutions have been introduced in aircraft manufacturing, some phases of the process still require a significant amount of manual work. One of the reasons is the challenging accessibility of several areas inside an aircraft, such as the cargo hold. The EU-project *COMANOID* addressed the use case of gluing brackets (for holding cables or pipes) onto the inner hull of an aircraft. As an airplane can contain several ten thousands of these support elements, the task is repetitive and monotone, but requires high precision due to the quality requirements in aviation industries. Many of the brackets must be placed in rather unergonomic locations that require workers to kneel or bend down, for instance. Thus, the task can be classified as a monotone low-added-value task that subjects the worker to physical stress with the risk of repetitive strain injury.

The addressed scenario involved mounting the brackets on the cargo deck of an *Airbus A350 XWB*. To conduct the final demonstrations under realistic conditions, *Airbus Group Innovations* provided an exact replica of the cargo deck of the aircraft at the production site in *Saint-Nazaire* (see Fig. 9.11a, c). Due to the limited height of the cargo deck,[2] the workers (and humanoid robots) can only operate on the cargo deck in a rather crouched posture, which again emphasizes the need for automation. The humanoid robot and the human worker are supposed to share the same workspace, which demands for robust and compliant balancing algorithms to ensure safety, such as the control frameworks presented in Chaps. 6 and 7. Furthermore, the robot must

---

[1] *European Union Research and Innovation Programme Horizon 2020*, grant agreement number H2020-ICT-645097 (*COMANOID*).

[2] *Airbus S.A.S* specifies a height of 1.76 m for the door of the forward cargo hold of the *A350-900* and *A350-1000* (AIRBUS S.A.S. 2005).

(a) Picking up the bracket from a dispenser on a table.

(b) Positioning the bracket using the right hand as additional contact to enlarge the support polygon.

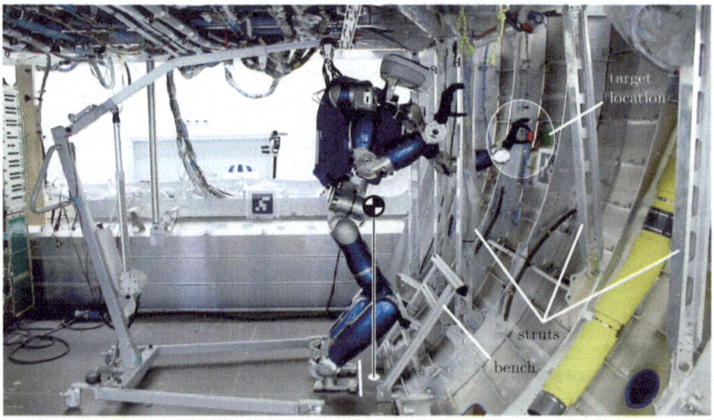

(c) Positioning the bracket using both knees as additional contacts to enlarge the support polygon.

Fig. 9.11 Use case of the EU-project *COMANOID*

be able to enter the shop floor via the same stairs as the workers. As the shop floor can be covered with tools, cables, or other equipment, the robot must also be able to walk over rough terrain. Thus, the task requires a high level of versatility and agility as provided by a legged humanoid robot. In particular, the scenario involved the following steps:

1. Enter the shop floor via stairs.
2. Pick up a bracket from a dispenser on a table.
3. Enter the cargo deck of the aircraft and proceed to the final working position.
4. Position the bracket at the predefined location on the fuselage.
5. Exit the working area.

The humanoid robot *TORO* performed the stair climbing in the Lab of *DLR* using the MCB balancer from Sect. 6.1. The negotiated stairs feature a height of 18 cm and a step length of 20 cm, which corresponds to the stairs in the mockup at *Saint-Nazaire*. The results are reported in Sect. 9.1.2. The second step (picking up the bracket) was demonstrated at *Saint-Nazaire*, as shown in Fig. 9.11a. Again the MCB balancer was used to coordinate grasping with the support and the balancing task during the whole-body motion. The location of the bracket on the table was provided by the perception system of *TORO* using *AprilTags* (Olson 2011). The locomotion inside the aircraft utilized a walking controller based on the MCB+ balancer (Mesesan et al. 2019). Here, it is important that the robot keeps a crouched body-posture to respect the limited height of the cargo deck of the aircraft.

The last step addresses the main focus of the EU-project *COMANOID*, which is on multi-contact planning and control. As soon as the robot reaches the final working position, it is supposed to glue the bracket at a predefined position on the inner hull of the aircraft. However, the distance to the hull is so large that even human workers cannot reach it directly. Instead, they have to lean forward using their hands for additional support. In order to increase the reach of the arms, the robot follows the human example by including additional contacts into the balancing problem, such that the support polygon is extended to the front. This in turn allows the robot to shift the CoM and thereby the torso to the front, which increases the workspace of the arms such that the target location on the hull can be reached. Note that the robot is not allowed to step on the hull to avoid damage to the aircraft. However, the cargo hold features a set of struts on both sides for reinforcing the structure of the aircraft. As they are of a robust design, the robot is allowed to utilize the struts for gaining additional support.

In the context of the project, we developed two different strategies for mastering these multi-contact tasks, which are both shown in Fig. 9.11. The first strategy was demonstrated in the lab at *DLR* in a functional mockup of the *A350 XWB* cargo hold (see Fig. 9.11b). The robot *TORO* grasps one of the struts with the right hand and leans to the front to place the bracket with the left hand on the hull of the aircraft. Here, the challenge is that the contact configuration is asymmetric, which results in an asymmetric support polygon. For this reason, the robot moves the center of mass not only to the front but also to the right, such that the CoM always stays on top of the support polygon. Note that the CoM is shifted in front of the feet, which means that the robot would fall over without the additional hand contact. The robot is operated using the multi-contact balancing controller (MCB) presented in Sect. 6.1. A video of the demonstration can be found on the *YouTube* channel of the *Institute of Robotics and Mechatronics* at *DLR* (RMC 2019).

The second strategy was demonstrated at *Saint-Nazaire* using the more sophisticated mockup of the *A350 XWB* cargo deck (see Fig. 9.11c). Here, the robot employs the knees as additional contacts to extend the support polygon to the front. For this purpose, *TORO* kneels on a portable bench similar to the assistive devices (e. g. stools and ladders) used by human workers. Again, the robot shifts the center of mass in front of the feet, which means that the robot would fall over without the additional knee contacts. In order to account for the kinematic constraints invoked by the knees, the hierarchical multi-contact balancer (HMCB) presented in Chap. 7 is used in combination with the task hierarchy for confined spaces (see Table 7.2). Note that the cargo deck represents a perfect example for balancing in confined spaces, as *TORO* cannot move into a fully upright posture due to the limited height of the ceiling.

In summary, we were able to prove that the whole-body control algorithms presented in Chaps. 6 and 7 can be used to robustly and compliantly operate humanoid robots in multi-contact scenarios under realistic conditions, as in this case in aircraft manufacturing.

References

Abi-Farraj, F., Henze, B., Werner, A., Panzirsch, M., Ott, C., Roa, M.A.: Humanoid teleoperation using task-relevant haptic feedback. In: IEEE/RSJ International Conference on Intelligent Robots and Systems, pp. 5010–5017 (2018)

AIRBUS SAS: A350 Aircraft Characteristics - Airport and Maintanance Planning. Nov. 01/16 edn (2005)

Berenson, D., Srinivasa, S.S., Ferguson, D., Kuffner, J.J.: Manipulation planning on constraint manifolds. In: IEEE International Conference on Robotics and Automation, pp. 625–632 (2009)

Büttner, S., Márton, Z.C., Hertkorn, K.: Automatic scene parsing for generic object descriptions using shape primitives. Robot. Auton. Syst. **76**, 93–112 (2016)

Comanoid: Website of the EU-project COMANOID. www.comanoid.cnrs.fr (2019), accessed 13 Sep 2019

DIN eV: Barrierefreies Planen und Bauen, DIN-Taschenbuch, vol. 199, 7th edn, Beuth Verlag, Berlin, chap DIN 18065:2015-03 Gebäudetreppen - Begriffe, Messregeln, Hauptmaße (2016)

Englsberger, J., Ott, C., Roa, M.A., Albu-Schäffer, A., Hirzinger, G.: Bipedal walking control based on capture point dynamics. In: IEEE/RSJ International Conference on Intelligent Robots and Systems, pp. 4420–4427 (2011)

Gijsberts, A., Bohra, R., González, D.S., Werner, A., Nowak, M., Caputo, B., Roa, M.A., Castellini, C.: Stable myoelectric control of a hand prosthesis using non-linear incremental learning. Front. Neurorobotics **8**, 1–15 (2014)

Henze, B., Roa, M.A., Werner, A., Dietrich, A., Ott, C., Albu-Schäffer, A.: Experiments with human-inspired behaviors in a humanoid robot: quasi-static balancing using toe-off motion and stretched knees. In: IEEE International Conference on Robotics and Automation, pp. 2510–2516 (2019)

Kheddar, A., Caron, S., Gergondet, P., Comport, A., Tanguy, A., Ott, C., Henze, B., Mesesan, G., Englsberger, J., Roa, M.A., Wieber, P.B., Chaumette, F., Spindler, F., Oriolo, G., Lanari, L.,

Escande, A., Chappellet, K., Kanehiro, F., Rabaté, P.: Humanoid robots in aircraft manufacturing - the Airbus use-cases. IEEE Robot. Autom. Mag. **26**(4), 30–45 (2019)

Mesesan, G., Englsberger, J., Ott, C., Albu-Schäffer, A.: Dynamic walking on compliant and uneven terrain using DCM and passivity-based whole-body control. In: IEEE-RAS International Conference on Humanoid Robots, pp. 25–32 (2019)

Mur-Artal, R., Tardós, J.D.: ORB-SLAM2: an open-source SLAM system for monocular, stereo, and RGB-D cameras. IEEE Trans. Robot. **33**(5), 1255–1262 (2017)

Olson, E.: AprilTag: a robust and flexible visual fiducial system. In: IEEE International Conference on Robotics and Automation, pp. 3400–3407 (2011)

Porges, O., Connan, M., Henze, B., Gigli, A., Castellini, C., Roa, M.A.: A wearable, ultralight interface for bimanual teleoperation of a compliant, whole-body-controlled humanoid robot. In: IEEE International Conference on Robotics and Automation. www.youtube.com/watch?v=YLEUBFu5qgI (2019), accessed 13 Sep 2019

Porges, O., Stouraitis, T., Borst, C., Roa, M.A.: ROBOT2013: First Iberian Robotics Conference, Advances in Intelligent Systems and Computing, vol. 235, pp. 703–718. Springer International Publishing, Cham, chap Reachability and Capability Analysis for Manipulation Tasks (2014)

RMC: YouTube channel of the Institute of Robotics and Mechatronics at the German Aerospace Center (DLR). www.youtube.com/user/DLRRMC (2019), accessed 13 Sep 2019

Sagardia, M., Hulin, T., Hertkorn, K., Kremer, P., Schätzle, S.: A platform for bimanual virtual assembly training with haptic feedback in large multi-object environments. In: ACM Conference on Virtual Reality Software and Technology, pp. 153–162 (2016)

Sundaram, A.M., Henze, B., Porges, O., Marton, Z.C., Roa, M.A.: Autonomous bipedal humanoid grasping with base repositioning and whole-body control. In: IEEE-RAS International Conference on Humanoid Robots, pp. 395–402 (2018)

Werner, A., Henze, B., Rodriguez, D.A., Gabaret, J., Porges, O., Roa, M.A.: Multi-contact planning and control for a torque-controlled humanoid robot. In: IEEE/RSJ International Conference on Intelligent Robots and Systems, pp. 5708–5715 (2016)

Chapter 10
Discussion and Conclusion

This work aims at providing algorithms for balance control of legged, torque-controlled humanoid robots. A humanoid robot normally uses the feet for locomotion. This work extends this paradigm by addressing the challenge of multi-contact balancing, which allows a humanoid robot to exploit an arbitrary number of contacts to support itself. One of the main aspects is the solution of the wrench distribution problem, which arises from the resulting closed kinematic chain. Using multiple contacts for support increases the size of the support polygon, which in turn leads to an increased robustness of the stance and to an increased kinematic workspace of the robot. Both are important features for facilitating a transition of humanoid robots from science to real-world applications, where they are confronted with several challenging scenarios, such as climbing stairs and ladders, traversing debris, handling heavy loads, or working in confined spaces.

The developed framework also addresses the challenge of whole-body control by allowing a humanoid robot not only to maintain balance but also to interact with its environment, as e.g. by carrying or manipulating objects. Of course, the forces and torques arising from the interaction task must be considered by the balancing and/or support task to avoid falling.

The whole-body control framework is generalized by combining the techniques for multi-contact balancing with multi-objective control, which allows for a more generic task definition. Kinematic and dynamic conflicts between the tasks are resolved via a prioritization. For instance, the generalization allows for support contacts not only at the end-effectors (hands and feet) but at arbitrary locations on the robot's body, such as the knees, pelvis, backpack, shoulders, or the elbows. This ability is essential for operating a humanoid robot in confined spaces, where there might be not enough space to completely "unfold" the limbs to rely on hands and feet only.

The developed control framework implements a compliant balancing behavior in order to ensure safety in case of accidental collisions between robot and environment

B. Henze, *Whole-Body Control for Multi-Contact Balancing of Humanoid Robots*, Springer Tracts in Advanced Robotics 143,
https://doi.org/10.1007/978-3-030-87212-0_10

or between robot and human. Furthermore, the framework employs passivity-based methods to achieve robustness with respect to external disturbances.

In summary, the framework advances the state of the art by addressing the aspects listed below:

- Extension of the work by Ott et al. (2011) into a passivity-based framework (MCB)
  - for whole-body control by incorporating interaction tasks,
  - and for multi-contact balancing by allowing for an arbitrary number of contacts.
- Incorporation of feedforward terms to allow for dynamic trajectories (MCB+).
- Passivity control for balancing on movable and deformable surfaces (MCB-PC).
- Automatic CoM task for performing high-force interaction tasks (MCB-ACT).
- Generalization of the whole-body control framework by integrating hierarchical multi-objective control (HMCB). The latter allows for
  - a dynamic decoupling of the tasks via a task prioritization,
  - and balancing in confined spaces by allowing for arbitrary contact locations on the body of the robot.
- Extensive experimental evaluation of the whole-body control framework.
- Demonstration of the performance and versatility of the whole-body control framework in various applications including planning, teleoperation, and industrial manufacturing.

One of the above-mentioned applications was the EU-project *COMANOID* (Comanoid 2019; Kheddar et al. 2019), which addressed the challenge of introducing humanoid robots into aircraft manufacturing. Other means of benchmarking the balancing and locomotion capabilities of humanoid robots are provided by the EU-project *EUROBENCH* (Eurobench 2019): Torricelli et al. (2015) proposed a set of skills that are relevant for bipedal balancing and locomotion. Figure 10.1 shows an excerpt of the skills that are relevant for the scope of the developed whole-body control framework. For this purpose, some of the categories are summarized—in particular, the ones addressing surface inclinations and compliant terrain. Furthermore, we summarized the categories for surface tilts and translations, such that they comprise continuous as well as sudden surface motions. However, the category "High Force Interaction" was added to the skill set to account for the MCB-ACT balancing controller, presented in Sect. 6.5. Torricelli et al. (2015) proposed a classification of the categories into two groups based on whether the environment is static or in motion. We suggest a further subdivision of the latter group into passive and active environments. If the environment is passive, then the motion of the robot causes a motion of the environment, such as in the category "Seesaw". In case of an active environment, the causality is reverted: Here, the environment inflicts external forces and torques on the robot that in turn cause a motion of the robot, such as in the category "Pushes". Note that the categories "Body Sway Referenced Platform", "Bearing Weight", and "High Force Interaction" form another subgroup, where the robot generates forces and torques for the purpose of deliberately manipulating the environment, in contrast to the other categories, where the robot exploits the environment to gain support.

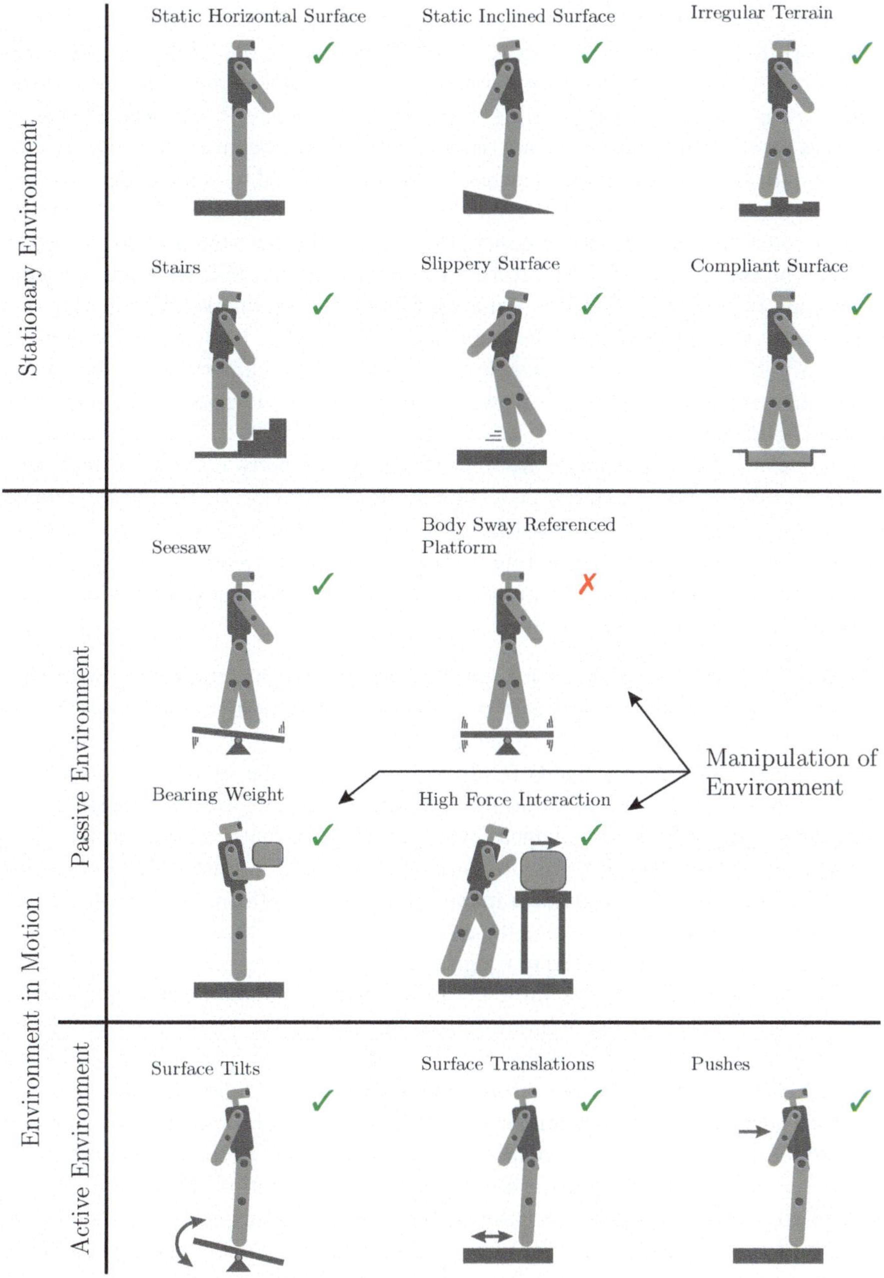

Fig. 10.1 Categories for benchmarking the balance performance of humanoid robots (based on Torricelli et al. 2015)

In the context of this work, we have demonstrated that the developed control framework for whole-body balancing is able to master 12 out of the 13 categories given in Fig. 10.1, including irregular terrain, stairs, and surface tilts and translations, for instance. The only missing scenario "Body Sway Referenced Platform" demands that the robot maintains not only its own balance but also the balance of the tiltable platform. This particular scenario does not only deal with the underactuation of the robot's base, but also with the underactuation of the platform. Thus, the control strategy must be extended, for instance, by utilizing the horizontal CoM motion to regulate the orientation of the platform. The presented whole-body control framework could be a potential starting point for this extension, thanks to its feature of a general task description. Due to the challenging nature of a "Body Sway Referenced Platform", this category is an interesting method for benchmarking the balancing capabilities of a humanoid robot. However, the practical relevance of this particular scenario is limited.

In conclusion, the presented framework for whole-body control of humanoid robots in multi-contact scenarios covers all the relevant categories of the benchmark given in Fig. 10.1. Therefore, it facilitates the assignment of humanoid robots to challenging and demanding tasks ranging from disaster scenarios and space and service robotics to industrial manufacturing. This work also revealed several recommendations for future research activities:

- In order to exploit the hands for multi-contact balancing, the robot kinematics must allow the hands to adjust to the environment. The humanoid robot *TORO* only features two joints in the wrist and forearm (see Chap. 5), which limits its capabilities to rotate the hands freely in Cartesian space. A mechanical design that incorporates three DoFs into the forearm (preferably with intersecting axes for mimicking a ball joint) appears to be in clear advantage over the two-DoF design of *TORO*. Besides this, the strength of *TORO*'s arms is limited by the elbow joint, which can provide a maximum torque of 40 Nm. Given a lever arm of approximately 0.4 m between the elbow and the hand, the maximum force that the arms can deliver is limited to roughly 100 N depending on the direction of the force. Although this range is sufficient for many manipulation tasks, it represents one of the limiting factors in multi-contact balancing. The arms can be employed to increase the robustness of the stance as demonstrated multiple times in this work. However, they are not strong enough to carry a relevant amount of the robot's weight, which disqualifies *TORO* for several multi-contact tasks, such as climbing. For these reasons, we recommend revising the mechanical design of the arms regarding strength and kinematics for the next generation of humanoid robots.
- In Chap. 7, the tasks are manually defined, of which several parts require the knowledge of an expert. In order to provide humanoid robotics to the general public, more generic methods for defining and switching tasks need to be investigated. Online task transitions especially are still a challenging field from a theoretical point of view.
- Thanks to Sect. 6.3 (MCB+), the framework enables humanoid robots to follow dynamic trajectories. The next logical step is to exploit this feature for dynamic walking, which is already under investigation by *DLR*.

- Many whole-body control frameworks require an explicit definition of the CoM task, e.g. by defining a constant setpoint for the CoM location. However, restricting the CoM to a static position reduces the kinematic capabilities of the robot (e.g. the reach of the arms during grasping). Therefore, we suggest introducing automatic CoM tasks into humanoid robotics that facilitate a suitable motion of the CoM, such as a free motion within the boundaries of the support polygon. This work presented a potential approach in the form of Sect. 6.5 (MCB-ACT), where an automatic CoM task was developed to account for high interaction forces. Another possibility is to follow the teleoperation setup presented in Sect. 9.2.2, where a potential is used to move the CoM in conjunction with the hands to avoid a singular arm configuration.
- Many whole-body control frameworks assume that the supporting contacts do not unintentionally detach from the environment, which is supposed to be prevented by respecting the contact model. However, practical experience has shown that the performance of the linear contact model presented in Sect. 3.4 can be unsatisfactory under low perpendicular forces ($f_{i,z} \approx f_{i,z}^{\min}$). Thus, we suggest investigating more sophisticated models for describing the contact properties. This work also provides several methods and procedures for increasing the robustness of the supporting contacts against lifting off, tilting, or sliding (see Sects. 6.7.1 and 9.1.2). However, these methods can only increase the robustness but are not able to stabilize the robot if an unintentional detaching already occurred. As a consequence, we recommend investigating further methods for detecting and reacting to misplaced or detached contacts. Furthermore, the aspect of detached contacts is linked to the problem of step recovery in multi-contact scenarios. Here, the challenge lies in the number of contacts as well as in the complexity of the cluttered environment. Similar problems have already been addressed in offline contact planning, but not in online step recovery.

Although there are still many challenges in humanoid robotics, this work advanced the state of the art by developing a whole-body control framework, which enables the robust and compliant operation of humanoid robots in multi-contact scenarios. It contributed to the field both with a thorough investigation of several theoretical aspects and with an extensive experimental evaluation. The research effort culminated in the EU-project *Comanoid* (Comanoid 2019), where we were able to demonstrate the potential of introducing humanoid robotics into aircraft manufacturing, which underlines the versatility and maturity of the developed framework.

References

Comanoid: Website of the EU-project COMANOID. www.comanoid.cnrs.fr. Accessed September 13th, 2019 (2019)

Eurobench: Website of the EU-project EUROBENCH. www.eurobench2020.eu. Accessed September 13th, 2019 (2019)

Kheddar, A., Caron, S., Gergondet, P., Comport, A., Tanguy, A., Ott, C., Henze, B., Mesesan, G., Englsberger, J., Roa, M.A., Wieber, P.B., Chaumette, F., Spindler, F., Oriolo, G., Lanari, L.,

Escande, A., Chappellet, K., Kanehiro, F., Rabaté, P.: Humanoid robots in aircraft manufacturing - the Airbus use-cases. IEEE Robot. Autom. Mag. **26**(4), 30–45 (2019)

Ott, C., Roa, M.A., Hirzinger, G.: Posture and balance control for biped robots based on contact force optimization. In: IEEE-RAS International Conference on Humanoid Robots, pp. 26–33 (2011)

Torricelli, D., González-Vargas, J., Veneman, J.F., Mombaur, K., Tsagarakis, N., del Ama, A.J., Gil-Agudo, A., Moreno, J.C., Pons, J.L.: Benchmarking bipedal locomotion: a unified scheme for humanoids, wearable robots, and humans. IEEE Robot. Autom. Mag. **22**(3), 103–115 (2015)

Appendix A
Positive Definiteness of Matrix Potential

The procedure for converting the potential from the matrix into the quaternion formulation is based on Zhang and Fasse (2000): Let η and $\boldsymbol{\eta}$ be the scalar and vector part of the unit quaternion $\boldsymbol{\eta}$ representing the rotation error $\boldsymbol{R}_{DA}$. According to Shuster (1993), $\boldsymbol{R}_{DA}$ is given by $\boldsymbol{R}_{DA} = \left(\eta^2 - \|\boldsymbol{\eta}\|\right) \boldsymbol{I} + 2\boldsymbol{\eta}\boldsymbol{\eta}^T + 2\eta \boldsymbol{S}(\boldsymbol{\eta})$, which can be converted into

$$\boldsymbol{R}_{DA} = \boldsymbol{I} - 2\boldsymbol{\eta}^T\boldsymbol{\eta}\boldsymbol{I} + 2\boldsymbol{\eta}\boldsymbol{\eta}^T + 2\eta\boldsymbol{S}(\boldsymbol{\eta}) \tag{A.1}$$

exploiting the length of the unit quaternion $1 = \eta^2 + \boldsymbol{\eta}^T\boldsymbol{\eta}$. Inserting $\boldsymbol{R}_{DA}$ into (4.22) yields

$$\begin{aligned} V^{K,\mathrm{rot}} &= -\operatorname{tr}\left((\boldsymbol{I} - 2\boldsymbol{\eta}^T\boldsymbol{\eta}\boldsymbol{I} + 2\boldsymbol{\eta}\boldsymbol{\eta}^T + 2\eta\boldsymbol{S}(\boldsymbol{\eta})\,(\boldsymbol{R}_D^T\,\boldsymbol{R}_{\mathrm{ref}}\,\tilde{\boldsymbol{K}}\,\boldsymbol{R}_{\mathrm{ref}}^T\,\boldsymbol{R}_D)\right) + \operatorname{tr}(\tilde{\boldsymbol{K}}) \\ &= -\operatorname{tr}\left(\boldsymbol{R}_D^T\,\boldsymbol{R}_{\mathrm{ref}}\,\tilde{\boldsymbol{K}}\,\boldsymbol{R}_{\mathrm{ref}}^T\,\boldsymbol{R}_D\right) + \operatorname{tr}(\tilde{\boldsymbol{K}}) \\ &\quad + 2\operatorname{tr}\left((\boldsymbol{\eta}^T\boldsymbol{\eta} - \boldsymbol{\eta}\boldsymbol{\eta}^T - \eta\boldsymbol{S}(\boldsymbol{\eta}))\,(\boldsymbol{R}_D^T\,\boldsymbol{R}_{\mathrm{ref}}\,\tilde{\boldsymbol{K}}\,\boldsymbol{R}_{\mathrm{ref}}^T\,\boldsymbol{R}_D)\right). \end{aligned} \tag{A.2}$$

Considering $\operatorname{tr}(\boldsymbol{S}(\boldsymbol{\eta})(\boldsymbol{R}_D^T\,\boldsymbol{R}_{\mathrm{ref}}\,\tilde{\boldsymbol{K}}\,\boldsymbol{R}_{\mathrm{ref}}^T\,\boldsymbol{R}_D)) = 0$ due to Theorems 4.3 and 4.1 leads to

$$\begin{aligned} V^{K,\mathrm{rot}} &= +\,2\boldsymbol{\eta}^T\boldsymbol{\eta}\operatorname{tr}(\tilde{\boldsymbol{K}}) \\ &\quad - 2\operatorname{tr}\left(\boldsymbol{\eta}\boldsymbol{\eta}^T\,(\boldsymbol{R}_D^T\,\boldsymbol{R}_{\mathrm{ref}}\,\tilde{\boldsymbol{K}}\,\boldsymbol{R}_{\mathrm{ref}}^T\,\boldsymbol{R}_D)\right). \end{aligned} \tag{A.3}$$

It can be easily verified that

$$\operatorname{tr}\left(\boldsymbol{\eta}\boldsymbol{\eta}^T\,(\boldsymbol{R}_D^T\,\boldsymbol{R}_{\mathrm{ref}}\,\tilde{\boldsymbol{K}}\,\boldsymbol{R}_{\mathrm{ref}}^T\,\boldsymbol{R}_D)\right) = \boldsymbol{\eta}^T(\boldsymbol{R}_D^T\,\boldsymbol{R}_{\mathrm{ref}}\,\tilde{\boldsymbol{K}}\,\boldsymbol{R}_{\mathrm{ref}}^T\,\boldsymbol{R}_D)\,\boldsymbol{\eta} \tag{A.4}$$

holds using Theorem 4.2, which leads to

B. Henze, *Whole-Body Control for Multi-Contact Balancing of Humanoid Robots*, Springer Tracts in Advanced Robotics 143,
https://doi.org/10.1007/978-3-030-87212-0

$$V^{K,\mathrm{rot}} = 2\eta^T \left(\mathrm{tr}(\tilde{\boldsymbol{K}}) - (\boldsymbol{R}_D^T\, \boldsymbol{R}_{\mathrm{ref}}\, \tilde{\boldsymbol{K}}\, \boldsymbol{R}_{\mathrm{ref}}^T\, \boldsymbol{R}_D)\right) \boldsymbol{\eta}. \tag{A.5}$$

Inserting the definition of the co-stiffness matrix (see Sect. 4.3) simplifies $V^{K,\mathrm{rot}}$ to

$$V^{K,\mathrm{rot}} = 2\eta^T\, \boldsymbol{R}_D^T\, \boldsymbol{R}_{\mathrm{ref}}\, \boldsymbol{K}_{\mathrm{rot}}\, \boldsymbol{R}_{\mathrm{ref}}^T\, \boldsymbol{R}_D\, \boldsymbol{\eta}, \tag{A.6}$$

which is positive definite with respect to the vector part $\boldsymbol{\eta}$. Thus, $V^{K,\mathrm{rot}}$ is positive definite with respect to the rotation matrix $\boldsymbol{R}_{DA}$:

$$V^{K,\mathrm{rot}} \begin{cases} = 0 \text{ if } \boldsymbol{R}_{DA} = \boldsymbol{I} \\ > 0 \text{ if } \boldsymbol{R}_{DA} \neq \boldsymbol{I} \end{cases}. \tag{A.7}$$

Appendix B
Averaging of Rotation Matrices

In Gramkow (2001), two methods are compared for computing the average of rotations based on a representation via unit quaternions and via ration matrices. According to Gramkow (2001) both methods yield a sufficient accuracy, but the one based on quaternions is simpler to compute than the one using rotation matrices. For this reason, we propose averaging rotation matrices by first converting them into quaternions, computing the average in quaternion space, and then converting the result back into a rotation matrix.

Let $\boldsymbol{\eta}_i$ be the unit quaternion representation of a set of general rotations $\boldsymbol{R}_i \in \mathcal{SO}(3)$ with $i = 1 \ldots n$. The weighted average of the rotations can be computed according to

$$\boldsymbol{\eta} = \frac{\sum_{i=1}^{n} \alpha_i \boldsymbol{\eta}_i}{\left\| \sum_{i=1}^{n} \alpha_i \boldsymbol{\eta}_i \right\|} \tag{B.1}$$

based on the method presented in Gramkow (2001). The weighting coefficients are represented by $\alpha_i \in [0; 1]$ satisfying $\sum_{i=1}^{n} \alpha_i = 1$.

Note that Gramkow (2001) states that the method can potentially fail if the quaternions are distributed all over the unit sphere of quaternion space. However, this can be avoided by (a) exploiting the fact that $\boldsymbol{\eta}_i$ and $-\boldsymbol{\eta}_i$ represent the same rotation in order to restrict the quaternions to one half of the unit sphere and by (b) assuming that the orientations are relatively close to each other. The latter assumption holds in our case, as we average several estimates of the orientation $\boldsymbol{R}_b$, which probably show only a moderate variation.

References

Gramkow, C.: On averaging rotations. Int. J. Comput. Vis. **42**(1/2), 7–16 (2001)
Shuster, M.D.: A survey of attitude representations. J. Astronaut. Sci. **41**(4), 439–517 (1993)
Zhang, S., Fasse, E.D.: Spatial compliance modeling using a quaternion-based potential function method. Multibody Syst. Dyn. **4**(1), 75–101 (2000)

B. Henze, *Whole-Body Control for Multi-Contact Balancing of Humanoid Robots*, Springer Tracts in Advanced Robotics 143,
https://doi.org/10.1007/978-3-030-87212-0

The manufacturer's authorised representative in the EU is Springer Nature Customer Service Centre GmbH, Europaplatz 3, 69115 Heidelberg, Germany. If you have any concerns regarding our products, please contact ProductSafety@springernature.com

Printed and bound by CPI Group (UK) Ltd, Croydon, CR0 4YY
30/09/2026
02225694-0004